INTRODUCTION TO THE EXPLICIT FINITE ELEMENT METHOD FOR NONLINEAR TRANSIENT DYNAMICS

INTRODUCTION TO THE EXPLICIT FINITE ELEMENT METHOD FOR NONLINEAR TRANSIENT DYNAMICS

SHEN R. WU and LEI GU

WILEY

A JOHN WILEY & SONS, INC., PUBLICATION

Published by John Wiley & Sons, Inc., Hoboken, New Jersey
Published simultaneously in Canada

Library of Congress Cataloging-in-Publication Data

Wu, Shen R., 1945–
Introduction to the explicit finite element method for nonlinear transient dynamics / Shen R. Wu, Lei Gu.
p. cm.
Includes index.
ISBN 978-0-470-57237-5 (hardback)
1. Finite element method. 2. Numerical analysis. I. Gu, Lei, 1959– II. Title.
QA297.W83 2012
518′.25–dc23

2012007291

ISBN: 9780470572375

10 9 8 7 6 5 4 3 2 1

MIX
Paper from responsible sources
FSC® C013604

To our families and friends for their love and support.

CONTENTS

PREFACE xv

PART I FUNDAMENTALS 1

1 INTRODUCTION 3

1.1 Era of Simulation and Computer Aided Engineering 3

1.1.1 A World of Simulation 3

1.1.2 Evolution of Explicit Finite Element Method 4

1.1.3 Computer Aided Engineering (CAE)—Opportunities and Challenges 5

1.2 Preliminaries 6

1.2.1 Notations 6

1.2.2 Constitutive Relations of Elasticity 8

2 FRAMEWORK OF EXPLICIT FINITE ELEMENT METHOD FOR NONLINEAR TRANSIENT DYNAMICS 11

2.1 Transient Structural Dynamics 11

2.2 Variational Principles for Transient Dynamics 13

2.2.1 Hamilton's Principle 13

2.2.2 Galerkin Method 15

2.3 Finite Element Equations and the Explicit Procedures 15
2.3.1 Discretization in Space by Finite Element 16
2.3.2 System of Semidiscretization 19
2.3.3 Discretization in Time by Finite Difference 19
2.3.4 Procedure of the Explicit Finite Element Method 20
2.4 Main Features of the Explicit Finite Element Method 21
2.4.1 Stability Condition and Time Step Size 22
2.4.2 Diagonal Mass Matrix 23
2.4.3 Corotational Stress 24
2.5 Assessment of Explicit Finite Element Method 24
2.5.1 About the Solution of the Elastodynamics 24
2.5.2 A Priori Error Estimate of Explicit Finite Element Method for Elastodynamics 25
2.5.3 About the Diagonal Mass Matrix 30

PART II ELEMENT TECHNOLOGY 37

3 FOUR-NODE SHELL ELEMENT (REISSNER–MINDLIN PLATE THEORY) 39

3.1 Fundamentals of Plates and Shells 40
3.1.1 Characteristics of Thin-walled Structures 40
3.1.2 Resultant Equations 42
3.1.3 Applications to Linear Elasticity 44
3.1.4 Kirchhoff–Love Theory 46
3.1.5 Reissner–Mindlin Plate Theory 47
3.2 Linear Theory of R-M Plate 47
3.2.1 Helmholtz Decomposition for R-M Plate 48
3.2.2 Load Scaling for Static Problem of R-M Plate 48
3.2.3 Load Scaling and Mass Scaling for Dynamic Problem of R-M Plate 49
3.2.4 Relation between R-M Theory and K-L Theory 50
3.3 Interpolation for Four-node R-M Plate Element 52
3.3.1 Variational Equations for R-M Plate 52
3.3.2 Bilinear Interpolations 52
3.3.3 Shear Locking Issues of R-M Plate Element 55
3.4 Reduced Integration and Selective Reduced Integration 56
3.4.1 Reduced Integration 56

3.4.2 Selective Reduced Integration 57
3.4.3 Nonlinear Application of Selective Reduced Integration—Hughes–Liu Element 58
3.5 Perturbation Hourglass Control—Belytschko–Tsay Element 60
3.5.1 Concept of Hourglass Control 61
3.5.2 Four-node Belytschko–Tsay Shell Element—Perturbation Hourglass Control 63
3.5.3 Improvement of Belytschko–Tsay Shell Element 68
3.5.4 About Convergence of Element using Reduced Integration 70
3.6 Physical Hourglass Control—Belytschko–Leviathan (QPH) Element 71
3.6.1 Constant and Nonconstant Contributions 71
3.6.2 Projection of Shear Strain 72
3.6.3 Physical Hourglass Control by One-point Integration 73
3.6.4 Drill Projection 74
3.6.5 Improvement of B-L (QPH) Element 76
3.7 Shear Projection Method—Bathe–Dvorkin Element 76
3.7.1 Projection of Transverse Shear Strain 76
3.7.2 Convergence of B-D Element 78
3.8 Assessment of Four-node R-M Plate Element 80
3.8.1 Evaluations with Warped Mesh and Reduced Thickness 80
3.8.2 About the Locking-free Low Order Four-node R-M Plate Element 85

4 THREE-NODE SHELL ELEMENT (REISSNER–MINDLIN PLATE THEORY) 88

4.1 Fundamentals of a Three-node C^0 Element 89
4.1.1 Transformation and Jacobian 89
4.1.2 Numerical Quadrature for In-plane Integration 91
4.1.3 Shear Locking with C^0 Triangular Element 91
4.2 Decomposition Method for C^0 Triangular Element with One-point Integration 92
4.2.1 A C^0 Element with Decomposition of Deflection 92
4.2.2 A C^0 Element with Decomposition of Rotations 96
4.3 Discrete Kirchhoff Triangular Element 97
4.4 Assessment of Three-node R-M Plate Element 102
4.4.1 Evaluations with Warped Mesh and Reduced Thickness 102
4.4.2 About the Locking-free Low Order Three-node R-M Plate Element 105

5 EIGHT-NODE SOLID ELEMENT 107

5.1 Trilinear Interpolation for the Eight-node Hexahedron Element 107
5.2 Locking Issues of the Eight-node Solid Element 111
5.3 One-point Reduced Integration and the Perturbed Hourglass Control 113
5.4 Assumed Strain Method and Selective/Reduced Integration 115
5.5 Assumed Deviatoric Strain 118
5.6 An Enhanced Assumed Strain Method 118
5.7 Taylor Expansion of Assumed Strain about the Element Center 120
5.8 Evaluation of Eight-node Solid Element 123

6 TWO-NODE ELEMENT 128

6.1 Truss and Rod Element 128
6.2 Timoshenko Beam Element 129
6.3 Spring Element 131
6.3.1 One Degree of Freedom Spring Element 131
6.3.2 Six Degrees of Freedom Spring Element 132
6.3.3 Three-node Spring Element 133
6.4 Spot Weld Element 134
6.4.1 Description of Spot Weld Separation 134
6.4.2 Failure Criterion 135
6.4.3 Finite Element Representation of Spot Weld 137

PART III MATERIAL MODELS 139

7 MATERIAL MODEL OF PLASTICITY 141

7.1 Fundamentals of Plasticity 142
7.1.1 Tensile Test 142
7.1.2 Hardening 144
7.1.3 Yield Surface 145
7.1.4 Normality Condition 150
7.1.5 Strain Rate Effect/Viscoplasticity 152
7.2 Constitutive Equations 153
7.2.1 Relations between Stress Increments and Strain Increments 153
7.2.2 Constitutive Equations for Mises Criterion 157
7.2.3 Application to Kinematic Hardening 158

7.3 Software Implementation 159
7.3.1 Explicit Finite Element Procedure with Plasticity 160
7.3.2 Normal (Radial) Return Scheme 160
7.3.3 A Generalized Plane Stress Model 163
7.3.4 Stress Resultant Approach 164
7.4 Evaluation of Shell Elements with Plastic Deformation 169

8 CONTINUUM MECHANICS MODEL OF DUCTILE DAMAGE 175

8.1 Concept of Damage Mechanics 175
8.2 Gurson's Model 177
8.2.1 Damage Variables and Yield Function 178
8.2.2 Constitutive Equation and Damage Growth 179
8.3 Chow's Isotropic Model of Continuum Damage Mechanics 180
8.3.1 Damage Effect Tensor 181
8.3.2 Yield Function and Constitutive Equation 183
8.3.3 Damage Growth 185
8.3.4 Application to Plates and Shells 187
8.3.5 Determination of Parameters 188
8.4 Chow's Anisotropic Model of Continuum Damage Mechanics 189

9 MODELS OF NONLINEAR MATERIALS 192

9.1 Viscoelasticity 192
9.1.1 Spring–Damper Model 192
9.1.2 A General Three-dimensional Viscoelasticity Model 196
9.2 Polymer and Engineering Plastics 197
9.2.1 Fundamental Mechanical Properties of Polymer Materials 197
9.2.2 A Temperature, Strain Rate, and Pressure Dependent Constitutive Relation 198
9.2.3 A Nonlinear Viscoelastic Model of Polymer Materials 199
9.3 Rubber 200
9.3.1 Mooney–Rivlin Model of Rubber Material 200
9.3.2 Blatz–Ko Model 202
9.3.3 Ogden Model 203
9.4 Foam 203
9.4.1 A Cap Model Combining Volumetric Plasticity and Pressure Dependent Deviatoric Plasticity 205
9.4.2 A Model Consisting of Polymer Skeleton and Air 205
9.4.3 A Phenomenological Uniaxial Model 207

9.4.4 Hysteresis Behavior 208
9.4.5 Dynamic Behavior 209
9.5 Honeycomb 209
9.5.1 Structure of Hexagonal Honeycomb 210
9.5.2 Critical Buckling Load 210
9.5.3 A Phenomenological Material Model of Honeycomb 211
9.5.4 Behavior of Honeycomb under Complex Loading Conditions 213
9.6 Laminated Glazing 214
9.6.1 Application of J-integral 214
9.6.2 Application of Anisotropic Damage Model 215
9.6.3 A Simplified Model with Shell Element for the Laminated Glass 216

PART IV CONTACT AND CONSTRAINT CONDITIONS 219

10 THREE-DIMENSIONAL SURFACE CONTACT 221

10.1 Examples of Contact Problems 221
10.1.1 Uniformly Loaded String with a Flat Rigid Obstacle 222
10.1.2 Hertz Contact Problem 225
10.1.3 Elastic Impact of Two Balls 226
10.1.4 Impact of an Elastic Rod on the Flat Rigid Obstacle 228
10.1.5 Impact of a Vibrating String to the Flat Rigid Obstacle 231
10.2 Description of Contact Conditions 233
10.2.1 Contact with a Smooth Rigid Obstacle—Signorini's Problem 233
10.2.2 Contact between Two Smooth Deformable Bodies 237
10.2.3 Coulomb's Law of Friction 240
10.2.4 Conditions for "In Contact" 242
10.2.5 Domain Contact 242
10.3 Variational Principle for the Dynamic Contact Problem 243
10.3.1 Variational Formulation for Frictionless Dynamic Contact Problem 243
10.3.2 Variational Formulation for Frictional Dynamic Contact Problem 247
10.3.3 Variational Formulation for Domain Contact 250

10.4 Penalty Method and the Regularization of Variational Inequality 252
10.4.1 Concept of Penalty Method 252
10.4.2 Penalty Method for Nonlinear Dynamic Contact Problem 256
10.4.3 Explicit Finite Element Procedure with Penalty Method for Dynamic Contact 258

11 NUMERICAL PROCEDURES FOR THREE-DIMENSIONAL SURFACE CONTACT 261

11.1 A Contact Algorithm with Slave Node Searching Master Segment 262
11.1.1 Global Search 263
11.1.2 Bucket Sorting Method 264
11.1.3 Local Search 266
11.1.4 Penalty Contact Force 268
11.2 A Contact Algorithm with Master Segment Searching Slave Node 272
11.2.1 Global Search with Bucket Sorting Based on Segment's Capture Box 272
11.2.2 Local Search with the Projection of Slave Point 273
11.3 Method of Contact Territory and Defense Node 273
11.3.1 Global Search with Bucket Sorting Based on Segment's Territory 274
11.3.2 Local Search in the Territory 274
11.3.3 Defense Node and Contact Force 275
11.4 Pinball Contact Algorithm 277
11.4.1 The Pinball Hierarchy 277
11.4.2 Penalty Contact Force 278
11.5 Edge (Line Segment) Contact 279
11.5.1 Search for Line Contact 279
11.5.2 Penalty Contact Force of Edge-to-Edge Contact 281
11.6 Evaluation of Contact Algorithm with Penalty Method 282

12 KINEMATIC CONSTRAINT CONDITIONS 289

12.1 Rigid Wall 289
12.1.1 A Stationary Flat Rigid Wall 290
12.1.2 A Moving Flat Rigid Wall 291
12.1.3 Rigid Wall with a Curved Surface 293

12.2 Rigid Body 296
12.3 Explicit Finite Element Procedure with Constraint Conditions 298
12.4 Application Examples with Constraint Conditions 300

REFERENCES **305**

INDEX **325**

PREFACE

This book pertains to the use of the explicit finite element method in simulations of nonlinear transient dynamics problem. The explicit finite element method has developed into a useful tool in solving the large deformation transient dynamics problem. With a better solution to the difficult problems that are commonly encountered when using the implicit finite element method, we are able to widen the application to various contact/impact engineering problems.

Many theories and computational methods of the explicit finite element method have been widely discussed in various technical journals. However, the authors feel that there is a lack of a more systematic and comprehensive reference book on this particular subject. On the basis of years of experience in the application and research of the explicit finite element method, the authors feel that there is a need for such a reference book. We hope that this book can bring some useful insights and a better understanding to researchers and engineers in their study in this area.

The book is organized in four parts containing 12 chapters. Part I describes the fundamentals of the explicit finite element method for nonlinear transient dynamics. Part II describes the finite element technologies. Part III discusses material models. Part IV devotes to contact algorithms and constraint conditions.

Chapter 1 gives an introduction to the explicit finite element method and a summary of elasticity in preparation for our discussions in later parts of the book. Chapter 2 covers the basic variational principle for transient dynamics of large deformation and the formulation of explicit finite element equations. The convergence and accuracy assessment for applications of linear elastodynamics are also introduced. Chapter 3 describes the four-node shell elements based on Reissner–Mindlin plate theory. Various methods, such as reduced integration and projection method, to avoid or control shear locking are discussed. The convergence theory for Bathe–Dvorkin element is briefly introduced. Chapter 4 describes the three-node shell element (based on Reissner–Mindlin plate theory). Techniques, such as decomposition and discrete Kirchhoff theory, to control shear locking are discussed. Chapter 5 covers the eight-node solid element with several methods to control shear locking. Two numerical examples of elastodynamics problems are used to evaluate the elements discussed in Chapters 3–5. Chapter 6 introduces several two-node elements, which can be used for modeling a special feature of structural connection, including spring elements, spot weld, etc. Chapter 7 provides a review of the plasticity theory and the discussion of its material models for software implementation in the explicit finite element. Chapter 8 gives a brief discussion of material failure models based on continuum damage mechanics. It covers Gurson's micromechanics model and Chow's phenomenological models. Chapter 9 describes models of other nonlinear materials, including viscoelasticity, polymer, rubber, foam, honeycomb, and laminated glazing. Chapter 10 discusses contact problems. Several example problems with analytical solutions are introduced. The variational inequality for large deformation of transient dynamics is derived. The penalty method is introduced to regularize the variational inequality. Chapter 11 introduces the numerical procedures for three-dimensional contact based on the penalty method discussed in Chapter 10. Chapter 12 introduces several kinematic constraint conditions used to model certain features of nonlinear transient dynamics, including rigid wall and rigid body.

Unlike the implicit finite element method, the explicit finite element method is still in its developing stages. The above introduction only serves as a foundation for further discussion and exploration. We are primarily concerned about the large deformation of transient dynamics. The explicit finite element method faces similar challenges as the implicit finite element method in regard to certain basic theories and computational methods. The authors hope that this book will generate more interest among fellow researchers to collaborate and study these topics to bring about further improvement in engineering applications.

We wish to express our sincere appreciation to our colleagues and friends who have shown their enthusiastic encouragement in this book project. We would like to express our sincere gratitude to Professor J.T. Oden of University of Texas at Austin and Professor T. Belytschko of Northwestern University for their valuable inputs. We want to extend our deep appreciation to the team at Wiley. We specially thank Dr. Priya Prasad for his continuous encouragement and support in the publication of this book. Finally, we would like to thank our families for their love and support, for without them our work in this book would not have been possible.

Shen R. Wu and Lei Gu

PART I

FUNDAMENTALS

CHAPTER 1

INTRODUCTION

1.1 ERA OF SIMULATION AND COMPUTER AIDED ENGINEERING

1.1.1 A World of Simulation

"Computer simulation" has become a popular terminology in almost all disciplines of science and engineering today. Successful stories of computer simulation on various research projects have been reported in many professional conferences and events. In recent years, many technical journals have emerged dedicating to theories, techniques, and applications of simulations. Simulation shines in almost every aspect of research.

In its final report of 2006, the Blue Ribbon Panel on simulation-based Engineering Science of US National Science Foundation claimed the critical importance of simulation technology in the twenty-first century and considered it as the national priority for tomorrow's engineering and science (available at http://www.nsf.gov/pubs/reports/sbes_final_report.pdf). The working group of scientists of computational mechanics, applied mathematics, and other disciplines has envisioned revolutionizing engineering science through simulation. Simulation is essentially the computational science and engineering. It involves heavily the use of finite element method and other numerical approaches. In the past half century, finite element methods have been used for many engineering applications with the advances of high-speed computing power and software functionality. The evolution of finite element technology has also stimulated the development of computer architectures and technologies. As many physical events are too costly for any type of failure, computer simulation has become a

Introduction to the Explicit Finite Element Method for Nonlinear Transient Dynamics, First Edition.
Shen R. Wu and Lei Gu.

highly desired tool to evaluate the process before carrying out the actual procedure. For example, medical doctors can first perform computer simulation on a bypass surgery procedure for treating disease in aorta and iliac artery to assess the potential results without subjecting any human life to danger. These scientific and engineering applications have placed additional importance on numerical simulation to provide precise and accurate information.

Approximate solutions to the differential and/or integral equations from various engineering problems have been in demand for a long time due to the difficulty in obtaining analytical solutions. Courant (1943) constructed the approximate solution to St. Venant torsion problem by triangulation with linear approximation for the minimum potential energy and the Ritz method. In fact, Courant (1943) demonstrated all the basic concepts of the finite element method. In the mid-1950s, Argyris (1954, 1957) and his colleagues extensively developed certain generalization of the linear theory of structures and presented procedures for analyzing complex discrete structures. Turner et al. (1956) analyzed classical elasticity equation and illustrated the triangular element properties for plane stress. Clough (1960) named such an approximation method the "finite element," for the first time. Since then, work and research on the finite element method has grown extensively. While many algorithms and applications of linear problems were still under development, nonlinear analysis has been developed at a significantly faster pace. Oden (1972) among others demonstrated significant achievements in nonlinear applications and provided the basic concepts and algorithms of nonlinear finite element methods.

Following the development of the fundamentals, finite element software was quickly commercialized and further propelled engineering applications. The first software program was delivered by Ed Wilson. The subsequent development became SAP and NONSAP. The first nonlinear commercial software MARC led by Pedro Marcal and ADINA led by Jürgen Bathe were among the early software developed for nonlinear structural dynamics. Finite element then started to be introduced into universities' colloquiums. It is critical that, in accompanying the development of numerical methods and engineering applications, the mathematical theories about interpolation, convergence, and error estimation of the finite element methods have also been heavily developed to provide strong support for the finite element method. The monograph edited by Ciarlet and Lions (1991) is an excellent collection of the mathematical achievements. We agree with the statement by Belytschko (1996) that extending from linear static analysis to nonlinear dynamic analysis greatly increases the level of difficulty. The generally adopted solver for nonlinear problems has been basically a Newton–Raphson procedure or a modified one. These numerical schemes are the foundation of successful engineering applications. For strongly nonlinear problems, however, the Newton–Raphson iteration can fail to converge. The algorithm to obtain a convergent solution within reasonably short time has been a focal point of finite element researches.

1.1.2 Evolution of Explicit Finite Element Method

The explicit finite element method has been successfully applied to various situations of nonlinear transient dynamics in the past decades. It is now widely adopted in the manufacturing process as well as in the research activity. As reported in journals and conferences, many problems have been solved by using explicit finite element. The applications involve various industries and manufacturing processes. The following references are just a few examples: Anghileri et al. (2005) and Ho and Smith (2006) for bird strike at the airplane; Xue and Schmid (2005) for train collision; Saha et al. (1995) and Houssini (2006) for automobile crashworthiness; Neumayer et al. (2006) for package drop; Chow and Tai (2000) for sheet metal stamping; Lu and Wu (2006) for forging and extrusion processes; Medvedev (2002) for welding. The list of applications goes on and on. These examples have a common feature, that is, dynamic contact or say impact. We name these applications as impact engineering for late reference.

The structural analysis for the impact engineering such as above is a class of transient dynamics. It is a highly nonlinear system including large deformation, large rotation, nonlinear material, contact, impact, etc. For such a system, usually only numerical solution can be expected. Even with numerical approach, engineers have seen substantial challenges from large deformation in dynamic buckling and postbuckling mode. In this area, the traditional (implicit) approach had not achieved much satisfaction until the explicit finite element method emerged as a powerful tool. The explicit approach provides an alternative problem-solving procedure. It is essentially an incremental method. Apart from the traditional implicit method, explicit approach basically does not form the system stiffness matrix and does not need to invert the large matrix. Hence, the explicit method has avoided certain difficulties of nonlinear programming that the implicit method has.

As described in Belytschko et al. (2000), the explicit finite element software was originated in the United States. Several groups of scientists had worked on the concept of explicit integration for nonlinear transient dynamics. Wilkins (1964) was among the earliest publications on explicit finite element methods. As reported by Constantino (1967), the first explicit software was built in 1964.

Other early developments include HONDO and later PRONTO led by Sam Key; SADCAT, WHAMS, and Super WHAMS led by Ted Belytschko, and DYNA-2D/3D led by John Hallquist. The commercial software boomed in the mid-1980s. We have seen PAMCRASH in the market first, followed by RADIOSS, DYTRAN, and ABAQUS-explicit. In later 1980s, headed by John Hallquist, LS-DYNA was commercialized. In fact, the fast development and implementation of many modern numerical technologies make LS-DYNA distinguished from the pack of commercial software.

1.1.3 Computer Aided Engineering (CAE)—Opportunities and Challenges

As Moore's rule predicted, the computing power increases tenfolds every 5 years. The CAE engineers have witnessed and enjoyed the great advances in computer architectures and software functionalities.

With growing computing power, expectations for more accurate predictive analysis (by the project management) have also risen. Simulation as an important design tool has been built into the manufacturing process. This brings a tough challenge to engineers as they try to assess the reliability of the results predicted by the computer simulation, even before the prototype test is conducted. Being over confident and overly reliant on simulation results have at times led to wrong and costly decisions. In recent years, the concept of verification and validation (V & V) has been proposed; see Oden et al. (2003) and Babuska and Oden (2004) for basic concepts and theories, also Oberkampf and Barone (2004) for engineering practices. Verification and validation is critical for certain types of simulation, whose errors could lead to major disasters. The essential point is how to systematically justify the numerical solutions.

From our years of engineering experiences, the authors strongly feel that it would be helpful for engineers to have a deep understanding of the "back bones" of the software. One of the main objectives of this book is to introduce the related theory and technology for the explicit finite element method. This book can also serve as a textbook for related disciplines in graduate level work and studies. This book identifies certain unresolved issues currently existing in finite element formulation and its implementation in software. It is also the authors' intent to assist researchers to find interesting and challenging topics for their studies that will eventually help engineers make better computer simulations.

1.2 PRELIMINARIES

1.2.1 Notations

Several aspects of applied mechanics, applied mathematics, and numerical methods are involved in this book. Due to the complexity of the course, many physical variables and parameters will be employed. Many of them have components in three-dimensional (3D) space and are time dependent. Notations commonly seen in engineering literatures will be used to identify these variables, in a consistent manner. In case if same symbol is used for different variables in different discipline, we will choose an alternative definition. The following is a partial list of the most important variables in the text:

u	displacement
v	velocity
a	acceleration
ε	strain
σ	stress
t	time
ζ	thickness of plate/shell
h	element size
E	Young's modulus
G	shear modulus
ν	Poisson ratio
λ, μ	Lamé elasticity constants
ρ	mass density
ξ, η	coordinates of reference system

Φ shape functions of finite element
$(\dot{f})$ first-order time derivative of function f
$(\ddot{f})$ second-order time derivative of function f
$f_{,x}$ a (partial) derivative $\partial f/\partial x$

Exceptions will accompany additional explanations whenever it is necessary.

Both indices and bold faces will be used to represent the vectors, matrices, and tensors. Indices will be used for the components of vector variables, for example, f_j indicates the j-component of variable f. Regarding coordinates, usually 1, 2, and 3 are for the x-, y-, and z-directions, respectively. Indices are also used for matrices, tensors, and other variables. For example, u_j^N will be used later for x_j-component of displacement of node N. To avoid any possible confusion with the sequence of matrix multiplications, or the multiplication with tensors of order 3 and higher or variables with multiindices, the index notation will be used more often. Bold-faced variables will also be used, when their number of components is easy to understand and their operation will not be confused.

The lower-case indices are most likely used for spatial components, with Latin indices for 3D variables and Greek indices for two-dimensional (2D) variables. Capital Latin indices are often used for nodal variables of finite elements.

Simple tensor operations will be used for shorthand writing purposes, which should be easily understood by readers without extensive knowledge of tensor analysis. Cartesian coordinate system will be used exclusively, except in special situations where additional explanations are provided. Hence, there is essentially no difference for superscripts and subscripts or contravariant and covariant components of the tensors. In particular, $u_{i,j}$ simply means a partial derivative $\partial u_i/\partial x_j$.

The commonly used convention of summation on repeated indices is adopted. The convention of summation only applies to paired variables with the same indices. Summation of tripled or more variables will use the traditional notation Σ. This convention is also extended to summation involving nodal values of finite elements. For instance,

$u_j v_j = \sum_j u_j v_j$: a dot product of two vectors $\boldsymbol{u} \cdot \boldsymbol{v}$.
$a_{ij} b_j = \sum_j a_{ij} b_j$: a multiplication of a matrix with a vector $\boldsymbol{Ab}$.
$u_N \Phi_N = \sum_N u_N \Phi_N$: interpolation formula with finite element nodal values and the shape functions.

Note that the number of components in above examples is not critical and easy to understand. The pair of indices in the summation is called dummy index, which can be replaced by any character. This is a necessary practice when an index would appear to be triple or more but summation is really acting on two variables only. Some differential operators can be expressed using the convention of summation:

$u_{j,j} = \nabla \bullet \boldsymbol{u}$: divergence of vector $\boldsymbol{u}$.
$w_{,jj} = \nabla^2 w$: Laplacian of function w.

Special tensors and their functionalities are adopted:

$$\delta_{ij} = \begin{cases} 0 & \text{if } i \neq j \\ 1 & \text{if } i = j \end{cases} : \quad \text{Kronecker delta,}$$

$$\in_{ijk} = \begin{cases} 0 & \text{if any two indices are equal} \\ 1 & \text{if } i, j, k = 1, 2, 3 \text{ or } 2, 3, 1 \text{ or } 3, 1, 2 : \\ -1 & \text{if } i, j, k = 3, 2, 1 \text{ or } 2, 1, 3 \text{ or } 1, 3, 2 \end{cases} \quad \text{3D permutation tensor.}$$

Part of their operational functionalities is listed below for later reference:

$$\delta_{ij} u_j = u_i,$$

$$\delta_{jj} = 3, \quad \delta_{ij}\delta_{jk} = \delta_{ik}, \quad \delta_{ij}\delta_{ij} = \delta_{jj} = 3,$$

$$\in_{ijk}\in_{imn} = \delta_{jm}\delta_{kn} - \delta_{jn}\delta_{km}, \quad \in_{ijk}\in_{ijn} = 2\delta_{kn},$$

$$c_k = \in_{ijk} a_i b_j : \mathbf{c} = \boldsymbol{a} \times \mathbf{b} : \quad \text{vector product of two vectors in 3D space,}$$

$$\in_{ijk} a_i b_j c_k = \boldsymbol{a} \times \boldsymbol{b} \bullet \boldsymbol{c} : \quad \text{mixed product of three vectors in 3D space.}$$

This threefold summation represents a mixed product of three vectors, which is equivalent to the volume framed by the vectors ***a***, ***b***, and ***c***.

The 2D Kronecker delta and permutation tensor are defined with α and β ranging from 1 to 2:

$$\delta_{\alpha\beta} = \begin{cases} 0 & \text{if } \alpha \neq \beta \\ 1 & \text{if } \alpha = \beta \end{cases} : \quad \text{2D Kronecker delta,}$$

$$\in_{\alpha\beta} = \begin{cases} 0 & \text{if } \alpha = \beta \\ 1 & \text{if } \alpha, \beta = 1, 2 : \\ -1 & \text{if } \alpha, \beta = 2, 1 \end{cases} \quad \text{2D permutation tensor.}$$

The related properties are, for example,

$$\delta_{\beta\beta} = 2,$$

$$\in_{\alpha\beta} a_\alpha b_\beta = a_1 b_2 - a_2 b_1 = \begin{vmatrix} a_1 & a_2 \\ b_1 & b_2 \end{vmatrix} : \quad \text{2D determinant,}$$

$$\in_{\alpha\beta} \psi_{,\beta} : (\psi_{,2}, -\psi_{,1}) : \quad \text{differential operator of curl on a scalar function.}$$

1.2.2 Constitutive Relations of Elasticity

Elasticity is the foundation of structural mechanics. Here we summarize the constitutive relations for later reference, but we would not provide detailed review for elasticity as we focus on nonlinear problems. For 3D solid material, let u_i be the components of displacement. The strain of small deformation is

$$\varepsilon_{ij} = (u_{i,j} + u_{j,i})/2. \tag{1.1}$$

The corresponding stresses are determined by the generalized Hooke's law:

$$\sigma_{ij} = E_{ijkl}\varepsilon_{kl}. \tag{1.2}$$

Here, $\boldsymbol{E}$ is called the elasticity tensor. The inverse relation is expressed with the compliance tensor $\boldsymbol{C}$:

$$\varepsilon_{ij} = C_{ijkl}\sigma_{kl}. \tag{1.3}$$

For the general elasticity, both $\boldsymbol{E}$ and $\boldsymbol{C}$ are symmetric with

$$E_{ijkl} = E_{klij} = E_{jikl} = E_{ijlk},\; C_{ijkl} = C_{klij} = C_{jikl} = C_{ijlk}. \tag{1.4}$$

For isotropic elastic materials, there are only two independent material parameters. The elasticity tensor can be expressed with Young's modulus E and Poisson ratio ν, or using Lamé elasticity constants λ and μ. We have

$$E_{ijkl} = \frac{E\nu}{(1+\nu)(1-2\nu)}\delta_{ij}\delta_{kl} + \frac{E}{1+\nu}\delta_{ik}\delta_{jl} = \lambda\delta_{ij}\delta_{kl} + 2\mu\delta_{ik}\delta_{jl}, \tag{1.5a}$$

$$C_{ijkl} = -\frac{\nu}{E}\delta_{ij}\delta_{kl} + \frac{1+\nu}{E}\delta_{ik}\delta_{jl} = -\frac{\lambda}{2\mu(3\lambda+2\mu)}\delta_{ij}\delta_{kl} + \frac{1}{2\mu}\delta_{ik}\delta_{jl}, \tag{1.5b}$$

$$\begin{aligned} \sigma_{ij} &= \lambda\delta_{ij}\varepsilon_{kk} + 2\mu\varepsilon_{ij}, \\ \varepsilon_{ij} &= -\frac{\nu}{E}\delta_{ij}\sigma_{kk} + \frac{1}{2\mu}\sigma_{ij}. \end{aligned} \tag{1.6}$$

The elasticity constants are related with the following formulae:

$$\lambda = \frac{E\nu}{(1+\nu)(1-2\nu)},\; \mu = \frac{E}{2(1+\nu)}. \tag{1.7}$$

We also use $G = \mu$, called shear modulus. Besides, we define the bulk modulus K with

$$\begin{aligned} K &= \frac{E}{3(1-2\nu)} = \frac{2(1+\nu)\mu}{3(1-2\nu)} = \frac{3\lambda+2\mu}{3}, \\ \sigma_{jj} &= 3K\varepsilon_{jj}. \end{aligned} \tag{1.8}$$

Plane stress or the generalized plane stress state is of particular interest, where $\sigma_{33} = 0$, and other stress components are independent of the thickness. We have

$$\varepsilon_{33} = -\frac{\lambda}{\lambda + 2\mu}\varepsilon_{\delta\delta}, \tag{1.9}$$

$$\begin{aligned} \sigma_{\alpha\beta} &= \frac{E}{1-\nu^2}(\nu\delta_{\alpha\beta}\varepsilon_{\eta\eta} + (1-\nu)\varepsilon_{\alpha\beta}), \\ \varepsilon_{\alpha\beta} &= -\frac{\nu}{E}\delta_{\alpha\beta}\sigma_{\eta\eta} + \frac{1}{2\mu}\sigma_{\alpha\beta}, \end{aligned} \tag{1.10}$$

$$\begin{aligned} \sigma_{11} &= E_1(\varepsilon_{11} + \nu\varepsilon_{22}),\ \varepsilon_{11} = (\sigma_{11} - \nu\sigma_{22})/E, \\ \sigma_{22} &= E_1(\varepsilon_{22} + \nu\varepsilon_{11}),\ \varepsilon_{22} = (\sigma_{22} - \nu\sigma_{11})/E, \\ \sigma_{ij} &= 2\mu\varepsilon_{ij}, i \neq j,\ \varepsilon_{ij} = \sigma_{ij}/2\mu, \quad i \neq j, \end{aligned} \tag{1.11}$$

$$E_1 = \frac{E}{1-\nu^2}. \tag{1.12}$$

CHAPTER 2

FRAMEWORK OF EXPLICIT FINITE ELEMENT METHOD FOR NONLINEAR TRANSIENT DYNAMICS

As a starting point, this chapter describes the system of governing equations, the variational principles, and the formulation of explicit finite element. It also serves as the framework for the development of later chapters.

2.1 TRANSIENT STRUCTURAL DYNAMICS

As an example, Figure 2.1 depicts the large deformation of a structural component due to impact. It presents several challenging aspects of the solid mechanics. There are large deformation and large rotation, usually named geometrical nonlinearity; and plasticity, usually named material nonlinearity. There are also cases of material contact/impact, which occur at the a priori unknown location. In addition, there are various constraint conditions defined by nonlinear equations or inequalities.

For a general case illustrated in Figure 2.2, the system consists of the following equations and inequalities, cf. Wu (2009):

$$\rho \ddot{u}_i - \sigma_{ij,j} = f_i(t, \boldsymbol{x}) \text{ in } \Omega, \quad i, j = 1, 2, 3, \tag{2.1a}$$

$$u_i(0,\boldsymbol{x}) = U_i^0(\boldsymbol{x}), \quad \dot{u}_i(0,\boldsymbol{x}) = U_i^1(\boldsymbol{x}) \text{ in } \Omega, \tag{2.1b}$$

$$u_i = U_i(t) \text{ on } \Gamma_u, \tag{2.1c}$$

Introduction to the Explicit Finite Element Method for Nonlinear Transient Dynamics, First Edition.
Shen R. Wu and Lei Gu.

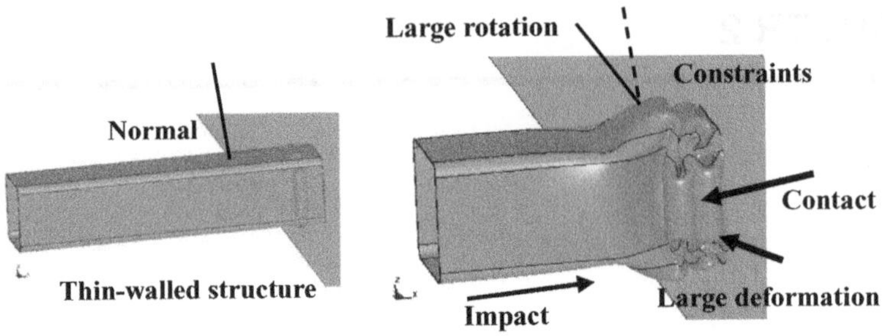

FIGURE 2.1 The large deformation of a component.

$$\sigma_{ij}N_j = g_i(t, \boldsymbol{x}) \text{ on } \Gamma_s, \tag{2.1d}$$

$$\begin{cases} \dot{u}_N \leq 0, \quad \sigma_N \leq 0, \quad \sigma_N \dot{u}_N = 0, \\ |\boldsymbol{\sigma}_T| \leq \mu |\sigma_N|, \\ \quad \text{if } |\boldsymbol{\sigma}_T| < \mu|\sigma_N| \text{ then } \dot{u}_{Ti} = 0, \\ \quad \text{if } |\boldsymbol{\sigma}_T| = \mu|\sigma_N| \text{ then } \exists\lambda > 0, \dot{u}_{Ti} = -\lambda\sigma_{Ti}, \end{cases} \quad \text{on } \Gamma c, \tag{2.1e}$$

$$\dot{\sigma}_{ij} = \dot{\sigma}_{ij}(E, \nu, E_t, x_i, u_i, \dot{\varepsilon}_{ij}, \ldots), \tag{2.1f}$$

$$\dot{\varepsilon}_{ij} = (\dot{u}_{i,j} + \dot{u}_{j,i})/2, \tag{2.1g}$$

where ρ is the mass density and f_i represents the components of the body force. U_i^0 and U_i^1 are the given initial displacement and initial velocity, respectively. $U_i(t)$ is the prescribed displacement on part of boundary, Γ_u. g_i represents the components of the tractions on part of the boundary, Γ_s, and N_i represents the components of the

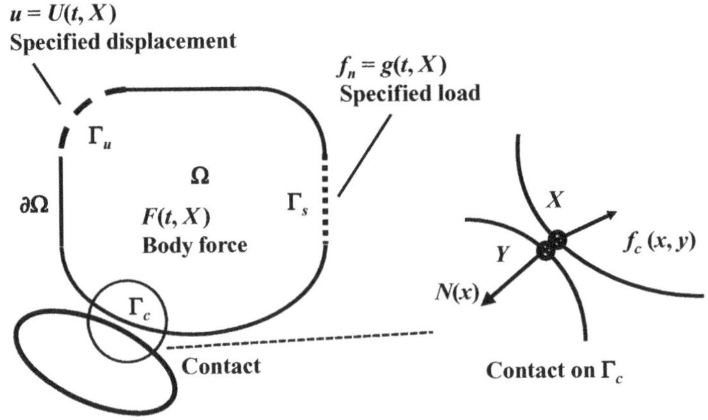

FIGURE 2.2 System of transient dynamics.

unit outward normal vector of Γ. The convention of summation on repeated indices is adopted. The boundary is composed of three disjoint portions:

$$\begin{aligned} \partial\Omega &= \bar{\Gamma}_u \cup \bar{\Gamma}_s \cup \bar{\Gamma}_c, \\ \Gamma_u \cup \Gamma_s &= \Gamma_s \cup \Gamma_c = \Gamma_u \cup \Gamma_c = \varphi. \end{aligned} \tag{2.1h}$$

As usual, displacement is prescribed on Γ_u and traction is prescribed on Γ_s. We assume that contact will happen on part of Γ_c or whole of Γ_c itself but not anywhere else. Here, $\sigma_N = \sigma_{ij}N_iN_j$ and $(\boldsymbol{\sigma}_T)_i = \sigma_{ij}N_j - \sigma_N N_i$ in (2.1e) represent the normal and tangential components of the traction on Γ_c. $\dot{u}_N$ and $\dot{u}_{Ti}$ are the normal and tangential components of velocity. We will discuss details about the contact conditions (2.1e) in Chapters 10 and 11. The constitutive relation (2.1f) in rate form is employed for the nonlinear analysis. E and ν are the Young's modulus and Poisson ratio of elasticity. E_t is the tangent modulus for nonlinear material.

For linear elasticity, strain–displacement relation can use the whole quantity instead of incremental form (2.1g). The constitutive relation (2.1f) is simplified to (1.5)–(1.7). Other equations and inequalities remain the same. Part 3 will provide detailed discussions for nonlinear materials. Part 4 will discuss contact and constraints. We will lay down the foundation of the explicit finite element method for the system in this chapter.

2.2 VARIATIONAL PRINCIPLES FOR TRANSIENT DYNAMICS

A basic system without contact and constraint will be discussed in this section. Chapters 10 and 11 will extend from here to describe details about variational principle with contact conditions.

2.2.1 Hamilton's Principle

In the classical mechanics course, the Hamilton's principle states that the action integral $\mathscr{L} = \int_{t_1}^{t_2} L\mathrm{d}t$ of the Lagrangian potential $L = T - V$ takes stationary value for the true motion among all motions that carry a conservative system from t_1 to t_2, that is, $\delta\mathscr{L} = 0$. Here, T is the kinetic energy and V is the potential energy. For a nonlinear system, the potential energy is accumulated in time, and is usually not expressed explicitly as a potential function of displacement or velocity. Alternatively, we can use the incremental form applicable to nonconservative system with $\delta V = \int \boldsymbol{Q}\delta\boldsymbol{q}\mathrm{d}\Omega$. Here, $\boldsymbol{Q}$ is for the generalized forces and $\boldsymbol{q}$ is for the generalized coordinates (displacement, rotation, etc.). Then, we have

$$\delta\mathscr{L} = \int_{t_1}^{t_2} (\delta T - \delta V)\mathrm{d}t = 0. \tag{2.2}$$

For a system with the specified displacement U_i on a portion of boundary, Γ_u, and the prescribed traction g_i on Γ_s, the rest portion of boundary, and subjected to the body force f_i, we can write

$$\delta V = \int_\Omega \sigma_{ij}\delta\varepsilon_{ij}\mathrm{d}\Omega - \int_\Omega f_i\delta u_i\mathrm{d}\Omega - \int_{\Gamma_s} g_i\delta u_i\mathrm{d}\Gamma. \tag{2.3}$$

The kinetic energy takes the usual form

$$\begin{aligned} T &= \int_\Omega \rho\dot{u}_i\dot{u}_i\mathrm{d}\Omega/2, \\ \delta T &= \int_\Omega \rho\dot{u}_i\delta\dot{u}_i\mathrm{d}\Omega. \end{aligned} \tag{2.4}$$

Here, δu_i is viewed as the variation of displacement. Integrating (2.2) with (2.4) by parts with respect to t, using $\delta u_i|_{t_1} = \delta u_i|_{t_2} = 0$ specified by Hamilton's principle, we obtain

$$\int_{t_1}^{t_2} \delta T\mathrm{d}t = \int_{t_1}^{t_2}\left(\int_\Omega -\rho\ddot{u}_i\delta u_i\mathrm{d}\Omega\right)\mathrm{d}t. \tag{2.5}$$

Substituting (2.3) and (2.5) into (2.2), we obtain

$$\int_{t_1}^{t_2}\left(\int_\Omega (-\rho\ddot{u}_i\delta u_i - \sigma_{ij}\delta\varepsilon_{ij} + f_i\delta u_i)\mathrm{d}\Omega + \int_{\Gamma_s} g_i\delta u_i\mathrm{d}\Gamma\right)\mathrm{d}t = 0. \tag{2.6}$$

Since the displacement condition (2.1c) is specified, the variation on Γ_u is not allowed, that is, $\delta u_i|_{\Gamma_u} = 0$. Using Gauss–Green theorem and (2.1g) (for small variation), the integration by parts for (2.3) leads to

$$\delta V = \int_\Omega -\sigma_{ij,j}\delta u_i\mathrm{d}\Omega + \int_{\Gamma_s}\sigma_{ij}n_j\delta u_i\mathrm{d}\Gamma - \int_\Omega f_i\delta u_i\mathrm{d}\Omega - \int_{\Gamma_s} g_i\delta u_i\mathrm{d}\Gamma. \tag{2.7}$$

By applying (2.5) and (2.7), (2.2) is reduced to

$$\begin{aligned} \int_{t_1}^{t_2}\Bigg(\int_\Omega -\rho\ddot{u}_i\delta u_i\mathrm{d}\Omega &+ \int_\Omega \sigma_{ij,j}\delta u_i\mathrm{d}\Omega - \int_{\Gamma_s}\sigma_{ij}n_j\delta u_i\mathrm{d}\Gamma \\ &+ \int_\Omega f_i\delta u_i\mathrm{d}\Omega + \int_{\Gamma_s} g_i\delta u_i\mathrm{d}\Gamma\Bigg)\mathrm{d}t = 0. \end{aligned} \tag{2.8}$$

If all the functions are smooth enough, the arbitrary variation δu_i results in the equation of motion (2.1a) and the force boundary conditions (2.1d). To see the details, we first let δu_i be zero on the boundary Γ_s and be arbitrary in Ω. This gives (2.1a) by the classical argument of calculus. Hence, the body integrals in (2.8) drop off. The rest are integrals on Γ_s. With arbitrary δu_i, it results in (2.1d). This is a classical method, which shows that Hamilton's principle and system (2.1) are essentially equivalent.

2.2.2 Galerkin Method

Various approximation methods for solving (2.1) are developed. The variational principle discussed in Section 2.2.1 can lead to the formulation of the energy method, cf. Meirovitch (1980). We now consider another concept of constructing the approximation—the Galerkin method. We use the test functions v_i, satisfying the homogeneous displacement boundary conditions like the variation. Multiplying v_i to the motion equation (2.1a) and the force boundary condition (2.1d), integrating the products in space domain Ω and on boundary Γ_s, respectively, and summing them together, we obtain the weighted residual

$$\int_\Omega (\rho \ddot{u}_i - \sigma_{ij,j}) v_i \mathrm{d}\Omega + \int_{\Gamma_s} \sigma_{ij} n_j v_i \mathrm{d}\Gamma = \int_\Omega f_i v_i \mathrm{d}\Omega + \int_{\Gamma_s} g_i v_i \mathrm{d}\Gamma. \tag{2.9}$$

Note that $v_i = 0$ on Γ_u. Using Gauss–Green formula for the second term on the left-hand side of (2.9), we have the variational equation

$$\int_\Omega (\rho \ddot{u}_i v_i + \sigma_{ij} v_{i,j}) \mathrm{d}\Omega = \int_\Omega f_i v_i \mathrm{d}\Omega + \int_{\Gamma_s} g_i v_i \mathrm{d}\Gamma. \tag{2.10}$$

This is essentially the same as (2.6). The equivalence of (2.10) and (2.1) can be established in the same way as described in Section 2.2.1.

Conceptually, δu_i in (2.6) is considered a type of virtual displacement. The test function v_i in (2.10) can also be considered as a type of virtual displacement. In this scenario, (2.6) and (2.10) are viewed as the energy form. The kinetic energy is viewed as a transform of integration by parts in time domain. On the other hand, the test function v_i can be considered as virtual velocity. Then, (2.10) has the characteristics of energy rate or virtual power.

2.3 FINITE ELEMENT EQUATIONS AND THE EXPLICIT PROCEDURES

So far, we have discussed the fundamental variational principles and their equivalence to the system of partial differential equations (2.1). Besides the method of weighted residual used to derive the variational equation (2.10) in the average sense of integration, only stress but not its gradient appears in (2.10). These features illustrate the meaning of "weak solution," which represent the nature of finite element method.

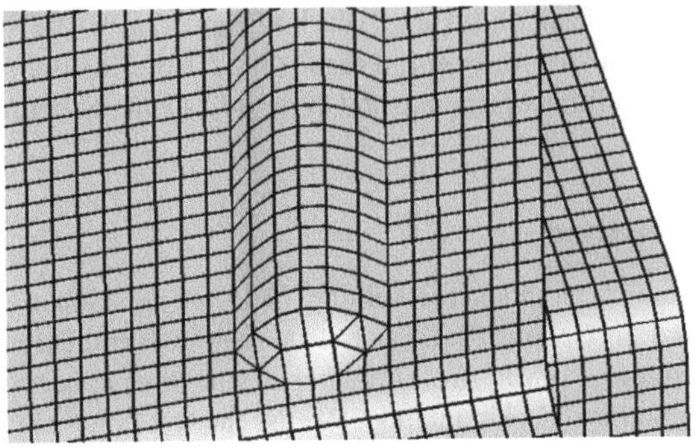

FIGURE 2.3 Mesh.

2.3.1 Discretization in Space by Finite Element

Our finite element method starts with grading (meshing) the material domain into a set of finite elements, for example, the four-node quadrilaterals and/or the three-node triangles for the thin-walled structures. As an illustration, a portion of the mesh for the component shown in Figure 2.1 is presented in Figure 2.3:

$$\Omega = \cup \Omega_e, \quad \Omega_e \cap \Omega_f = \varphi, \quad \text{if } e \neq f. \tag{2.11}$$

The geometry is then approximated by a set of nodes and finite elements. The original boundary, which may be a curve, is approximated by the collection of the exterior element boundaries. In addition, if the physical domain is a thin-walled curved surface meshed by using four-node quadrilaterals or three-node triangles, the curvature of the structure is also approximated, similar to the curved surface boundary of a solid body. The material domain is thus discretized and approximated by the finite elements with the mesh. In our discussion, we focus on the linear elements of Lagrangian family.

For each node (N) of the mesh we define a base function Φ_N, such that Φ_N has value one at node N and zero at all other nodes. By examining a patch of the elements connected to node N, as shown in Figure 2.4, we find that Φ_N has zero value at the boundary of the patch of these elements. This is why these base functions are named hat functions in some textbooks.

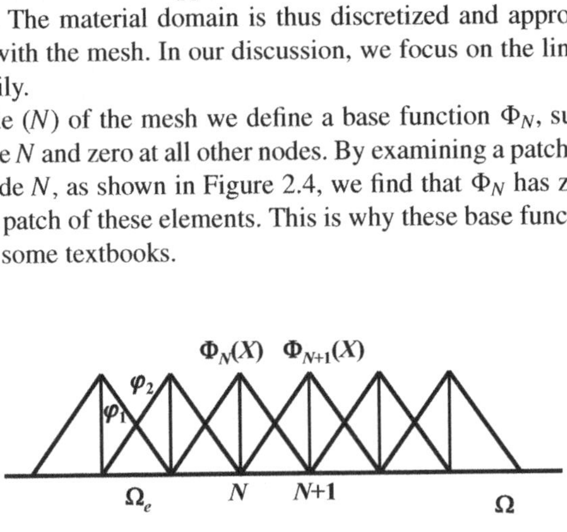

FIGURE 2.4 Shape function.

The base functions have the following unity property, which can be verified easily:

$$\begin{aligned} &\Phi_N(x^M) = \delta_{MN}, \\ &\sum\nolimits_N \Phi_N(x) \equiv 1. \end{aligned} \tag{2.12}$$

In one of the neighboring elements of node N, for example, Ω_e, its Jth node $n_J = N$. The restriction of the base function Φ_N to Ω_e is called a shape function of element Ω_e, associated with node n_J, denoted by φ_J. These shape functions φ_J, corresponding to all nodes of the specific element Ω_e, form the set of shape functions of the element Ω_e. On the other hand, the assembly of the element shape functions associated with node N from all the connected elements forms the base function $\Phi_N = \sum_{e|N\in\Omega_e} \varphi_J$. Obviously the unity property (Eq. 2.12) holds true for the element shape functions too.

We now process the approximation of a variable defined in the material domain by the finite elements. The values of the variable can be approximated by interpolation by using the nodal values of the variable and shape functions. For example, $\tilde{f}(x) = \sum_{J=1} f^J \varphi_J(x)$ is an approximation of $f(x)$ viewed in element Ω_e. This approximation is exact at the nodes. The interpolation of a one-dimensional (1D) example is illustrated in Figure 2.5. The terminology shape function is named after the geometry representation of the interpolation. When assembling the elements in the mesh, the representation of the variable becomes

$$\tilde{f}(X) = \sum_e \sum_{J=1}^{N_e} f^J \varphi_J(x) = \sum_N f^N \Phi_N(x). \tag{2.13}$$

It is exact at the nodes but approximated at the rest points of the domain. Equation 2.13 forms an interpolation for $f(X)$ in Ω.

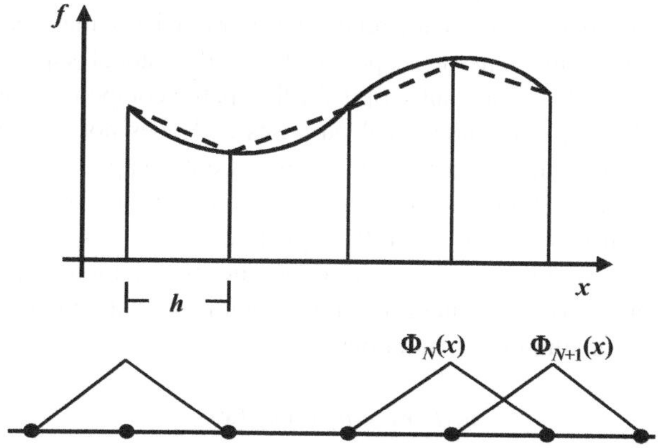

FIGURE 2.5 Interpolation of a function using base functions.

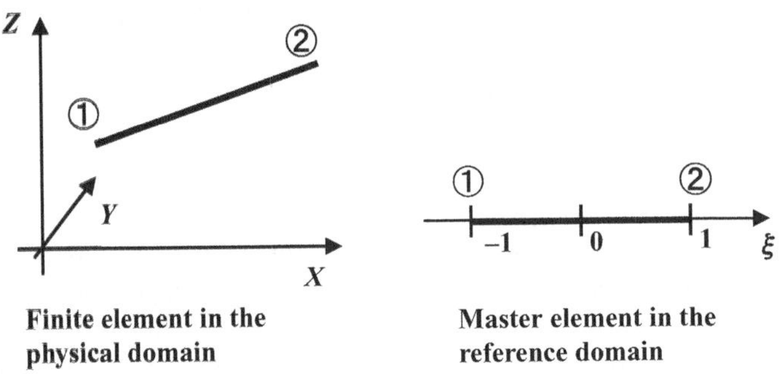

FIGURE 2.6 Element transform.

The calculation related to the variables is basically the operations in each of the elements with these shape functions. The assembly of all the element calculations forms the calculation of the whole domain. Usually the element calculation is based on an element local coordinate system. A transform from the physical domain Ω_e to a reference domain $\hat{\Omega}_e$ is needed, at least for simple illustration. For example, we can use shape functions to construct the transform, as illustrated in Figure 2.6 for a two-node linear element:

$$x = \sum_{J=1}^{N_e} x^J \varphi_J(\xi), \tag{2.14}$$

where N_e is the total number of nodes of the element Ω_e, x represents a generic particle in the element of physical domain, and ξ represents the particle in the reference element $\hat{\Omega}_e$, also called the master element. x^J represents node J in the physical system. We generally do not need to distinguish the functional notation when using the physical parameter x or the reference parameter ξ. Here, both the transform and the interpolation use the same set of functions and the same formula, this type of element is called the isoparametric element.

From here onward, under normal circumstances, we will use the capital subscripts to indicate the base functions or shape functions; the capital superscript for the nodal values; and the lowercase subscripts for the spatial components of mechanical variables. This is simply for notational purposes and does not bear any meaning of tensor's operation with covariant and contravariant components. Here, we use Cartesian systems only. Latin indices from 1 to 3 are used for 3D applications and Greek indices from 1 to 2 are used for 2D applications.

For the transient dynamic system, we introduce the interpolation as approximations for the displacement, velocity, and acceleration, with summation convention on the number of nodes for simplifying notations:

$$\begin{aligned} u_i(t, x) &= u_i^N(t)\Phi_N(x), \\ v_i(t, x) &= v_i^N(t)\Phi_N(x), \\ a_i(t, x) &= a_i^N(t)\Phi_N(x). \end{aligned} \tag{2.15}$$

The approximation assumes the form of separation of variables, which is commonly used in the classical mechanics. The interpolation for these variables use the same base functions and shape functions. The relations of displacement, velocity, and acceleration with time derivatives now come to the nodal values as functions of time:

$$\begin{aligned} v_i^N(t) &= \dot{u}_i^N(t), \\ a_i^N(t) &= \dot{v}_i^N(t) = \ddot{u}_i^N(t). \end{aligned} \tag{2.16}$$

2.3.2 System of Semidiscretization

We also use the interpolation (2.13) for test functions $v_i = v_i^N \Phi_N$, but without time-dependence. We plug these interpolations into the variational equation (2.10) and obtain the discrete form of variational equation:

$$\int_\Omega \left(\rho \Phi_M \Phi_N \ddot{u}_i^N + \sigma_{ij} \Phi_{M,j}\right) v_i^M \mathrm{d}\Omega = \int_\Omega f_i \Phi_M v_i^M \mathrm{d}\Omega + \int_{\Gamma_s} g_i \Phi_M v_i^M \mathrm{d}\Gamma. \tag{2.17}$$

As usual, we drop the arbitrary variables $v_i{}^m$ and derive from (2.17), the equation of motion in the discrete weak form. At any time point, we can rewrite it in a form of the ordinary differential equations in time domain:

$$\begin{aligned} &\sum\nolimits_N \int_\Omega (\rho \Phi_M \Phi_N) \mathrm{d}\Omega\, \ddot{u}_i^N(t) \\ &\quad = -\int_\Omega (\sigma_{ij} \Phi_{M,j}) \mathrm{d}\Omega + \int_\Omega f_i \Phi_M \mathrm{d}\Omega + \int_{\Gamma_s} g_i \Phi_M \mathrm{d}\Gamma. \end{aligned} \tag{2.18}$$

Note that the stress σ_{ij} is a function of the deformation and material properties. Therefore, it contains unknowns u_i and possibly $\dot{u}_i$ but not $\ddot{u}_i$. Equation 2.18 is recognized as one in the form of Newton's second law, written in a matrix form:

$$\begin{aligned} \boldsymbol{F} &= \boldsymbol{M}A, \\ F_i^M &= -\int_\Omega (\sigma_{ij} \Phi_{M,j}) \mathrm{d}\Omega + \int_\Omega f_i \Phi_M \mathrm{d}\Omega + \int_{\Gamma_s} g_i \Phi_M \mathrm{d}\Gamma, \\ M_{MN} &= \int_\Omega (\rho \Phi_M \Phi_N) \mathrm{d}\Omega, \\ A_i^N &= \ddot{u}_i^N(t). \end{aligned} \tag{2.19}$$

It is understood that domain and boundary integrations are decomposed to integrations on the elements.

2.3.3 Discretization in Time by Finite Difference

System (2.19) is a system of second-order ordinary differential equations in time, whether linear or nonlinear. Among various approaches developed for solving this system, here we are only interested in explicit scheme, which uses central difference to approximate the acceleration; see Newmark (1959) for more general discussions. For simplicity, assume that the time domain $[0, T]$ is uniformly divided into N equal subintervals $[t_n, t_{n+1}]$, with $0 = t_0 < t_1 < \cdots < t_N = T$, $t_{n+1} - t_n = \Delta t = T/N$. The velocity and acceleration as time derivatives are approximated by the finite difference method, expressed in the vector form as:

$$\begin{cases} \partial_t \boldsymbol{u}_{n+1/2}^h = \left(\boldsymbol{u}_{n+1}^h - \boldsymbol{u}_n^h\right)/\Delta t \sim \dot{\boldsymbol{u}}_{n+1/2}^h, \\ \partial_{t^2} \boldsymbol{u}_n^h = \left(\partial_t \boldsymbol{u}_{n+1/2}^h - \partial_t \boldsymbol{u}_{n-1/2}^h\right)/\Delta t = \left(\boldsymbol{u}_{n+1}^h - 2\boldsymbol{u}_n^h + \boldsymbol{u}_{n-1}^h\right)/\Delta t^2 \\ \sim \ddot{\boldsymbol{u}}_n^h \sim \left(\dot{\boldsymbol{u}}_{n+1/2}^h - \dot{\boldsymbol{u}}_{n-1/2}^h\right)/\Delta t, \end{cases} \tag{2.20a}$$

$$\begin{cases} \partial_t \boldsymbol{u}_{n+1/2}^h = \partial_t \boldsymbol{u}_{n-1/2}^h + \partial_{t^2} \boldsymbol{u}_n^h \Delta t, \\ \boldsymbol{u}_{n+1}^h = \boldsymbol{u}_n^h + \partial_t \boldsymbol{u}_{n+1/2}^h \Delta t. \end{cases} \tag{2.20b}$$

When the acceleration is approximated by the central difference defined in (2.20), the finite element equation is reduced from (2.19):

$$\partial_{t^2} \boldsymbol{u}_n^h = \boldsymbol{M}^{-1} \boldsymbol{F}_n. \tag{2.21}$$

2.3.4 Procedure of the Explicit Finite Element Method

Conceptually, we can use (2.19) in the form of Newton's second law with current information available at t_n to calculate acceleration explicitly. The external force contributed by the given load conditions in (2.19) is calculated directly. The internal force contributed by stress is calculated by the formulation built in the element and material models. With the central difference method, the approximate acceleration is obtained from (2.21). The velocity and the nodal positions (or displacement) are then calculated by simple algebraic operations described in (2.20). The incremental deformation brings in the increments of strain and stress. It follows that the internal force is obtained. Meanwhile the external force at t_{n+1} is calculated by the prescribed loading conditions. Then the acceleration at time t_{n+1} is subsequently calculated. The schedule of these events is depicted in Figure 2.7. These steps form a cyclic procedure.

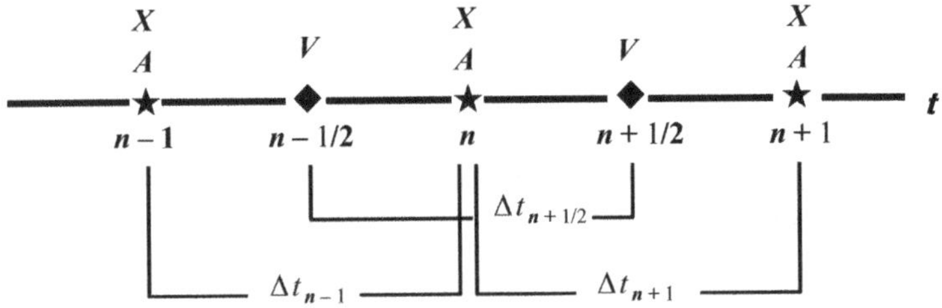

FIGURE 2.7 Events of time integration. (S.R. Wu, A priori error estimates for explicit finite element for linear elasto-dynamics by Galerkin method and central difference method. Elsevier 2003.)

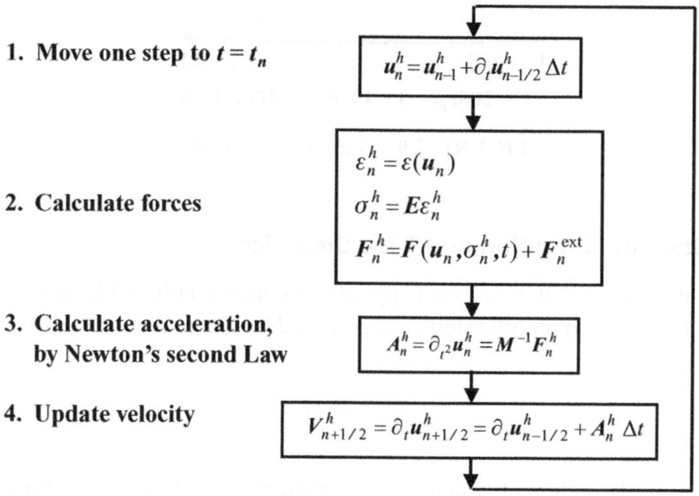

FIGURE 2.8 Explicit finite element procedure. (Reprinted from Wu and Qiu, 2009. Copyright (2008), with permission from Wiley-Blackwell.)

We can design the algorithm of explicit finite element method with the framework shown in Figure 2.8. Here, we use the elastic material model as a demonstration. When nonlinear materials are involved, the process for stress calculation will be updated. When we discuss contact and constraint conditions later in Part 4, the procedure will expand with additional modules. In fact, any of the steps in the cyclic process, illustrated in Figure 2.8, can be a starting point of the program. For example, we can choose to start with the motion due to initial velocity. Early development of explicit finite element method can be found in Belytschko et al. (1975); also in Hughes et al. (1979). Later advances in explicit finite element technology can be found in Belytschko et al. (2000).

2.4 MAIN FEATURES OF THE EXPLICIT FINITE ELEMENT METHOD

As discussed in Section 2.3, the explicit scheme has a simple and neat solution procedure. This procedure can produce a unique solution, as long as all the material models and constraint conditions are well defined. As a matter of fact, every step shown in Figure 2.8 does not have a chance to yield multiple solutions.

In this process, we have graded the physical domain and approximated the geometry by finite element mesh. We have used interpolation and finite difference to approximate the physical variables in space and time domain, respectively. Then, we have formulated the variational equation for weak solution to approximate the partial differential equations.

Here, we discuss few important features that are critical to the explicit scheme.

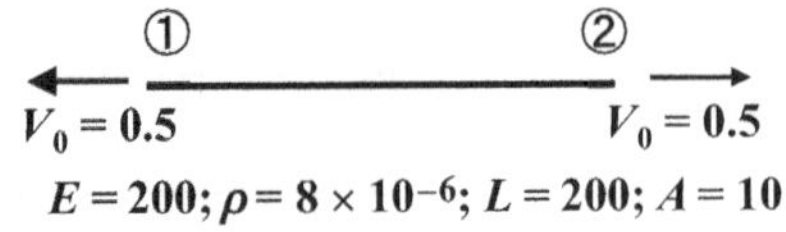

FIGURE 2.9 A truss/rod element.

2.4.1 Stability Condition and Time Step Size

It is well understood that the explicit approach is conditionally stable and has to meet the stability condition on time step size, proposed by Courant et al. (1928):

$$\Delta t < 2/\omega. \tag{2.22}$$

Here, ω is the natural frequency. The application of this rule is illustrated in Flanagan and Belytschko (1981a, 1981b).

We use a simple example, depicted in Figure 2.9, as a demonstration. Consider an elastic rod (truss) element with Young's modulus $E = 200$, Poisson ratio $\nu = 0$, mass density $\rho = 8 \times 10^{-6}$, length $L = 20$, cross-sectional area $A = 10$. The nodal mass $m = \rho AL/2$. Initial velocity $v_0 = 0.5$ is assigned to its two nodes in opposite directions. We can obtain the natural frequency of the element $\omega = 2\sqrt{E/\rho}/L = 500$ and the critical time step size $\Delta t = 2/\omega = 4 \times 10^{-3}$.

For this example, we can easily get the analytical solution $u_2(t) = v_0 \sin(\omega t)/\omega = -u_1(t)$. It is a vibration with magnitude $v_0/\omega = 10^{-3}$ and period $T = 2\pi/\omega$. The displacement u_2 calculated with the critical time step 4×10^{-3} is shown in Figure 2.10a. It diverges quickly. The numerical result with a time step size 4×10^{-4}, 10% of the critical value, is presented in Figure 2.10b. It is quite accurate. Note that the critical time step size is only for controlling the stability. For better accuracy, we may need to use smaller time steps.

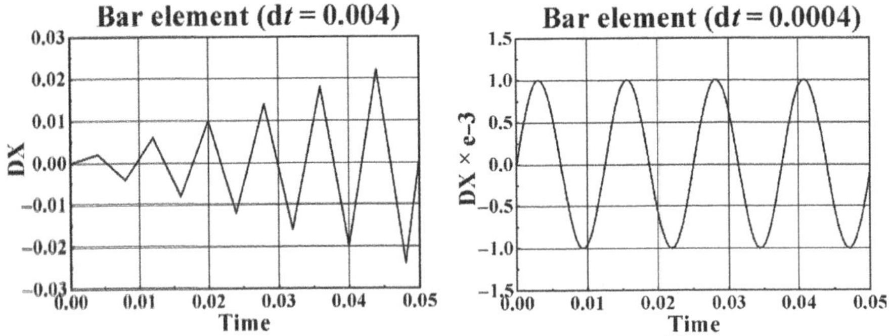

(a) (b)

FIGURE 2.10 Stability due to time step size: (a) $dt = 0.004$; (b) $dt = 0.0004$.

The computation based on the procedure in Figure 2.8 gives the following results:

$$u_1(t_1) = -v_0\Delta t, \quad u_2(t_1) = v_0\Delta t,$$
$$\varepsilon = (u_2(t_1) - u_1(t_1))/L = 2v_0\Delta t/L, \quad \sigma = E\varepsilon, \quad f = A\sigma = 2v_0\Delta tEA/L,$$
$$a_1(t_1) = f/m, \quad a_2(t_1) = -f/m,$$
$$v_2(t_{1.5}) = v_0 + a_2\Delta t = v_0(1 - 2\Delta t^2 EA/Lm) = v_0(1 - 4\Delta t^2 E/\rho L^2),$$
$$u_2(t_2) = u_2(t_1) + v_2(t_{1.5})\Delta t = u_2(t_1)(2 - \omega^2\Delta t^2).$$

If $\Delta t \geq 2/\omega$, then $u_2(t_2) \leq -2u_2(t_1)$. This means that during calculation, the displacement of moving one-step forward results in the displacement of moving more than two-steps backward in one cycle. The system diverges quickly, as shown in Figure 2.10a.

For a system model, the stability condition needs to evaluate the maximum frequency of the whole system. This is not practical for explicit procedure, not only due to the computational cost but also the lack of eigen solver in the explicit code. It is shown in Flanagan and Belytschko (1981a) that $\omega_{\max}^{\text{sys}} \leq \omega_{\max}^{\text{ele}}$, that is, $\Delta t_{\min}^{\text{ele}} \leq \Delta t_{\text{critical}}$. Therefore, if using the minimum element time step for the system computation, the stability condition will be satisfied on the conservative side. Several approaches can improve the efficiency of computation, such as subcycling technique, cf. Belytschko et al. (1979), Hulbert and Hughes (1988).

2.4.2 Diagonal Mass Matrix

The diagonal mass matrix is one of the important features that makes the explicit method efficient and practical. When using diagonal mass matrix, the step to calculate acceleration by applying Newton's second law in (2.21) is reduced to a simple division without the need of inverting the mass matrix. The diagonal mass matrix could be a direct engineering intuition of mass discretization by lumping to the nodal points. It can be formulated by row or column summation as an approximation, which is easy for software implementation. It can also be derived from using methods such as orthogonal basis functions, which is straightforward; under integration, cf. Andreev et al. (1992) and Cohen et al. (1994); lower order finite element interpolation in the calculation of kinetic energy, cf. Tong et al. (1971); and nonconforming piecewise constant interpolation, cf. Fujii (1972). While the diagonal mass matrix is optional to the eigen problems or the implicit schemes, it is essential to the explicit finite element method for transient dynamics. Without it, the explicit scheme would lose its efficiency and would not have enjoyed its success today.

Recall the mass matrix defined in (2.19), the row (or column) summation gives

$$(\boldsymbol{M}_d)_{MN} = \delta_{MN} \sum_J \int_\Omega (\rho\Phi_M\Phi_J)\mathrm{d}\Omega = m_M\delta_{MN}. \tag{2.23}$$

Here, δ_{MN} is the notation of Kronecher delta, and there is no summation on index M. The last step is due to the unity property (2.12) with

$$m_M = \int_\Omega \rho \Phi_M \mathrm{d}\Omega. \tag{2.24}$$

The matrix in (2.23) is named diagonal mass matrix, or lumped mass matrix. Correspondingly, the matrix in (2.19) is called consistent mass matrix denoted by $\boldsymbol{M}_c$, since it uses the same interpolations and quadrature as used for the force calculation. We will continue the discussion later in Section 2.5.3.

2.4.3 Corotational Stress

During large deformation, the element and the associated local coordinate system change their positions and orientations with respect to the global system. The theory of continuum mechanics requires stress objectivity, typically achieved by using transform with Jaumann or other stress rate. Argyris (1964, 1965) initiated a natural approach of corotational stress. The idea is to separate the rigid body motion from the deformation in stress increment calculation, which is always important for large deformation problems. In case if the element local system has negligible change with respect to the nodal positions, then the stress calculated in the element local system is equal to Cauchy stress and satisfies the objectivity. The transformation is therefore saved. Bending dominated deformation is one of the applications with stress objectivity satisfied reasonably. See Belytschko and Hsieh (1973) for more discussions. This approach, however, is not suitable when large shear deformation is expected.

2.5 ASSESSMENT OF EXPLICIT FINITE ELEMENT METHOD

In this section, we summarize the theoretical achievements on error estimation of explicit finite element methods.

2.5.1 About the Solution of the Elastodynamics

For the second-order hyperbolic equation, Lions and Magenes (1972) proved the existence and uniqueness of the solution for a general class of boundary conditions and initial conditions. To ensure uniqueness without rigid body motion, the displacement should be prescribed on part of the boundary with positive measure. Evans (1998) demonstrated the regularity and higher regularity of solution with requirement on smoothness and compatibility of data.

To study the elasticity problem, the ellipticity of the stress operator is a critical factor. This property has been established by using Kohn's inequality, cf. Ciarlet (1988). With these preliminaries, the existence and uniqueness statements can be extended to the general three-dimensional (3D) (including lower space) elastodynamics.

2.5.2 A Priori Error Estimate of Explicit Finite Element Method for Elastodynamics

Since Clough (1960) proposed the terminology of finite element method, engineering applications of finite element method have accelerated. Mathematicians have since participated and contributed to the technology by assessing its convergence behavior. ZlÁmal (1968) is recognized as the first study applicable to general analysis, which was developed within the same framework as the modern theory of error analysis. The theories regarding convergence and error estimate have quickly covered many aspects of elliptic problems and extended to parabolic and hyperbolic problems. Among vast publications, Strang and Fix (1973) and also Oden and Reddy (1976) represent the early development. Ciarlet and Lions (1991) is an excellent reference presenting a collection of the mathematical achievements for finite element methods. Ainsworth and Oden (2000), Babuska and Strouboulis (2001), and Brenner and Scott (2002) present more information about the new development. The mathematical developments have provided strong support for the successful engineering applications of the finite element methods.

Generally speaking, for elliptic problems, if certain conditions for the data are satisfied (an ideal situation) and also when mesh is refined, the error of the displacement will converge to zero. This is the minimal acceptance requirement for the numerical algorithm. Usually, it is expressed as the boundedness of error, which is the difference between the finite element solution u^h and the exact solution u, in terms of the power of element size h. The power index is named the rate of convergence, expressed below, for example,

$$||u^h - u||_{H^m} \leq Ch^{k+1-m}||u||_{H^{k+1}}. \tag{2.25}$$

where h is the maximum element size, m is the order of the differential operator in the variational equations (for elliptic problem, usually $2m$ represents the order of differential equations), and k is the highest degree of complete polynomials employed in the interpolation. The constant C in (2.25) is independent of the solution and the element size but dependent of the load and boundary conditions. The error is measured by using Sobolev norms, defined as

$$||u||_{H^m} = \sqrt{\int \left(u^2 + \sum_{j=1}^{p} \left(\frac{\partial^j u}{\partial x^{\alpha_x} \partial y^{\alpha_y} \partial z^{\alpha_z}} \right)^2 \right) \mathrm{d}\Omega}, \quad p = 1, \ldots, m(\geq 1)$$

$$\text{all combinations with } \alpha_x + \alpha_y + \alpha_z = j, \tag{2.26}$$

$$||u||_{L^2} = \sqrt{\int u^2 \mathrm{d}\Omega},$$

$$||u||_{L^\infty} = \sup(|u|).$$

For the second-order elliptic problem, whose corresponding variational equation contains the first derivatives only ($m = 1$), when linear elements are used ($k = 1$), from (2.25), the error is bounded by

$$\begin{aligned} ||u^h - u||_{H^1} &\leq Ch, \\ ||u^h - u||_{L^2} &\leq Ch^2. \end{aligned} \tag{2.27}$$

This is the same rate of convergence by piece-wise linear interpolation of (2.13), and is therefore the best we can expect. Hence, it is named the optimal convergence rate. The first estimate in (2.27) can be derived by using energy method. The second estimate in (2.27), measuring error in L^2-norm, is developed by using Aubin–Nitsche method. The error estimate theories for complex situations, such as curved boundary and crack, are also established. For details, we are referred to Ciarlet and Lions (1991) and the references quoted there.

In transient structural dynamics, the spatial differential operator corresponds to a static one of elliptic problem. Many results of elliptic problems are valuable resources for the transient dynamics. The a priori error estimate for transient dynamics was first reported by Dupont (1973). Newmark method was used for the time derivatives in the framework of implicit scheme and a static projection with the elliptic operator was introduced. The main conclusion is cited below:

Assessment 2.1

$$||u^h - u||_{L^\infty(L^2(\Omega))} + ||\dot{u}^h - \dot{u}||_{L^\infty(L^2(\Omega))} \leq C(h^{k+1} + \Delta t^2). \tag{2.28}$$

Note that the index k in Dupont (1973) corresponds to k+1 in our discussion here, where the following notations are used for time-dependent problems:

$$\begin{aligned} ||v(x,t)||_{L^\infty(0,\mathrm{T};L^2)} &= ||v(x,t)||_{L^\infty(L^2(\Omega))} = \sup_t \{||v(x,t)||_{L^2(\Omega)}\}, \\ ||v(x,t)||_{L^\infty(0,\mathrm{T};H^1)} &= ||v(x,t)||_{L^\infty(H^1(\Omega))} = \sup_t \{||v(x,t)||_{H^1(\Omega)}\}. \end{aligned} \tag{2.29}$$

For discretization in time, denote $||v(x,t)||_{\tilde{L}^\infty(0,\mathrm{T};H^1)}$ for $\sup_n\{||v(x,t_n)||_{H^1}\}$, etc. Optimal convergence rate can be obtained for displacement and velocity measured in $L^\infty(L^2(\Omega))$ norm, if the time step is small.

The static projection was also employed by Oden and Reddy (1976) with the central difference method in the framework of explicit integration. Similar to Dupont (1973)'s results, optimal convergence rate can be obtained,

Assessment 2.2

$$||u^h - u||_{\tilde{L}^\infty(L^2(\Omega))} + ||\dot{u}^h - \dot{u}||_{\tilde{L}^\infty(L^2(\Omega))} \leq C(\mathrm{err\_0} + h^{k+1} + \Delta t^2). \tag{2.30}$$

Here, err_0 is determined by the initial conditions and their interpolations. Optimal convergence rate is achievable when err_0 and Δt are small. Both investigations directly obtained results of displacement and velocity in L^2 norm but not for H^1 norm. This is different from the analysis for elliptic problems.

In light of Oden and Reddy (1976), the a priori error estimate for the explicit finite element method of 3D elastodynamics was developed by Wu (2003). The error estimates were also extended to H^1 norm for both displacement and velocity. For central difference approximation of the time derivatives, the error also depends on how the initial conditions are approximated. The error brought by the approximation propagates during the whole process. The results are summarized in the following,

Assessment 2.3 *Under certain conditions, the error of explicit finite element solution for the transient dynamics is bounded and described below:*

$$\begin{aligned}
||\boldsymbol{u}_n^h - \boldsymbol{u}_n||_{\tilde{L}^\infty(L^2(\Omega))} &\leq C(h^{\min(k+1,2)} + \Delta t^2), \\
||\boldsymbol{u}_n^h - \boldsymbol{u}_n||_{\tilde{L}^\infty(H^1(\Omega))} &\leq C(h^{\min(k,2)} + \Delta t^2), \\
||\partial_t \boldsymbol{u}_n^h - \dot{\boldsymbol{u}}_{n+1/2}||_{\tilde{L}^\infty(L^2(\Omega))} &\leq C(h^{\min(k+1,2)} + \Delta t^2), \\
||\partial_t \boldsymbol{u}_n^h - \dot{\boldsymbol{u}}_{n+1/2}||_{\tilde{L}^\infty(H^1(\Omega))} &\leq C(h^{\min(k+1,2)} + \Delta t^2)h^{-1}.
\end{aligned} \tag{2.31}$$

It is worth noting that in this book we do not discuss the details of required conditions and all the complex mathematical details about how to derive the theories. They can be found in the quoted references and may be too mathematical to be of interest to engineers. However, should special attention arises, additional explanations will be provided.

The conditions mentioned are related to the body force, boundary conditions, and initial conditions. For example of the impact problem, the material body moves with a constant initial velocity without initial displacement. Hence, there is no error due to the approximation for the initial conditions and boundary conditions except the contact conditions and certain type of constraint conditions, which will be discussed later. Another situation is a kind of metal forming, where the sheet metal rests on the die at the beginning, which may bring in little initial deformation with zero initial velocity but subjected to impact loading. Although these examples are all strongly nonlinear, in large varieties of engineering applications with explicit finite element, usually the conditions needed for Assessment 2.3 can be satisfied. It is worth mentioning that great efforts are involved for the challenging theoretical assessment for nonlinear transient dynamics. It may be wise to first work out the associated static equilibrium problems. Even for static analysis, there is, however, no general theory available yet on accuracy of finite element method for large deformation with nonlinear material.

Many engineering applications using explicit finite element software involve linear elements only. Correspondingly, in (2.31), the index $k = 1$. In addition, the consideration of stability requires reducing the time step when mesh is refined, usually and naturally, in proportion to the element size. Then the terms related to h and Δt in (2.31) are of the same order. The optimal rates of convergence can be achieved as following:

Assessment 2.4 *Under certain conditions,*

$$||\boldsymbol{u}_n^h - \boldsymbol{u}_n||_{\tilde{L}^\infty(L^2(\Omega))} + ||\partial_t \boldsymbol{u}_n^h - \dot{\boldsymbol{u}}_{n+1/2}||_{\tilde{L}^\infty(L^2(\Omega))} \leq Ch^2,$$
$$||\boldsymbol{u}_n^h - \boldsymbol{u}_n||_{\tilde{L}^\infty(H^1(\Omega))} + ||\partial_t \boldsymbol{u}_n^h - \dot{\boldsymbol{u}}_{n+1/2}||_{\tilde{L}^\infty(H^1(\Omega))} \leq Ch. \tag{2.31a}$$

To demonstrate the convergence behavior of the explicit finite element method, we investigate an example of the axial small vibration of a uniform elastic rod, which was studied by Wu (2003). The governing equation, boundary conditions, and initial conditions are defined below:

$$\begin{aligned}
\rho\ddot{u} - Eu'' &= 0,\\
u(t,0) &= 0,\\
u(t,a) &= 0,\\
u(0,x) &= 0,\\
\dot{u}(0,x) &= \psi(x) = v_0(xa^3 - 2ax^3 + x^4)/a^4
\end{aligned} \tag{2.32}$$

According to Evans (1998), the initial velocity is smooth enough to ensure a smooth solution:

$$\begin{aligned}
u &= \sum \frac{\psi_m}{\omega_m}\sin(\alpha_m x)\sin(\omega_m t),\\
\alpha_m &= m\pi/a,\\
\omega_m &= \alpha_m C, \quad C = \sqrt{E/\rho},\\
\psi_m &= (1-(-1)^m)\frac{48v_0}{(m\pi)^5}.
\end{aligned} \tag{2.33}$$

The solution satisfies the fundamental requirements, so that the error estimation (2.31) is valid. Linear truss elements with only axial response are used to model this 1D problem. The parameters are $E = 200$ (kN/mm^2), $\rho = 8.0 \times 10^{-6}$ (kg/mm^3), $a = 1000$ (mm), and $v_0 = 1.0$ (mm/ms). A set of uniform meshes are used for computation and the results are compared with the analytical solution (2.33). At the center point $x = 500$, the numerical error of the displacement from $t = 99$ to 100 (ms) is depicted in Figure 2.11. It is observed that when mesh is refined, the error reduces and tends to zero. Furthermore, with a refinement of one-to-two split, the element size reduces to half and the error of the displacement roughly reduces to 1/4. The errors of the displacement and velocity at $t = 9.96$, measured in L^2 and H^1 norms

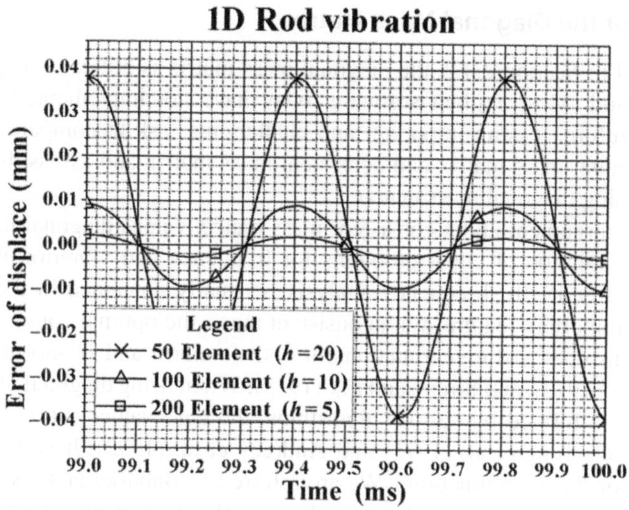

FIGURE 2.11 Numerical error of the displacement at mid-span of the rod. (S.R. Wu, A priori error estimates for explicit finite element for linear elasto-dynamics by Galerkin method and central difference method. Elsevier 2003.)

are shown in Figure 2.12. The slopes of the error curves plotted in the log–log scale match the rates of convergence by the error estimation (Eq. 2.31a) asymptotically. It is worth mentioning that for this example, high precision of data, including input and output, is needed to obtain results with high resolution.

The error estimation for certain types of nonlinear hyperbolic equations has also been derived. For example, see Oden and Fost (1973) and Yuan and Wang (1985).

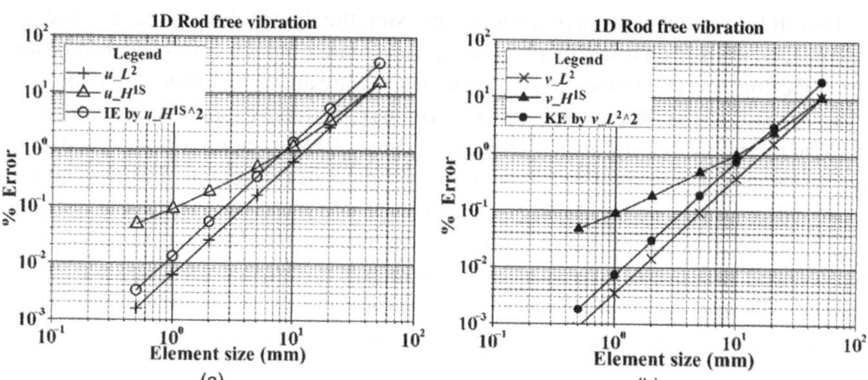

FIGURE 2.12 Approximation errors of the rod free vibration at 9.96 ms: (a) displacement and internal energy; (b) velocity and kinetic energy. (S.R. Wu, A priori error estimates for explicit finite element for linear elasto-dynamics by Galerkin method and central difference method. Elsevier 2003.)

2.5.3 About the Diagonal Mass Matrix

As mentioned in Section 2.4.2, the diagonal mass matrix is critical to explicit finite element method. Most explicit software today have not implemented the matrix inverter. Therefore, it is an important and challenging task to compare the results by using diagonal mass matrix and consistent mass matrix and assess the accuracy affected by using diagonal mass matrix.

In fact, the argument about consistent mass matrix versus diagonal mass matrix regarding the solution of eigen problems for frequency and vibration mode dates back to the early 1970s. Frequency analysis is usually contained in the structural design as an important feature. With consistent mass, the optimal convergence rates for both eigenvalues and eigenfunctions in both L^2-norm and H^1-norm have been proved, cf. Fix (1973), and Strang and Fix (1973). When using diagonal mass matrix, which might be formed by various algorithms, the eigenvalues can converge with optimal rate, cf. Tong et al. (1971) and Andreev et al. (1992). It is, however, not always true for the eigenfunctions. We are referred to Babuška and Osborn (1991) for more in-depth investigations. In a simple example of eigenvalue problem like the rod discussed in Section 2.5.2, it was shown in Babuška and Osborn (1991) that the methods using consistent mass and diagonal mass have the same convergence rates. In addition, the frequencies by consistent mass converge from above, whereas the frequencies by lumped mass converge from below.

For explicit finite element method for elastic transient dynamics, by investigating the contributions by the difference of the two mass matrices, Wu (2006) proved the following:

Assessment 2.5 *Under certain conditions, the displacement and velocity obtained by using diagonal mass matrix can have the same optimal convergence rate by using consistent mass matrix.*

Therefore, there is no loss of accuracy in using the diagonal mass matrix. The assessment was based on linear elements only. As a matter of fact, currently most (if not all) of the engineering applications of explicit software are using linear elements only.

To investigate the effect of diagonal mass matrix, an iterative approach of conjugate gradient method was employed by Song and Duan (1998) to approximate the inverse of consistent mass matrix. For axisymmetric problem of impact into a solid material, improved results by using consistent mass were observed. An iterative procedure based on the concept of matrix splitting was reported in Wu and Gu (2003) to approximate the inverse of consistent mass matrix or a mixed one. The process is illustrated below, and is applicable for comparing the results of general nonlinear transient analysis. The mixed mass matrix is defined with a parameter $\alpha \in [0, 1]$:

$$\boldsymbol{M}^{\text{mix}} = (1-\alpha)\boldsymbol{M}^{d} + \alpha\boldsymbol{M}^{c} = \boldsymbol{M}^{d} + \alpha\boldsymbol{M}^{f}. \tag{2.34}$$

Obviously, $\alpha = 0$ represents a diagonal matrix and $\alpha = 1$ represents a consistent matrix. The inverse of the mixed mass matrix and the nodal acceleration can be calculated by

$$\begin{aligned}(\boldsymbol{M}^{\text{mix}})^{-1} &= \left(\boldsymbol{I} + \alpha(\boldsymbol{M}^d)^{-1}\boldsymbol{M}^f\right)^{-1}(\boldsymbol{M}^d)^{-1} \\ &= \left(\boldsymbol{I} - \alpha(\boldsymbol{M}^d)^{-1}\boldsymbol{M}^f + (-\alpha(\boldsymbol{M}^d)^{-1}\boldsymbol{M}^f\right)^2 + \cdots)(\boldsymbol{M}^d)^{-1}, \\ \ddot{\boldsymbol{u}} &= (\boldsymbol{M}^{\text{mix}})^{-1}\boldsymbol{F} = \ddot{\boldsymbol{u}}_0 + \ddot{\boldsymbol{u}}_1 + \ddot{\boldsymbol{u}}_2 + \cdots,\end{aligned} \tag{2.35}$$

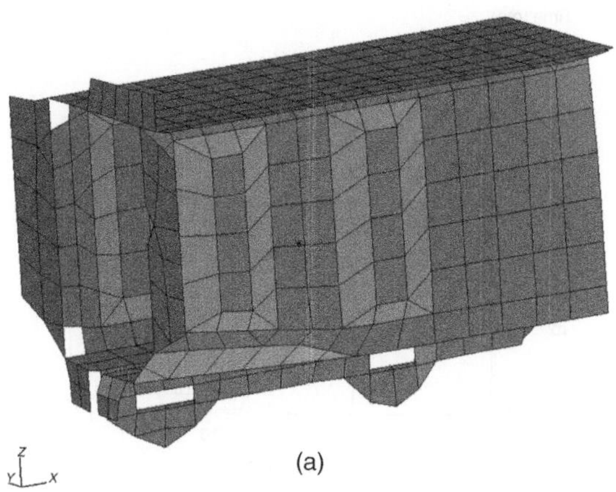

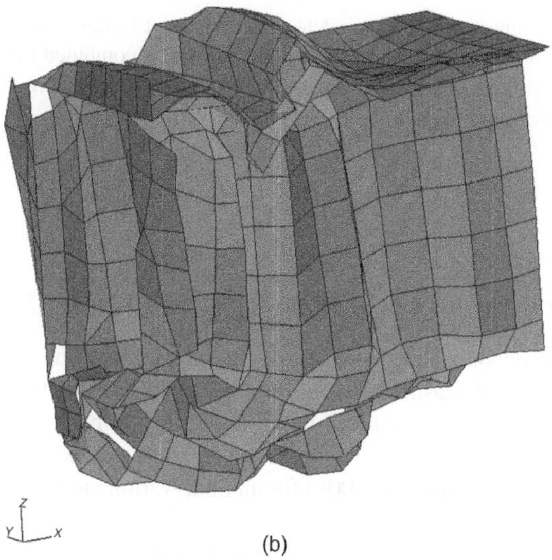

FIGURE 2.13 Crash can: (a) undeformed geometry; (b) deformation after 5 ms impact. (Reprinted from Wu and Qiu, 2009. Copyright (2008), with permission from Wiley-Blackwell.)

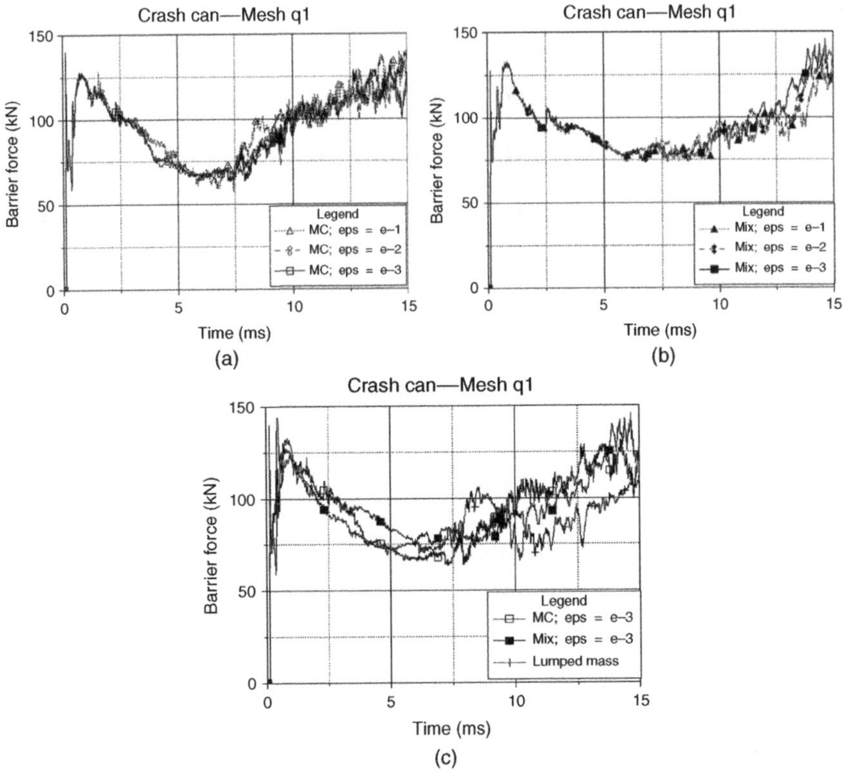

FIGURE 2.14 Component impact model—barrier impact forces calculated from a coarse mesh: (a) by consistent mass, with various tolerances; (b) by combined mass ($\alpha = 0.5$), with various tolerances; (c) results by various mass matrices. (Reprinted from Wu and Qiu, 2009. Copyright (2009), with permission from Wiley-Blackwell.)

where we recursively have: for $k = 1, 2, 3, \ldots,$

$$\ddot{\boldsymbol{u}}_0 = (\boldsymbol{M}^d)^{-1}\boldsymbol{F},\ \ddot{\boldsymbol{u}}_k = (-\alpha(\boldsymbol{M}^d)^{-1}\boldsymbol{M}^f)^k\ddot{\boldsymbol{u}}_0 = (-\alpha(\boldsymbol{M}^d)^{-1}\boldsymbol{M}^f)\ddot{\boldsymbol{u}}_{k-1}.$$

The iterative procedure discussed above was proved to converge by Wu and Qiu (2009), so that it could reasonably represent the inverse of consistent mass matrix or a mixed one. Before we discuss in-depth about the element technology, material model, and contact algorithm, we would like to present two examples studied in Wu and Gu (2003), also Wu and Qiu (2009) for an observation on the performance of the iterative procedure.

The first example is the impact of an aluminum structural component onto a rigid barrier. The aluminum component is about 300 mm long, with the cross section of the size about 150 mm × 110 mm, and is moving with a speed of 15 mm/ms and with

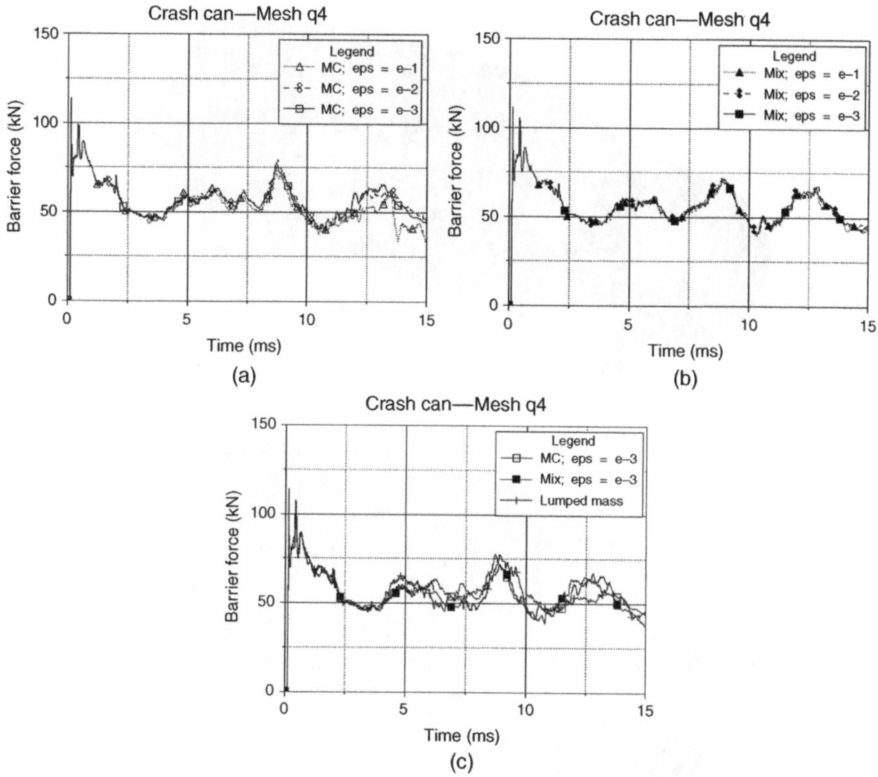

FIGURE 2.15 Component impact model—barrier impact forces calculated from a fine mesh: (a) by consistent mass, with various tolerances; (b) by combined mass ($\alpha = 0.5$), with various tolerances; (c) results by various mass matrices. (Reprinted from Wu and Qiu, 2009. Copyright (2008), with permission from Wiley-Blackwell.)

an attached mass of 500 kg. The component is modeled by shell elements, which will be discussed later. The geometry of the component before and after impact computed by using the diagonal mass matrix is depicted in Figure 2.13.

The iterative process uses a tolerance for controlling the relative difference on iterative results. The barrier impact force calculated by the approximate consistent mass matrix and mixed mass matrix with $\alpha = 0.5$ using a coarse mesh is shown in Figure 2.14. The results by using tolerance range from 0.001 to 0.1 are close in both cases. The iterative process is believed to work properly. The difference in results by using different mass matrices however is noticeable. A refined mesh is then used for further investigation. It is observed from Figure 2.15 that the difference in results by using different tolerance levels is also small, and the difference by using different matrix, including the usual diagonal mass matrix is also small.

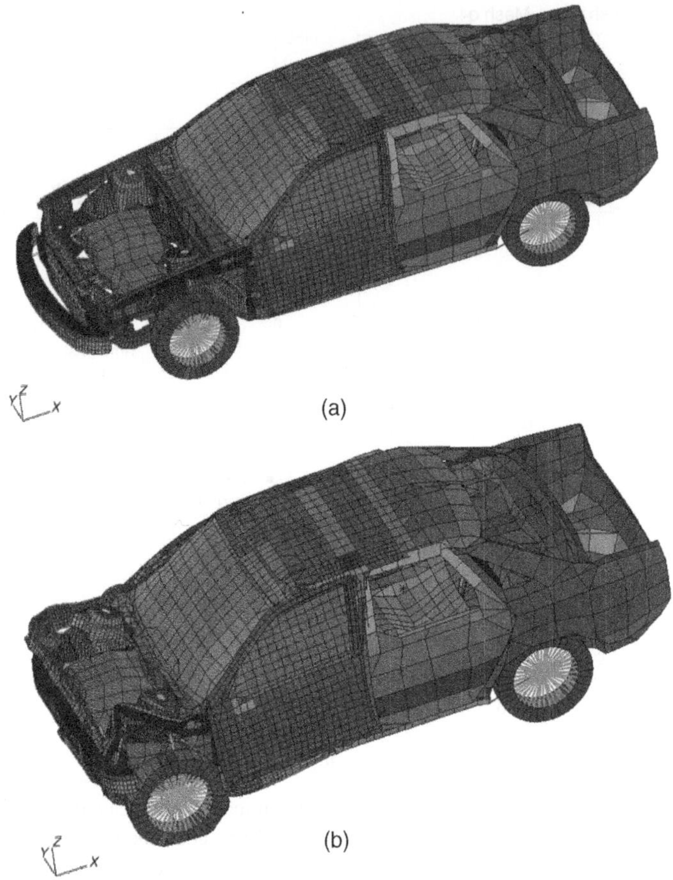

FIGURE 2.16 A vehicle model: (a) before impact; (b) after 100 ms impact. (Reprinted from Wu and Qiu, 2009. Copyright (2008), with permission from Wiley-Blackwell.)

The second example is a vehicle front impact. The geometry and deformation after impact are presented in Figure 2.16. The results produced by using various mass matrices are shown in Figure 2.17. For the iterative process, only the results produced by using tolerance = 0.001 are used for comparison. Some differences in results are observed. Since refined mesh is not available for this example, the comparison is not yet conclusive. For applications of impact engineering, a type of transient dynamics with large deformation, this amount of difference can be considered as noncritical, particularly when only a quick evaluation is needed for a concept design with many details not yet available.

The numerical experiments show that when the same tolerance is used, the iterative process for a mixed mass matrix converges faster than the consistent mass matrix.

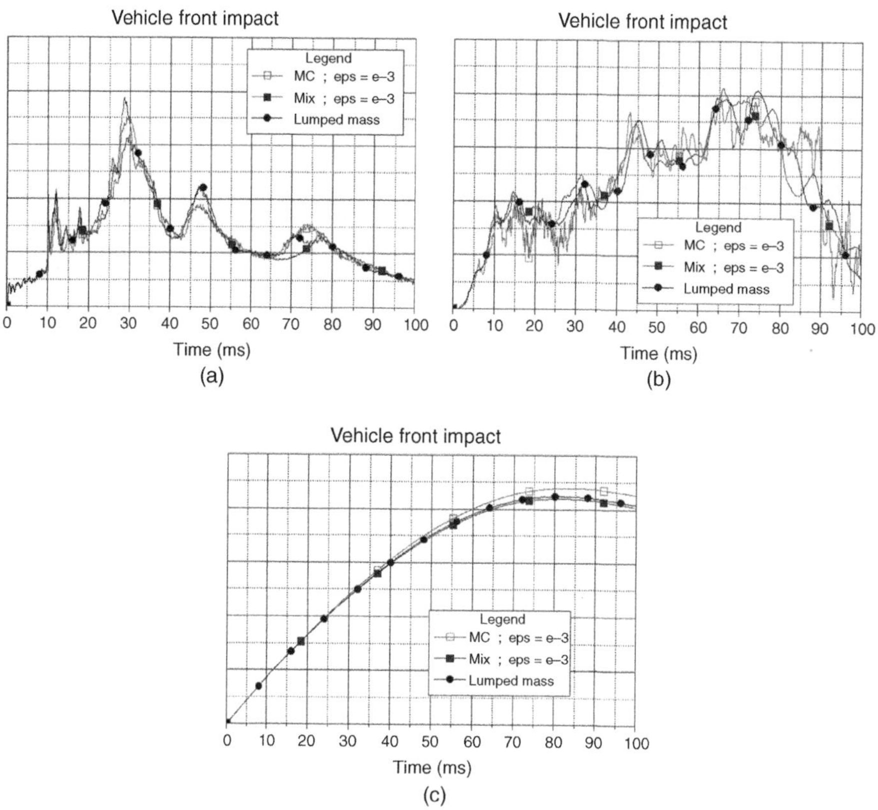

FIGURE 2.17 Vehicle impact model—results by using various mass matrices: (a) barrier impact forces; (b) deceleration at B-pillar; (c) displacement at B-pillar. (Reprinted from Wu and Qiu, 2009. Copyright (2008), with permission from Wiley-Blackwell.)

With tolerance = 0.001 used in these two examples, the computing time needed for consistent mass matrix and mixed mass matrix is nearly ten times and three times of that by using diagonal mass matrix, respectively. It is also found that for stability requirement and the control of oscillatory noises, these mass matrices need smaller time steps than what the diagonal mass matrix needs.

PART II

ELEMENT TECHNOLOGY

CHAPTER 3

FOUR-NODE SHELL ELEMENT (REISSNER–MINDLIN PLATE THEORY)

Many structures are of thin-walled type, named plates or shells, whose thickness is much smaller than the lateral spans. It usually has the advantages of high load capacity versus weight ratio or high load capacity versus the quantity of structural material used, particularly in the applications with complex geometry. The body structures of airplanes, automobiles, buildings, and ships are just a few examples. In applications, the shell structures are designed to carry sustainable loads and do not allow large deformation. The research focuses on strength and stability. In many of these cases, the shells' ability of supporting load is mainly determined by the bending stiffness. On the other hand, in some applications, large bending deformation is the desired mode and contributes to the main part of strain energy.

The development of shell theory and numerical method for shell structures has become one of the central themes of research. The complexity and variety of shell/plate theories also bring challenges to the numerical methods. The shell theory is basically a mathematical model of two-dimensional (2D) configuration to characterize the behavior of thin-walled three-dimensional (3D) structures. However, various difficulties and challenges are embedded in the shell model. Various theories for shell structures have been proposed in the long history of development, cf. Volmir (1956), Timoshenko and Woinowsky-Krieger (1959), and Flügge (1960). Various finite element methods for plates and shells have also been developed. Argyris (1964), Gallagher (1975), Bathe and Wilson (1976), Zienkiewicz (1977), Bathe (1982), Hughes (1987), and Belytschko et al. (2000) are good source for references.

Introduction to the Explicit Finite Element Method for Nonlinear Transient Dynamics, First Edition.
Shen R. Wu and Lei Gu.

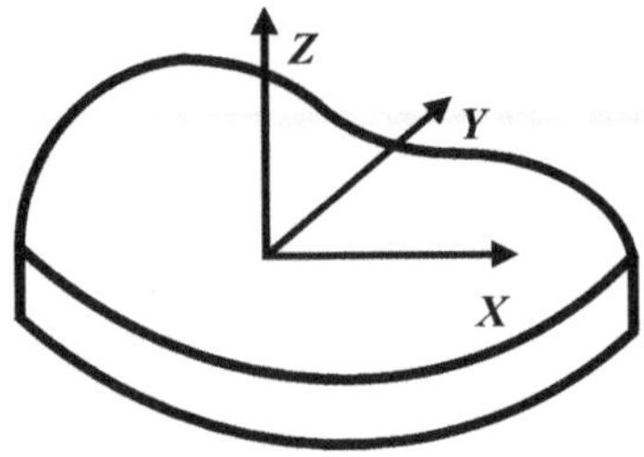

FIGURE 3.1 Configuration of a plate.

For historical commentary, we refer to Stolarski et al. (1995) and many references quoted there. For theories of partial differential equations about plates and shells, and mathematical theory on convergence and error estimates of the plate and shell elements, we refer to Bernadou (1993, 1996), Ciarlet (2000), and Chapelle and Bathe (2003). In this chapter, we will discuss a few types of shell elements for applications of explicit finite element method. Note that the shell elements implemented in the commercial explicit finite element software are essentially based on plate theory, without curvatures as the variables. But these elements can be used to analyze the curved shell structures, provided that the geometry and deformation are modeled appropriately.

3.1 FUNDAMENTALS OF PLATES AND SHELLS

3.1.1 Characteristics of Thin-walled Structures

For simplicity, consider a flat plate and establish a Cartesian coordinate system for it. Let X_1 and X_2 be parallel to the plate surface, and X_3 be perpendicular to the surface, depicted in Figure 3.1. When bending happens, rotation is involved in the material deformation. Consider a cross section of the plate, shown in Figure 3.2.

The displacement of a generic point with distance z to the mid-surface (as a reference surface) can be considered as the combination of the displacement of the

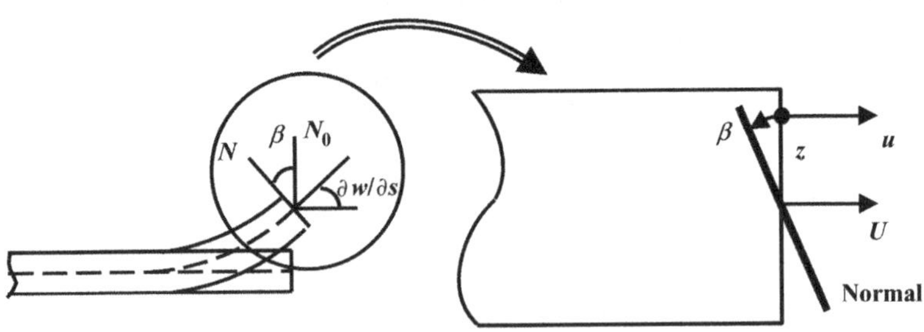

FIGURE 3.2 Deformation and rotation of the normal at a cross section of the plate.

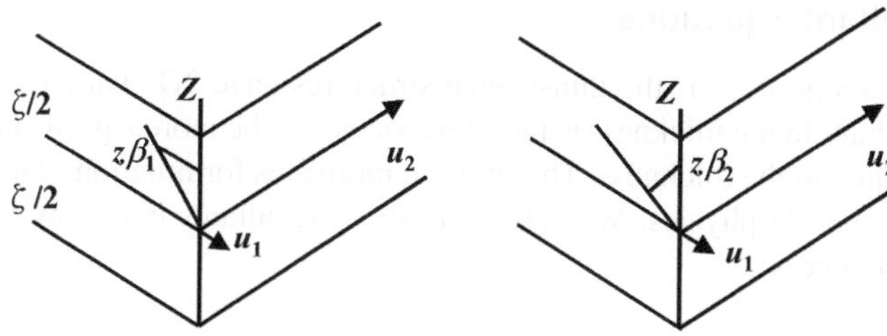

FIGURE 3.3 Displacement of a generic point. (S.R. Wu, A priori error estimation of a 4-node Reissner-Mindlin plate element for elasto-dynamics. Elsevier 2005.)

mid-surface and the rotation of the normal of the mid-surface:

$$U = u - z\beta. \tag{3.1}$$

Here, u represents the displacement at mid-surface and β represents the rotation of the normal, following the bending action of the plate.

For bending dominated problem, it is observed that the deformation in the thickness is small comparing to the thickness. In small deformation theory, it is further assumed that the deflection is small comparing to the thickness. As an approximation to the displacements, we assume that the normal is kept straight during deformation. Then, extending from (3.1), the displacement of a generic point is expressed as

$$\begin{aligned} U_\alpha &= u_\alpha - z\beta_\alpha, \quad \alpha = 1, 2, \\ U_3 &= w. \end{aligned} \tag{3.2}$$

Here, u_α, w, and β_α are functions of X_1 and X_2, independent of $X_3 = z$, illustrated in Figure 3.3. Sometimes, this type of geometry model is considered as first-order approximation of plates and shells, due to the linear form in thickness.

For large deformation, a rate form of (3.2) applicable to the velocity and acceleration replaces (3.2) for displacement:

$$\begin{aligned} &V_\alpha = \dot{U}_\alpha = \dot{u}_\alpha - z\dot{\beta}_\alpha, \quad A_\alpha = \ddot{U}_\alpha = \ddot{u}_\alpha - z\ddot{\beta}_\alpha, \quad \alpha = 1, 2, \\ &V_3 = \dot{U}_3 = \dot{w}, \quad A_3 = \ddot{U}_3 = \ddot{w}. \end{aligned} \tag{3.2a}$$

When investigating the load mainly in the nature of pressure or weight, we assume that the loading is perpendicular to the surface. It is also commonly assumed that the stress in the thickness direction is smaller than that in the other two directions. Hence, we consider the stress state with

$$\begin{aligned} &\sigma_{33}\Big|_{-\zeta/2}^{+\zeta/2} = q\Big|_{-}^{+} = q, \\ &\sigma_{\alpha 3}|_{\pm\zeta/2} = 0, \quad \alpha = 1, 2, \\ &\sigma_{33} << 1. \end{aligned} \tag{3.3}$$

Here, ζ represents the thickness, with the system set at the mid-surface.

3.1.2 Resultant Equations

As briefed in Section 3.1.1, the thin-walled structures have 2D characteristics. The action of resultant in the thickness is therefore viewed to be more significant than the action distributed in the thickness. Therefore, it motivates formulation of a 2D model reduced from the 3D physics. We define the stress resultants below by integrating through the thickness:

$$
\begin{aligned}
N_{\alpha\beta} &= \int_{-\zeta/2}^{\zeta/2} \sigma_{\alpha\beta}\mathrm{d}z, \\
M_{\alpha\beta} &= \int_{-\zeta/2}^{\zeta/2} z\sigma_{\alpha\beta}\mathrm{d}z, \qquad \alpha, \beta = 1, 2, \\
Q_{\alpha} &= \int_{-\zeta/2}^{\zeta/2} \sigma_{\alpha 3}\mathrm{d}z.
\end{aligned}
\tag{3.4}
$$

Correspondingly, we need to integrate the motion equation (2.1a) through the thickness. Besides three equations for translational motions, we have two more equations formulated with rotation of the normal and moments. With (3.2a), we obtain (with U_α for the generic displacement in (2.1a) and $U_3 = w$ for the deflection)

$$
\begin{aligned}
&\int_{-\zeta/2}^{\zeta/2} \rho \ddot{U}_\alpha \mathrm{d}z = \rho\zeta \ddot{u}_\alpha, \\
&\int_{-\zeta/2}^{\zeta/2} \rho \ddot{U}_3 \mathrm{d}z = \rho\zeta \ddot{w}, \\
&\int_{-\zeta/2}^{\zeta/2} z\rho \ddot{U}_\alpha \mathrm{d}z = -\rho\frac{\zeta^3}{12}\ddot{\beta}_\alpha.
\end{aligned}
\tag{3.5}
$$

With (3.3), we have

$$
\begin{aligned}
\int_{-\zeta/2}^{\zeta/2} \sigma_{\alpha j,j}\mathrm{d}z &= \int_{-\zeta/2}^{\zeta/2} (\sigma_{\alpha\beta,\beta} + \sigma_{\alpha 3,3})\mathrm{d}z \\
&= N_{\alpha\beta,\beta} + \sigma_{\alpha 3}\big|_{-\zeta/2}^{\zeta/2} = N_{\alpha\beta,\beta}, \\
\int_{-\zeta/2}^{\zeta/2} \sigma_{3j,j}\mathrm{d}z &= \int_{-\zeta/2}^{\zeta/2} (\sigma_{3\alpha,\alpha} + \sigma_{33,3})\mathrm{d}z \\
&= Q_{\alpha,\alpha} + \sigma_{33}\big|_{-\zeta/2}^{\zeta/2} = Q_{\alpha,\alpha} + q, \\
\int_{-\zeta/2}^{\zeta/2} z\sigma_{\alpha j,j}\mathrm{d}z &= \int_{-\zeta/2}^{\zeta/2} z(\sigma_{\alpha\beta,\beta} + \sigma_{\alpha 3,3})\mathrm{d}z \\
&= M_{\alpha\beta,\beta} + z\sigma_{\alpha 3}\big|_{-\zeta/2}^{\zeta/2} - \int_{-\zeta/2}^{\zeta/2} \sigma_{\alpha 3}\mathrm{d}z = M_{\alpha\beta,\beta} - Q_\alpha.
\end{aligned}
\tag{3.6}
$$

The body forces can also be treated in the resultant form:

$$\int_{-\zeta/2}^{\zeta/2} f_\alpha \mathrm{d}z = F_\alpha,$$

$$\int_{-\zeta/2}^{\zeta/2} f_3 \mathrm{d}z = F_3, \tag{3.7}$$

$$\int_{-\zeta/2}^{\zeta/2} z f_\alpha \mathrm{d}z = M_\alpha.$$

The motion equations in terms of resultants are obtained with (3.5)–(3.7):

$$\begin{aligned} \rho\zeta \ddot{u}_\alpha - N_{\alpha\beta,\beta} &= F_\alpha, \\ \rho\zeta \ddot{w} - Q_{\alpha,\alpha} &= F_3 + q = \bar{F}_3, \\ -\rho\frac{\zeta^3}{12}\ddot{\beta}_\alpha - M_{\alpha\beta,\beta} + Q_\alpha &= M_\alpha. \end{aligned} \tag{3.8}$$

The membrane action is separated from the bending and shearing actions. The bending and shearing are coupled and of great interest. Dropping the inertia terms, we obtain the static equilibrium equations:

$$\begin{aligned} -N_{\alpha\beta,\beta} &= F_\alpha, \\ -Q_{\alpha,\alpha} &= F_3 + q = \bar{F}_3, \\ -M_{\alpha\beta,\beta} + Q_\alpha &= M_\alpha. \end{aligned} \tag{3.8a}$$

We can also write the resultant form for the boundary conditions in a similar way. We have assumed pressure loading without shear forces on the top and bottom surfaces, where the normal $\boldsymbol{n} = \boldsymbol{e}_3$, described in (3.3). In fact, as the reduced 2D model of plates and shells, we are interested in the in-plane boundary conditions, prescribed on the side surfaces only, where the normal component $n_3 = 0$. For the loading given on Γ_s, we have the following in terms of stress resultants:

$$\int_{-\zeta/2}^{\zeta/2} \sigma_{\alpha\beta} n_\beta \mathrm{d}z|_{\Gamma_S} = N_{\alpha\beta} n_\beta|_{\Gamma_S} = g_\alpha,$$

$$\int_{-\zeta/2}^{\zeta/2} z\sigma_{\alpha\beta} n_\beta \mathrm{d}z|_{\Gamma_S} = M_{\alpha\beta} n_\beta|_{\Gamma_S} = m_\alpha. \tag{3.9}$$

On the other part of the boundary where the displacement is prescribed, we have

$$\begin{aligned} u_\alpha|_{\Gamma_u} &= \bar{u}_\alpha, \\ w|_{\Gamma_u} &= \bar{w}, \\ \beta_\alpha|_{\Gamma_u} &= \bar{\beta}_\alpha. \end{aligned} \tag{3.10}$$

The initial conditions are similarly defined as

$$
\begin{aligned}
u_\alpha(0, x_1, x_2) &= u_\alpha^0(x_1, x_2), \quad \dot{u}_\alpha(0, x_1, x_2) = u_\alpha^1(x_1, x_2), \\
w(0, x_1, x_2) &= w^0(x_1, x_2), \quad \dot{w}(0, x_1, x_2) = w^1(x_1, x_2), \\
\beta(0, x_1, x_2) &= \beta^0(x_1, x_2), \quad \dot{\beta}(0, x_1, x_2) = \beta^1(x_1, x_2).
\end{aligned}
\tag{3.11}
$$

The discussion so far is applicable to the general nonlinear case.

3.1.3 Applications to Linear Elasticity

Due to the assumption of (3.3), we adopt the generalized plane stress model for plates and shells with $\sigma_{33} = 0$. In view of the geometry relation (3.2), the components of strain for small deformation are

$$
\begin{aligned}
2\varepsilon_{\alpha\beta} &= u_{\alpha,\beta} + u_{\beta,\alpha} - z\beta_{\alpha,\beta} - z\beta_{\beta,\alpha}, \\
2\varepsilon_{\alpha 3} &= w_{,\alpha} - \beta_\alpha.
\end{aligned}
\tag{3.12}
$$

For linear elasticity, the stress components of plane stress model (1.10) are then expressed in terms of displacement and rotations of the normal:

$$
\begin{aligned}
\sigma_{\alpha\beta} &= \frac{E}{1-\nu^2}(\nu\delta_{\alpha\beta}\varepsilon_{\gamma\gamma} + (1-\nu)\varepsilon_{\alpha\beta}) \\
&= \frac{E}{1-\nu^2}(\nu\delta_{\alpha\beta}(u_{\gamma,\gamma} - z\beta_{\gamma,\gamma}) + \frac{1-\nu}{2}(u_{\alpha,\beta} + u_{\beta,\alpha} - z\beta_{\alpha,\beta} - z\beta_{\beta,\alpha})), \\
\sigma_{\alpha 3} &= 2\mu\varepsilon_{\alpha 3} = \mu(w_{,a} - \beta_a).
\end{aligned}
\tag{3.13}
$$

Substitution of these stress components in (3.4) leads to

$$
\begin{aligned}
N_{\alpha\beta} &= \frac{E\zeta}{1-\nu^2}\left(\nu\delta_{\alpha\beta}u_{\gamma,\gamma} + \frac{1-\nu}{2}(u_{\alpha,\beta} + u_{\beta,\alpha})\right), \\
M_{\alpha\beta} &= \frac{-E\zeta^3}{12(1-\nu^2)}\left(\nu\delta_{\alpha\beta}\beta_{\gamma,\gamma} + \frac{1-\nu}{2}(\beta_{\alpha,\beta} + \beta_{\beta,\alpha})\right), \\
Q_\alpha &= \kappa\mu\zeta(w_{,a} - \beta_a).
\end{aligned}
\tag{3.14}
$$

Here, a correction factor κ is introduced in the equation of shear force, to adjust the shear stress distribution due to the assumption of no shear force on the surface. When we consider bending deformation subjected to loading mainly in the normal direction, the transverse shear strains of (3.12) are constant and therefore are identically zeroes. It is true only for the Kirchhoff–Love (K-L) theory discussed in the following section; therefore, a correction is introduced—assuming that the shear strain is approximately quadratic through the thickness, and that we are only interested in their resultants instead of the detailed distribution:

$$\sigma_{\alpha 3} = c_\alpha(\zeta^2/4 - z^2),$$
$$Q_\alpha = \int_{-\zeta/2}^{\zeta/2} \sigma_{\alpha 3} \mathrm{d}z = \kappa \mu \zeta \gamma_{\alpha 3}. \tag{3.15}$$

Generally, $\kappa = 5/6$ is commonly adopted, see Babuška and Li (1992) for more discussion.

The corresponding derivatives of the stress resultants used in the motion equations are

$$-N_{\alpha\beta,\beta} = -\frac{E\zeta}{1-\nu^2}\left(\frac{1+\nu}{2}u_{\gamma,\gamma\alpha} + \frac{1-\nu}{2}u_{\alpha,\beta\beta}\right),$$
$$-Q_{\beta,\beta} = -\kappa\mu\zeta(w_{,\beta\beta} - \beta_{\beta,\beta}),$$
$$-M_{\alpha\beta,\beta} = \frac{E\zeta^3}{12(1-\nu^2)}\left(\frac{1+\nu}{2}\beta_{\gamma,\gamma\alpha} + \frac{1-\nu}{2}\beta_{\alpha,\beta\beta}\right).$$

The motion equations in (3.8) are reduced to

$$\rho\zeta\ddot{u}_\alpha - E_1\zeta\left(\frac{1+\nu}{2}u_{\gamma,\gamma\alpha} + \frac{1-\nu}{2}u_{\alpha,\gamma\gamma}\right) = F_\alpha,$$
$$\rho\zeta\ddot{w} - \kappa\mu\zeta(w_{,\gamma\gamma} - \beta_{\gamma,\gamma}) = \bar{F}_3, \tag{3.16}$$
$$-\rho\frac{\zeta^3}{12}\ddot{\beta}_\alpha + D\left(\frac{1+\nu}{2}\beta_{\gamma,\gamma\alpha} + \frac{1-\nu}{2}\beta_{\alpha,\gamma\gamma}\right) + \kappa\mu\zeta(w_{,a} - \beta_a) = M_\alpha.$$

Here, we introduce the following notations, with D as the bending stiffness:

$$E_1 = \frac{E}{1-\nu^2}, \quad D = \frac{E\zeta^3}{12(1-\nu^2)}. \tag{3.17}$$

The static equilibrium equation is simply derived by dropping the accelerations:

$$-E_1\zeta\left(\frac{1+\nu}{2}u_{\gamma,\gamma\alpha} + \frac{1-\nu}{2}u_{\alpha,\beta\beta}\right) = F_\alpha,$$
$$-\kappa\mu\zeta(w_{,\gamma\gamma} - \beta_{\gamma,\gamma}) = \bar{F}_3, \tag{3.16a}$$
$$D\left(\frac{1+\nu}{2}\beta_{\gamma,\gamma\alpha} + \frac{1-\nu}{2}\beta_{\alpha,\gamma\gamma}\right) + \kappa\mu\zeta(w_{,\gamma} - \beta_\gamma) = M_\alpha.$$

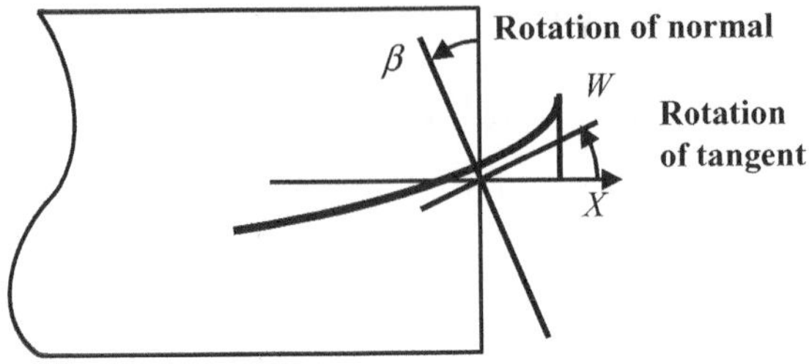

FIGURE 3.4 Rotation of the normal and rotation of the tangent.

3.1.4 Kirchhoff–Love Theory

In timely development of shell theories, K-L theory has played a particularly important role. The essential assumption of K-L theory, initially proposed by Kirchhoff (1850), is that the normal remains straight and normal, named as normality condition in the literatures. This is equivalent to the assumption that the rotation of the normal is the same as the rotation of the tangent plane, illustrated in Figure 3.4. For small deformation, this is expressed by

$$w_{,\alpha} = \beta_\alpha. \tag{3.18}$$

In the early stage of theoretical development for shell structures, many assumptions were introduced and some of which were found inconsistent. For the historical commentary on the small deflection theory of plate, see Love (1927). For the study on the systematic treatment of thin shell/thin plate, see Chien (1944).

For linear elasticity with assumption (3.18), the strain components (3.12) is reduced to

$$\begin{aligned} 2\varepsilon_{\alpha\beta} &= u_{\alpha,\beta} + u_{\beta,\alpha} - 2zw_{,\alpha\beta}, \\ 2\varepsilon_{\alpha 3} &= w_{,\alpha} - \beta_\alpha = 0. \end{aligned} \tag{3.19}$$

This means a pure bending deformation without transverse shear strains. At this point, we would like to revisit the resultant equations. Starting with the static equilibrium, we eliminate the contribution of shear strain from (3.8a) and obtain

$$\begin{aligned} -N_{\alpha\beta,\beta} &= F_\alpha, \\ -M_{\alpha\beta,\alpha\beta} &= M_{\alpha,\alpha} + \bar{F}_3 = \bar{q}. \end{aligned} \tag{3.20}$$

This is a system of two separate actions, the membrane and the bending.

On the other hand, with (3.18), the second equation of (3.14) becomes $M_{\alpha\beta} = -D(\nu\delta_{\alpha\beta}w_{,\gamma\gamma} + (1-\nu)w_{,\alpha\beta})$. This is for bending response due to deflection, which are more interested to us. The second equation of (3.20) is then reduced to an equation of deflection with fourth-order derivatives. Or alternatively, we can use

(3.16a) directly to obtain, also with the simple notation $\bar{q}$ for the total load:

$$-M_{\alpha\beta,\alpha\beta} = D\nabla^4 w = \bar{q}. \tag{3.21}$$

This is the well-known deflection equation. For dynamic problems, such as the lateral vibration or transient response, we can add a term of inertia force to obtain the equation of motion:

$$\rho\varsigma\ddot{w} + D\nabla^4 w = \bar{q}. \tag{3.22}$$

The inertia force in (3.22) is also in the sense of resultant through the thickness. Equation (3.22) is considered in the sense of per unit area of the plate.

Many static and dynamic cases have been studied and solved by using (3.21) and (3.22), which involve the biharmonic operator.

3.1.5 Reissner–Mindlin Plate Theory

The biharmonic operator involved in K-L theory needs C^1 interpolation in the finite element method. It is less attractive, although successful development has been achieved with great efforts. On the other hand, it is found that K-L theory is effective for thin structures but not satisfied for moderately thick structures. The biharmonic operator of the K-L theory is introduced from a differentiation on (3.8a). The purpose is to eliminate the shear force terms based on the assumption of a straight normal. Alternately, if the assumption on the normal is not enforced, then a theory of using only second-order derivatives is available. The later approach becomes more attractive when facing the challenges. One of the reasons is that the framework using C^0 interpolation for many other applications of finite element method can now be extended to plates and shells.

Reissner (1945, 1947) proposed a theory for moderately thick plate, assuming that the normal keeps straight but can rotate without remaining normal. The transverse shears are then retained in the equations. Mindlin (1951) developed the variational principle for the plate theory involving the transverse shear. This is commonly named as Reissner–Mindlin (R-M) theory.

As a matter of fact, the previous discussions in Sections 3.1.1–3.1.3 are essentially applicable to R-M theory. We will investigate more in the following sections.

3.2 LINEAR THEORY OF R-M PLATE

The motion equation (3.8), the boundary conditions (3.9) and (3.10), and the initial conditions (3.11) are in fact valid for general nonlinear applications. It is beneficial to explore the linear case at this point. Generally speaking, a robust theory or numerical algorithm for nonlinear analysis should perform well for the linear case as a reduced system. For linear elasticity, (3.16) derived in Section 3.1.3 is the system of governing equations for R-M plate theory. The system of equations can also be derived from

(2.1) by using the Hamilton principle through variational formulation, cf. Liew et al. (1998) or using the Galerkin method.

The membrane action is separated from the bending and shear actions. It forms a subsystem of membrane with 2D spatial variables that possesses the same form of a 2D linear elasticity. The general theory about transient dynamics discussed in Chapter 2 applies to this subsystem. Here, we focus on the bending and shear terms.

3.2.1 Helmholtz Decomposition for R-M Plate

For static problems, differentiating the equilibrium equation (3.8a) and eliminating the shear strains lead to K-L equation (3.12). When applying this approach to dynamic R-M equations (3.8) or (3.16), reducing to a separate system is not straightforward. For static problems and vibration problems without moment loading ($M_\alpha = 0$, which is usually negligible), Hu (1963) used the Helmholtz decomposition for the rotation $\boldsymbol{\beta}$ (as a 2D vector):

$$\begin{cases} \beta_1 = \psi_{,1} + \varphi_{,2}, \\ \beta_2 = \psi_{,2} - \varphi_{,1}. \end{cases} \tag{3.23}$$

Then, the second and third equations of (3.16) can be transformed to a separated system, briefly discussed below. Applications of cases with various boundary conditions can be found in Hu (1981) and the references cited there. We can define

$$w = \psi + L(\psi). \tag{3.24}$$

Here, $\boldsymbol{L}$ is a linear operator, introduced for facilitating the method of solving the equations. Substituting (3.23) and (3.24) into the second and third equations of (3.16) leads to two independent equations of ψ and φ:

$$\begin{cases} \rho\zeta(\ddot{\psi} + L(\ddot{\psi})) - \kappa\mu\zeta L(\nabla^2\psi) = \bar{F}_3, \\ -\dfrac{\rho\zeta^3}{12}\ddot{\varphi} + \dfrac{1-\nu}{2}D\nabla^2\varphi - \kappa\mu\zeta\varphi = 0. \end{cases} \tag{3.25}$$

3.2.2 Load Scaling for Static Problem of R-M Plate

For the static problem, Bathe and Brezzi (1985) introduced load scaling with the objective to investigate the effect of thickness to R-M plate and the relation between the R-M theory and the K-L theory. A division by ζ^3 at both sides was proposed to obtain a reference equation. We apply the method to the second and third equations of (3.16a), and obtain the scaled equilibrium system:

$$\begin{aligned} &-\kappa\mu\gamma_{\alpha,\alpha} = \hat{f}_3 \\ &\frac{E}{12(1-\nu^2)}\left(\frac{1+\nu}{2}\beta_{\gamma,\gamma\alpha} + \frac{1-\nu}{2}\beta_{\alpha,\gamma\gamma}\right) + \kappa\mu\gamma_\alpha = \hat{m}_\alpha. \end{aligned} \tag{3.26a}$$

Here, we define the scaled shear strain and loads by

$$\begin{aligned}\gamma_\alpha &= \zeta^{-2}(w_{,\alpha} - \beta_\alpha),\\ \hat{f}_3 &= \zeta^{-3}q,\\ \hat{m}_\alpha &= \zeta^{-3}M_\alpha.\end{aligned} \tag{3.26b}$$

Correspondingly, the scaled K-L problem is $D_0\nabla^4 w = \hat{f}_3$, $D_0 = E/12(1-\nu^2)$. It is independent of thickness and hence serves as a reference problem.

The Helmholtz decomposition was introduced by Brezzi and Fortin (1986) for the transverse shear strain:

$$\begin{aligned}\gamma_1 &= \psi_{,1} + p_{,2},\\ \gamma_2 &= \psi_{,2} - p_{,1}.\end{aligned} \tag{3.26c}$$

This formula is different from the one in Hu (1963) for rotations in (3.23). These two can be related if choosing $L(\psi) = \psi$. The mixed method was used to form a system with two additional scalar variables p and ψ. The existence of a unique solution of (3.26) was proved. Their result on regularity of the solution is briefly described below, see Arnold and Falk (1989a) for more discussions.

Assessment 3.1 *Under certain conditions, the solution of the scaled static problem* (3.26) *with boundary conditions* (3.10) *satisfies*

$$\|\boldsymbol{\beta}\|_2 + \|w\|_2 + \|\psi\|_2 + |p|_1 + \zeta\|p\|_2 + |\boldsymbol{\gamma}|_0 + \zeta\|\boldsymbol{\gamma}\|_1 \le C(\|g\|_0 + \|\bar{\boldsymbol{m}}\|_0) \tag{3.27}$$

with the generic constant C independent of thickness ζ.

3.2.3 Load Scaling and Mass Scaling for Dynamic Problem of R-M Plate

When extending to the dynamic problem, Brezzi (2003) suggested an additional mass scaling in the two equations of (3.16) for bending and shear, presented in Wu (2004):

$$\rho = \rho_0\zeta^2, \tag{3.28}$$

$$\begin{cases}\rho_0\ddot{w} - \kappa\mu\gamma_{\alpha,\alpha} = g\\[2mm] -\dfrac{\rho_0\zeta^2}{12}\ddot{\beta}_\alpha + \dfrac{E}{24(1-\nu^2)}E((1+\nu)\beta_{\eta,\eta\alpha} + (1-\nu)\beta_{\alpha,\eta\eta})\\ +\kappa\mu\gamma_\alpha = \bar{m}_\alpha.\end{cases} \tag{3.29}$$

The existence, uniqueness, and regularity of the solution to (3.29) were established in Wu (2004). The main results are described below.

Assessment 3.2 *For a clamped plate, under certain conditions, there exists a unique solution* (β_α, w) *of* (3.29) *with initial conditions* (3.11). *Moreover, there exists a* *constant $C > 0$, independent of the material parameters, such that*

$$\zeta||\dot{\boldsymbol{\beta}}||_0 + ||\dot{w}||_0 + ||\boldsymbol{\beta}||_1 + \zeta||\boldsymbol{\gamma}||_0$$
$$\leq C\big(\zeta||\boldsymbol{\beta}^1||_0 + ||w^1||_0 + ||\beta^0||_1 + \zeta||\gamma^0||_0 + \zeta^{-1}||\bar{\boldsymbol{m}}||_{L^2(L^2)} + ||g||_{L^2(L^2)}\big), \tag{3.30a}$$

$$\zeta||\ddot{\boldsymbol{\beta}}||_0 + ||\ddot{w}||_0 + ||\dot{\boldsymbol{\beta}}||_1 + \zeta||\dot{\boldsymbol{\gamma}}||_0$$
$$\leq C\,\big(\zeta^{-1}(||\boldsymbol{\beta}^0||_2 + ||\boldsymbol{\gamma}^0||_0 + ||\boldsymbol{m}(0)||_0) + ||\boldsymbol{\gamma}^0||_1 + ||g(0)||_0$$
$$+ ||\boldsymbol{\beta}^1||_1 + \zeta||\boldsymbol{\gamma}^1||_0 + \zeta^1||\dot{\boldsymbol{m}}||_{L^2(L^2)} + ||\dot{g}||_{L^2(L^2)}\big)\,, \tag{3.30b}$$

$$||\boldsymbol{\beta}||_2 \leq C\,\big(\zeta||\ddot{\boldsymbol{\beta}}||_0 + ||\boldsymbol{\gamma}||_0 + ||\boldsymbol{m}||_0\big)\,, \tag{3.30c}$$

$$||w||_2 \leq C\,\big(\zeta^2||\ddot{w}||_0 + ||\beta||_1 + \zeta^2||g||_0\big)\,, \tag{3.30d}$$

where $\boldsymbol{\gamma}^0 = \zeta^{-2}(\nabla w^0 - \boldsymbol{\beta}^0)$ and $\boldsymbol{\gamma}^1 = \zeta^{-2}(\nabla w^1 - \boldsymbol{\beta}^1)$. The bold-faced $\boldsymbol{\beta}$ and $\boldsymbol{\gamma}$ represent the 2D vector form of the corresponding components.

3.2.4 Relation between R-M Theory and K-L Theory

According to K-L theory, the thickness is assumed to be small, for example, with the aspect ratio $\zeta/L < 50$ (L is the characteristic length in the plane dimension). The R-M theory is designed to work for moderate thickness. A challenging question arises that the R-M theory is also expected to work for small thickness.

Consider the static bending problem described in Equation 3.26. Bathe and Brezzi (1985) provided a proof of the relation between the solutions of the R-M plate and the K-L plate, briefed below.

Assessment 3.3 *For a clamped plate, when the thickness approaches zero, the solution of the R-M plate approaches the solution of a K-L plate, in the following sense:*

$$\begin{aligned} &\beta_\alpha \to \bar{\beta}_\alpha, \quad w \to \bar{w}, \\ &\bar{\beta}_\alpha = \bar{w}_{,\alpha}, \\ &D_0 \nabla^4 \bar{w} = q, \end{aligned} \tag{3.31}$$

where $D_0 = E/12(1 - \nu^2)$. Note that the third equation of (3.31) is a scaled equation and is independent of thickness, hence serves as a reference problem. This feature is often used to verify the performance of shell element against the shear locking, which will be a focus point in the rest of this chapter. This relationship was extended to the elastodynamic problem by Wu (2004), briefly stated below,

Assessment 3.4 *With clamped boundary conditions and homogeneous initial conditions, the solution of Equation 3.29 converges in the following sense, as the thickness approaches zero:*

$$w \to \tilde{w}, \beta_\alpha \to \tilde{\beta}_\alpha, \dot{w} \to \dot{\tilde{w}}, \ddot{w} \to \ddot{\tilde{w}} \text{ weakly}$$

with

$$\tilde{\beta}_\alpha = \tilde{w}_{,\alpha}, \tag{3.32}$$

where $\tilde{w}$ is the solution of a reference K-L plate problem of elastodynamics with clamped boundary conditions and homogeneous initial conditions

$$\begin{aligned} \rho_0 \ddot{\tilde{w}} + D_0 \nabla^4 \tilde{w} &= g, \\ \tilde{w}\Big|_{\partial\Omega} = \frac{\partial \tilde{w}}{\partial n}\Big|_{\partial\Omega} &= 0, \\ \tilde{w}(0, \boldsymbol{x}) = \dot{\tilde{w}}(0, \boldsymbol{x}) &= 0. \end{aligned} \tag{3.33}$$

Note that due to the complexity and rich variety of the initial-boundary value problems, this relationship has only been proved to hold for certain types of boundary conditions and initial conditions. For example, consider a case with clamped boundary conditions and distributed initial conditions, without external loading, referring to Equation 3.11:

$$\begin{aligned} & w_1|_\Gamma = 0, \quad \partial w_1/\partial n|_\Gamma = 0, \\ & \beta_\alpha(0) = w(0) = 0, \\ & \dot{\beta}_\alpha(0) = 0, \quad \dot{w}(0, x, y) = w_1(x, y). \end{aligned}$$

If w_1 is not identically zero, the initial velocities of the transverse shear strains are not identically zero in the domain. The dynamic response will not stop the growth of the transverse shear. Therefore, when the thickness approaches zero, the solution of our R-M equation may not be expected to converge to a solution of a K-L plate, which requires zero transverse shear. If the boundary conditions do not promise zero shear strain, then it may also result in a situation that there is no convergence to the K-L problem.

For static problem, the boundary layer about the size of the thickness was identified for R-M plate, by Arnold and Falk (1989b, 1990). In this area, the solution behaves with $\partial^3 \boldsymbol{\beta}/\partial n^3 = O(\varsigma^{-1})$. Additional assessment can be found in Babuška and Li (1992). Häggblad and Bathe (1990) investigated the method of proper meshing in the boundary area. For dynamic problem, it seems more complicated and deserves significant effort of investigation.

3.3 INTERPOLATION FOR FOUR-NODE R-M PLATE ELEMENT

3.3.1 Variational Equations for R-M Plate

Recall the variational equation (Eq. 2.10) along with the concept of R-M plate (Eq. 3.2a). We use test functions in the form of velocity $V_\alpha = v_\alpha - z\eta_\alpha$, $V_3 = v_3$. The variational equation for R-M plate is derived from Equation 2.10:

$$\int_\Omega (\rho \ddot{U}_i V_i + \sigma_{ij} V_{i,j}) \mathrm{d}\boldsymbol{\Omega} = \int_\Omega f_i V_i \mathrm{d}\boldsymbol{\Omega} + \int_{\Gamma_s} g_i V_i \mathrm{d}\Gamma.$$

The volume integral is decomposed into the through-thickness integral and the in-plane integral. Using the resultants of stress, body force, and surface tractions defined in Equations 3.4, 3.7–3.9, it is straightforward to obtain

$$\int_A (\rho\zeta(\ddot{u}_\alpha v_\alpha + \ddot{w} v_3) + \rho\zeta^3 \ddot{\beta}_\alpha \eta_\alpha / 12 + N_{\alpha\beta} v_{\alpha,\beta} - M_{\alpha\beta}\eta_{\alpha,\beta} + Q_\alpha v_{3,\alpha}) \mathrm{d}A$$
$$= \int_A (F_\alpha v_\alpha + \bar{F}_3 v_3 - M_\alpha \eta_\alpha) \mathrm{d}A + \int_{\Gamma_s} (g_\alpha v_\alpha - m_\alpha \eta_\alpha) \mathrm{d}\Gamma. \tag{3.34}$$

This is equivalent to a system of variational equations

$$\begin{aligned}
&\int_A (\rho\zeta \ddot{u}_\alpha v_\alpha + N_{\alpha\beta} v_{\alpha,\beta}) \mathrm{d}A = \int_A F_\alpha v_\alpha \mathrm{d}A + \int_{\Gamma_s} g_\alpha v_\alpha \mathrm{d}\Gamma, \\
&\int_A (\rho\zeta \ddot{w} v_3 + Q_\alpha v_{3,\alpha}) \mathrm{d}A = \int_A \bar{F}_3 v_3 \mathrm{d}A, \\
&\int_A (\rho\zeta^3 \ddot{\beta}_\alpha \eta_\alpha / 12 - M_{\alpha\beta}\eta_{\alpha,\beta} - Q_\alpha \eta_\alpha) \mathrm{d}A \\
&= \int_A -M_\alpha \eta_\alpha \mathrm{d}A - \int_{\Gamma_s} m_\alpha \eta_\alpha \mathrm{d}\Gamma.
\end{aligned} \tag{3.34a}$$

3.3.2 Bilinear Interpolations

We only consider elements of the first order here as most applications with explicit software only use low order elements. For the bilinear isoparametric four-node element, four shape functions are used for interpolation. They are also used for coordinate transform from the material element in physical domain to a master element in

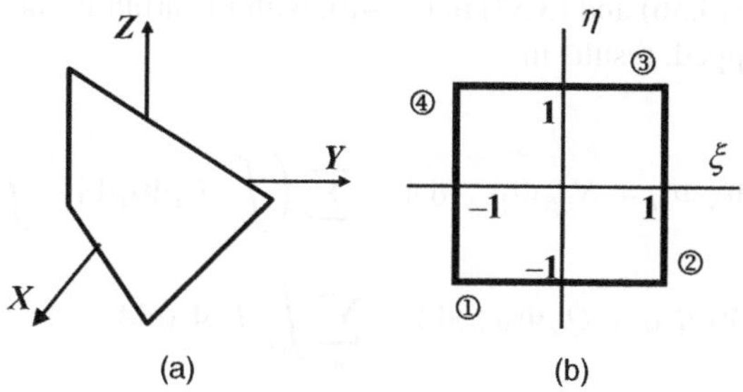

FIGURE 3.5 Element transformation from the physical domain to reference domain: (a) Material element in physical domain; (b) Master element in reference domain.

the reference domain, depicted in Figure 3.5. The shape functions are defined below

$$\Phi_N(\xi,\eta) = (1+\xi^N\xi)(1+\eta^N\eta)/4 = \left(\Sigma_N + \Lambda_N^1\xi + \Lambda_N^2\eta + \Gamma_N\xi\eta\right)/4, \quad (3.35)$$

where (ξ^N, η^N) are the nodal coordinates in the reference system. The coefficients in (3.35) are shown in Table 3.1.

In a simple and straightforward application of the interpolations, the displacement and the test functions in (3.34) are then discretized with the summation convention on index N from 1 to 4:

$$\begin{aligned} u_\alpha|_e &= u_\alpha^N(t)\Phi_N, \quad v_\alpha|_e = v_\alpha^M\Phi_M, \\ w|_e &= w_N(t)\Phi_N, \quad v_3|_e = v_3^M\Phi_M, \\ \beta_\alpha|_e &= \beta_\alpha^N(t)\Phi_N, \quad \eta_\alpha|_e = \eta_\alpha^M\Phi_M. \end{aligned} \quad (3.36)$$

The velocity and acceleration are discretized similar to (2.15):

$$\begin{aligned} \dot{u}_\alpha|_e &= \dot{u}_\alpha^N(t)\Phi_N, \quad \ddot{u}_\alpha|_e = \ddot{u}_\alpha^N(t)\Phi_N, \\ \dot{w}|_e &= \dot{w}_N(t)\Phi_N, \quad \ddot{w}|_e = \ddot{w}_N(t)\Phi_N, \\ \dot{\beta}_\alpha|_e &= \dot{\beta}_\alpha^N(t)\Phi_N, \quad \ddot{\beta}_\alpha|_e = \ddot{\beta}_\alpha^N(t)\Phi_N. \end{aligned} \quad (3.37)$$

TABLE 3.1 Coefficients of the Shape Functions for the Four-Node Element

Node	ξ	η	Σ_I	Λ_{1I}	Λ_{2I}	Γ_I
1	−1	−1	1	−1	−1	1
2	1	−1	1	1	−1	−1
3	1	1	1	1	1	1
4	−1	1	1	−1	1	−1

Source: Reprinted from Flanagan and Belytschko, 1981. Copyright (1981), with permission from Wiley-Blackwell.

Substitution of (3.36) and (3.37) in (3.34a), with the arbitrary nodal values of the test functions dropped, results in

$$\sum_e \int_{\Omega_e} (\rho\zeta \ddot{u}_\alpha^N \Phi_N \Phi_M + N_{\alpha\beta}\Phi_{M,\beta})\mathrm{d}A = \sum_e \left(\int_{\Omega_e} F_\alpha \Phi_M \mathrm{d}A + \int_{\Gamma_{se}} g_\alpha \Phi_M \mathrm{d}\Gamma \right),$$

$$\sum_e \int_{\Omega_e} (\rho\zeta \ddot{w}^N \Phi_N \Phi_M + Q_\alpha \Phi_{M,\alpha})\mathrm{d}A = \sum_e \int_{\Omega_e} \bar{F}_3 \Phi_M \mathrm{d}A,$$

$$\sum_e \int_{\Omega_e} (\rho\zeta^3 \ddot{\beta}_\alpha^N \Phi_N \Phi_M / 12 - M_{\alpha\beta}\Phi_{M,\beta} - Q_\alpha \Phi_M)\mathrm{d}A$$
$$= -\sum_e \left(\int_{\Omega_e} M_\alpha \Phi_M \mathrm{d}A + \int_{\Gamma_{se}} m_\alpha \Phi_M \mathrm{d}\Gamma \right).$$

Here, the in-plane domain integral takes into account the contributions from all the elements. The system yields the finite element equations

$$\begin{aligned} M_{MN}\ddot{u}_\alpha^N &= F^M_{(\mathrm{int})\alpha} + F^M_{(\mathrm{ext})\alpha}, \\ M_{MN}\ddot{w}^N &= F^M_{(\mathrm{int})3} + F^M_{(\mathrm{ext})3}, \\ I_{MN}\ddot{\beta}_\alpha^N &= M^M_{(\mathrm{int})\alpha} + M^M_{(\mathrm{ext})\alpha}. \end{aligned} \tag{3.38}$$

This is an extension of (2.19) to the R-M plate. Here, we denote

$$\begin{aligned} F^M_{(\mathrm{int})\alpha} &= -\sum_e \int_{\Omega_e} N_{\alpha\beta}\Phi_{M,\beta}\mathrm{d}A, \\ F^M_{(\mathrm{int})3} &= -\sum_e \int_{\Omega_e} Q_\alpha \Phi_{M,\alpha}\mathrm{d}A, \\ M^M_{(\mathrm{int})\alpha} &= \sum_e \int_{\Omega_e} (M_{\alpha\beta}\Phi_{M,\beta} + Q_\alpha \Phi_M)\mathrm{d}A, \end{aligned} \tag{3.39a}$$

$$\begin{aligned} F^M_{(\mathrm{ext})\alpha} &= \sum_e \left(\int_{\Omega_e} F_\alpha \Phi_M \mathrm{d}A + \int_{\Gamma_{se}} g_\alpha \Phi_M \mathrm{d}\Gamma \right), \\ F^M_{(\mathrm{ext})3} &= \sum_e \int_{\Omega_e} \bar{F}_3 \Phi_M \mathrm{d}A, \\ M^M_{(\mathrm{ext})\alpha} &= -\sum_e \left(\int_{\Omega_e} M_\alpha \Phi_M \mathrm{d}A + \int_{\Gamma_{se}} m_\alpha \Phi_M \mathrm{d}\Gamma \right), \end{aligned} \tag{3.39b}$$

$$M_{MN} = \sum_e \int_{\Omega_e} \rho\zeta \Phi_M \Phi_N \mathrm{d}A,$$
$$I_{MN} = \sum_e \int_{\Omega_e} \rho\zeta^3 \Phi_M \Phi_N \mathrm{d}A/12. \tag{3.39c}$$

Here, $\boldsymbol{I}$ is named the moment of inertia, related to the rotational motion of the nodes, brought from R-M theory.

It is worth noting that the displacement, velocity, and acceleration as the physical parameters are all included in the variational principle and the finite element equations. They are functions of time. But the test function is time independent due to the nature of the variational principle (or the weighted residual method) applied at a current time point with spatial actions only.

3.3.3 Shear Locking Issues of R-M Plate Element

The description in Section 3.3.2, which seems reasonable and straightforward, however, causes unexpected dissatisfaction known as locking, and particularly the shear locking for bending dominated applications. We provide a simple example to illustrate the issue of shear locking.

Examine a rectangular element that is aligned with the coordinate axes, so that

$$x^1 = x^4 = -1, x^2 = x^3 = 1.$$

For a case of "pure bending," with

$$w^N = \beta_y^N = 0, \beta_x^1 = \beta_x^4 = -\alpha, \text{ and } \beta_x^2 = \beta_x^3 = \alpha.$$

The transverse shear strain and stress are

$$\gamma_{xz} = -\beta_x = -\alpha x,$$
$$\tau_{xz} = -\alpha\mu x. \tag{3.40}$$

For pure bending, the shear strains are supposed to be zero. Its appearance in (3.40) is due to the application of bilinear interpolation (3.35). For elastic deformation with small thickness, for instance, the bending stiffness $D = E\zeta^3/12(1-\nu^2)$ will be much smaller than the shear stiffness $\mu\zeta$. Hence, the shear term will take too much energy and the bending is not represented appropriately. It becomes more severe when the thickness becomes smaller. As discussed in Section 3.2.4, under certain conditions, when the thickness approaches zero, the R-M solution should converge to the K-L solution. With locking issue, the transverse shear calculated by the finite element is large and the convergence to K-L solution is questionable. Therefore, such approximate solution is not reliable. For detailed discussion, see Stolarski et al. (1995), Suri et al. (1995), Arnold and Falk (1997), and the references included.

To address the shear locking issue mathematically, the concept of uniform convergence has been introduced and become a desired property of the robust R-M plate element. That means when mesh is refined, the finite element solution should not only converge but also converge with the rate independent of thickness. In another word, if the convergence rate deteriorates when the thickness is reduced, or the mesh has to be finer to obtain good convergence, then the element cannot be locking-free yet. The cases of R-M plate converging to K-L plate, as discussed in Section 3.2.2, often serve as a benchmark.

To alleviate or eliminate the locking issue has been the focus point of research on shell and plate finite elements. Various approaches, such as mixed method, nonconforming method, reduced integration, selective reduced integration, assumed strain, and discrete Kirchhoff method have been investigated in the past decades of rich development. Several elements have been developed successfully and proven to converge uniformly for static elasticity, see for example, Arnold and Falk (1989a), Brezzi et al. (1991), Durán and Liberman (1992), and Lyly et al. (1993). Also see Arnold et al. (2002) for a survey. On the other hand, deteriorate convergence or locking is found for some other elements, see Arnold and Falk (1997) for commentary.

A few of the R-M elements have been implemented in explicit finite element software for nonlinear applications of impact engineering. In the following sections, we will introduce some of these elements

3.4 REDUCED INTEGRATION AND SELECTIVE REDUCED INTEGRATION

3.4.1 Reduced Integration

To alleviate the shear locking with R-M plate element in thin shell/plate application, one of the early developments was to impose the Kirchhoff condition at several discrete points, introduced by Wempner et al. (1968). The method extended the classic K-L concept to R-M plate element, to be discussed later.

Another approach was the reduced integration, first introduced by Doherty et al. (1969) for axisymmetric solid mechanics analysis, using plane quadrilateral element. When lower order integration is used, the stiffness by numerical process is reduced. This is obvious from (3.40) and the parasitic stiffness of the shell/plate element is expected to diminish. If the strain energy is calculated using the reduced integration of one-point quadrature at the element center, instead of the usual 2×2 quadrature, the shear strain just becomes zero. The shear locking can thus be eliminated.

The approach applies to the volumetric locking too. Zienkiewicz et al. (1971) extended the concept to shell element. In the examples of simply supported square plate subject to uniform load and clamped circular plate under uniform load using the eight-node serendipity quadrilateral element, 2×2 Gauss quadrature was used for the transverse shear components. The usual 3×3 Gauss quadrature was used for the other components. Good results were obtained for both thick and thin situations. The lower order quadrature points for stress calculation are named Barlow points, according to the observation by Barlow (1976). We could obtain the same accuracy for stress at these locations as nodal displacement. For the more complex situation with the combination of membrane and bending, however, this approach did not show enough improvement from the full integration. With mesh refinement, slow convergence was observed. When 2×2 Gauss quadrature was used for all the components, good result was obtained even using only one or few elements. Implementation of the reduced integration for all terms also becomes easier. In addition, this saves computing time significantly. Hence, Zienkiewicz et al. (1971) recommended the all-around reduced integration for the R-M plate element.

3.4.2 Selective Reduced Integration

The approach of Doherty et al. (1969) using reduced integration selectively for certain terms only was called selective reduced integration. While Zienkiewicz et al. (1971) recommended all-around reduction of the order of quadrature, the selective reduced integration was extended to other elements. For instance, Hughes et al. (1977) utilized one-point reduced integration for the transverse shear strain of the bilinear four-node quadrilateral R-M plate bending element and the linear two-node Timoshenko beam element.

In the theoretical aspect, Malkus and Hughes (1978) proved that the selective reduced integration scheme was equivalent to certain class of mixed finite element method. On the other hand, Babuška et al. (1975) developed mathematical theory of mixed method, which in turn provided the support to selective reduced integration. Meanwhile the easy and simple implementation of selective reduced integration had shown advantage over the mixed method by avoiding complex formulation.

For constrained problems such as incompressible elasticity of beam and plate, the selective reduced integration scheme was also shown to be equivalent to a formula with penalty function applied to volume.

On the other hand, Kavanagh and Key (1972) commented that the stiffness matrix using selective reduced integration for eliminating the parasitic shear was not invariant with respect to coordinate rotation. In fact, the reduced integration resulted in rank deficiency of the stiffness matrix. In addition to the usual zero eigenvalues corresponding to rigid body motion, there were extra zero eigenvalues associated with an oscillatory keystone pattern of deformation, as illustrated in Figure 3.6. In finite element literatures, this was also called spurious, or kinematic, or zero energy mode, whereas "hourglass" was first used in finite difference literatures. Now, the word "hourglass" is popular in the applications of explicit finite element methods due

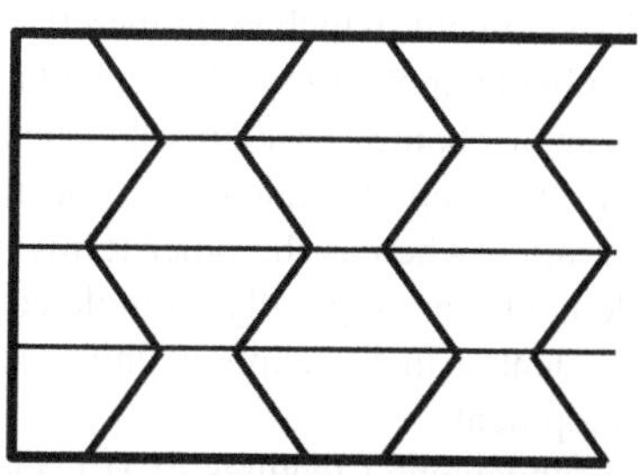

FIGURE 3.6 Keystone pattern of deformation.

to the application of one-point reduced integration, which we will discuss in the next sections.

An improved formulation of stiffness matrix was proposed in Kavanagh and Key (1972) as a combination of those calculated from low order integration. For example, denoted by *Q4(j)* for the stiffness matrix of a four-node quadrilateral element using *j* points Gauss quadrature, the new element stiffness matrix was formed as

$$\begin{cases} Q4X = \sum_{j=1}^{n} A_j Q4(j), \\ \sum A_j = 1. \end{cases} \tag{3.41}$$

The second equation was required to represent a constant strain. The coordinate invariance was preserved. The coefficients A_j could often be found for a particular application, but problem dependent.

The R-M plate element by Belytschko et al. (1981) extended this concept. The stiffness matrix was decomposed into two parts, bending and shear:

$$\boldsymbol{K} = \boldsymbol{K}_B + \boldsymbol{K}_S. \tag{3.42}$$

The bending portion used 2 × 2 quadrature. The shear portion used a mixed quadrature of one-point integration and 2 × 2 integration:

$$\begin{aligned} \boldsymbol{K}_S &= (1-\varepsilon)\boldsymbol{K}_S^{1\times 1} + \varepsilon \boldsymbol{K}_S^{2\times 2} = \boldsymbol{K}_S^{1\times 1} + \varepsilon \boldsymbol{K}_H, \\ \boldsymbol{K}_H &= \boldsymbol{K}_S^{2\times 2} - \boldsymbol{K}_S^{1\times 1}. \end{aligned} \tag{3.43}$$

$\boldsymbol{K}_H$ was used to stabilize the hourglass mode with a perturbation format. Where $\varepsilon = r\zeta^2/A$ was used with a scalar r, recommended to be 0.03–0.1. Here, ζ is the thickness and A is the element area.

3.4.3 Nonlinear Application of Selective Reduced Integration—Hughes–Liu Element

In the series of quadrilateral elements with selective reduced integration, Hughes and Liu (1981) developed an element, which combined the membrane and bending actions, and aimed at the nonlinear applications. Incremental formulation was implemented for general nonlinear mechanics problems. For small increment, the strain increment was calculated. In this step, the calculation for transverse shear strain increment used reduced integration whereas the other terms used regular integration. The Hughes–Liu (H-L) element was originally formulated with the implicit static framework, and then implemented in the explicit finite element software, such as LS-DYNA in its early development.

It is worth noting some fundamental features of H-L element. The thickness of the element is considered a variable, changing from node to node, defined by the

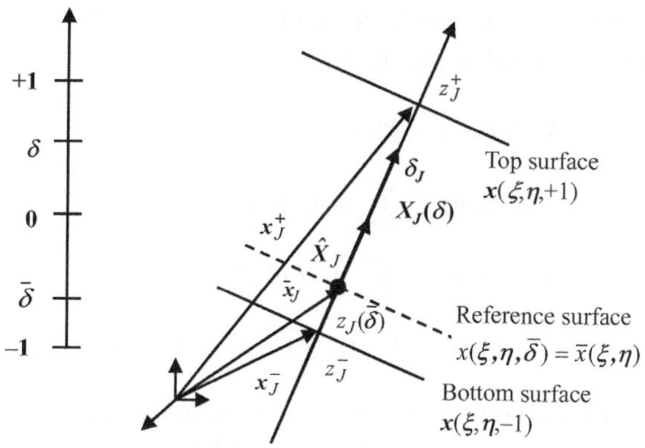

FIGURE 3.7 Geometry relation of a generic point to the reference surface in H-L element. (Reprinted from Hughes and Liu, 1981. Copyright (1981), with permission from Elsevier.)

coordinates of the corner points on top and bottom surfaces. The reference surface can be defined at any position in the thickness, and not necessarily the "mid-surface." The vectors at the nodes, emanating from the bottom surface to the top surface, are called fibers, as depicted in Figure 3.7. The fibers are not necessarily normal to the reference surfaces. The fiber length at node J and its directional unit vector are defined as

$$\begin{aligned} L_J &= \|\boldsymbol{x}_J^+ - \boldsymbol{x}_J^-\|, \\ \tilde{\boldsymbol{e}}_J &= (\boldsymbol{x}_J^+ - \boldsymbol{x}_J^-)/L_J. \end{aligned} \tag{3.44}$$

For a given parameter $\bar{\delta} \in [-1, 1]$, a reference point on the fiber is identified using interpolation in fiber direction:

$$\begin{aligned} \bar{\boldsymbol{x}}_J(\bar{\delta}) &= \boldsymbol{x}_J^+ \varphi_+(\bar{\delta}) + \boldsymbol{x}_J^- \varphi_-(\bar{\delta}) \\ \varphi_+(\bar{\delta}) &= (1 + \bar{\delta})/2 \\ \varphi_-(\bar{\delta}) &= (1 - \bar{\delta})/2. \end{aligned} \tag{3.45}$$

A reference surface is then generated by using $\bar{x}_J(\bar{\delta})$ with 2D interpolation:

$$\bar{\boldsymbol{x}}(\xi, \eta, \bar{\delta}) = \sum_J \bar{\boldsymbol{x}}_J(\bar{\delta}) \Phi_J(\xi, \eta). \tag{3.46}$$

In the nodal fiber direction, the distances from the top surface and bottom surface to the reference point $\bar{\boldsymbol{x}}_J(\bar{\delta})$ are described below

$$\begin{aligned} z_J^+ &= \|\boldsymbol{x}_J^+ - \boldsymbol{x}_J(\bar{\delta})\| = L_J \varphi_-(\bar{\delta}), \\ z_J^- &= -\|\boldsymbol{x}_J^- - \boldsymbol{x}_J(\bar{\delta})\| = -L_J \varphi_+(\bar{\delta}). \end{aligned} \tag{3.47}$$

With these preparations, a generic point inside the element can be described by using interpolations, depicted in Figure 3.7:

$$
\begin{aligned}
\boldsymbol{X}(\xi,\eta,\delta) &= \bar{\boldsymbol{x}}(\xi,\eta) + \hat{\boldsymbol{x}}(\xi,\eta,\delta),\\
\bar{\boldsymbol{x}}(\xi,\eta) &= \sum \bar{\boldsymbol{x}}_J(\bar{\delta})\Phi_J(\xi,\eta),\\
\hat{\boldsymbol{x}}(\xi,\eta,\delta) &= \sum \hat{\boldsymbol{x}}_J(\delta)\Phi_J(\xi,\eta),\\
\hat{\boldsymbol{x}}_J(\delta) &= (z_J^+\varphi_+(\delta) + z_J^-\varphi_-(\delta))\tilde{\boldsymbol{e}}_J.
\end{aligned}
\tag{3.48}
$$

The deformation is also represented by the deformation in the reference plane and the fiber. In incremental form similar to (3.48), we have

$$
\begin{aligned}
\Delta U(\xi,\eta,\delta) &= \Delta\boldsymbol{u}(\xi,\eta) + \Delta\hat{\boldsymbol{u}}(\xi,\eta,\delta),\\
\Delta\boldsymbol{u}(\xi,\eta) &= \sum \Delta\boldsymbol{u}_J(\bar{\delta})\Phi_J(\xi,\eta),\\
\Delta\hat{\boldsymbol{u}}(\xi,\eta,\delta) &= \sum \Delta\hat{\boldsymbol{u}}_J(\delta)\Phi_J(\xi,\eta),\\
\Delta\hat{\boldsymbol{u}}_J(\delta) &= (z_J^+\varphi_+(\zeta) + z_J^-\varphi_-(\zeta))\Delta\tilde{\boldsymbol{e}}_J.
\end{aligned}
\tag{3.49}
$$

Here, $\Delta\hat{\boldsymbol{u}}$ represents the fiber displacement. $\Delta\tilde{\boldsymbol{e}}_J$ is basically a small rotation increment with a constant fiber length at node J. The unit vector should be updated for the next time step calculation, $\tilde{\boldsymbol{e}}_J = (\tilde{\boldsymbol{e}}_J + \Delta\tilde{\boldsymbol{e}}_J)/|\tilde{\boldsymbol{e}}_J + \Delta\tilde{\boldsymbol{e}}_J|$.

The strain increments can be expressed by using the derivatives of the displacement increments, as usual. At this point, the reduced integration is used for the transverse shear strain components, whereas the other strain components use the regular integration. For instance, if 2×2 integration is used for the general terms of the element, then the one-point reduced integration is used for the transverse shear strain.

Other features of H-L element include the stress increment transformation and rotation matrix, which are for achieving the stress objectivity based on the theory of continuum mechanics.

3.5 PERTURBATION HOURGLASS CONTROL—BELYTSCHKO–TSAY ELEMENT

The discussion in Section 3.4 is emanating from the static applications. Several points are worth noting when dealing with nonlinear transient dynamics by explicit finite element.

It usually takes many time steps to simulate a transient dynamic event. The calculations repeated in every time step should be as few as possible in order to save computing time. As a comparison, quadratic element generally has better accuracy and converges with higher rate than what the linear element does. We could use quadratic element, whose size is twice that of the linear element. The model has the same (or nearly same) amount of nodes and 25% of the total number of linear elements. When using reduced integration, the quadratic element needs 2×2 quadrature points, but the linear element needs only one point. Therefore, both methods use almost the same amount of total integration points. Due to the complexity of formulation of the quadratic elements, the stress evaluation at an integration point takes more calculation than that of the linear element. On the other hand, the frequency of the quadratic element is higher than that of the linear element due to the higher order interpolation. As a result, the time step size of the quadratic element will be smaller than twice of the time step size of the corresponding linear element, which is half size of the quadratic element. So far the advantage of using quadratic element over linear element has not been verified for a general class of applications of nonlinear transient dynamics. Software implementation for explicit finite element is mainly for linear element. Thus, we focus on linear element only.

On the other hand, in transient dynamics, any error occurred at any time will possibly propagate into the future; and from neighboring elements to the farther region. Something that seems less important in the static analysis may become a major issue in transient dynamic analysis. Thus, the finite element method for the transient dynamics deserves more thoughtful development. One typical issue is the rank deficiency of the stiffness matrix with extra zero eigenvalue generated from the reduced integration scheme.

3.5.1 Concept of Hourglass Control

The rank deficiency of stiffness due to the reduced integration scheme has motivated people to devote years of continuous effort of research. Since shape functions are used for interpolation in finite element method, it is helpful to investigate the role of the shape functions in stress and strain calculation. For example, the x-velocity is expressed by (3.35 and 3.36)

$$\dot{u}_x = \dot{u}_x^N \Phi_N(\xi, \eta) = \dot{u}_x^N \left(\Sigma_N + \Lambda_N^1 \xi + \Lambda_N^2 \eta + \Gamma_N \xi \eta\right)/4. \tag{3.50}$$

At any point (ξ, η), $\dot{u}_x$ can be viewed as a four-dimensional (4D) vector formed by linear combination of vectors $(\boldsymbol{\Sigma}, \boldsymbol{\Lambda}_1, \boldsymbol{\Lambda}_2, \boldsymbol{\Gamma})$ with nodal values as the coefficients. These four vectors, listed in Table 3.1, form an orthogonal set and hence perform as the base vectors of the 4D vector space. The velocity fields corresponding to these vectors are depicted in Figure 3.8.

In finite difference method with the regular rectangular grid mesh, the displacement of hourglass shape produces no dilatation strain and zero stress. Artificial viscosity

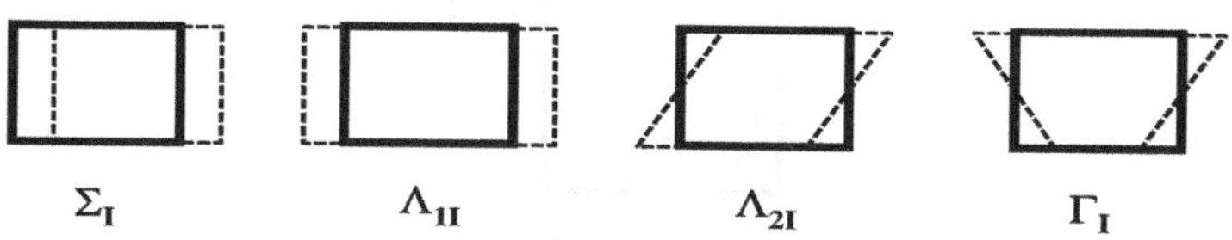

FIGURE 3.8 Basic deformation modes of the four-node shell element. (Reprinted from Flanagan and Belytschko, 1981. Copyright (1981), with permission from Wiley-Blackwell.)

was proposed by Maenchen and Sack (1964) to stabilize the stiffness matrix. However, the application has been limited. The viscosity was added to the nodes without distinguishing hourglass modes from either uniform strain or rigid body motion. It was found in Petschek and Hansen (1968) about the 2D elastic flow analysis that the missing bilinear terms in velocity field were responsible for the hourglass modes.

Similar study on finite element method was reported by Belytschko (1974). In the hydrodynamic analysis with Navier-Stokes equations using linear element, the assumption of constant pressure, which seemed to be attractive, resulted in no resistance to motion of hourglass mode. A linear pressure element, consistent with displacement field was then proposed with

$$P = P_0 + P_1 x + P_2 y. \tag{3.51}$$

The coefficients P_0, P_1, and P_2 were determined by using the equation of state.

Kosloff and Frazier (1978) proposed an hourglass stiffness matrix, superimposed to the singular stiffness matrix resulted from the one-point reduced integration. Consider a simple model of rectangular element subject to a linear distribution of stress σ_x along the sides $x = \pm a$, shown in Figure 3.9. For the plane stress case, the displacement, strain and stress are suggested below:

$$\begin{cases} u_x = \sigma_0 xy/E \\ u_y = \sigma_0(a^2 - x^2 - \nu y^2)/2E \end{cases} \quad \begin{cases} \varepsilon_x = \sigma_0 y/E \\ \varepsilon_y = -\nu\sigma_0 y/E \\ \gamma_{xy} = 0 \end{cases} \quad \begin{cases} \sigma_x = \sigma_0 y \\ \sigma_y = 0 \\ \gamma_{xy} = 0 \end{cases} . \tag{3.52}$$

The stiffness is then calculated from this model analytically (note that the original paper gave $\varepsilon_y = 0$ without the term $\nu\, y^2$ in u_y, which does not satisfy elasticity equation, but does not affect the strain energy). Similar step is taken for σ_y along the sides $y = \pm b$. The stiffness thus formed was named hourglass stiffness matrix. It was added to the calculated singular stiffness matrix, which was a result of using one-point reduced integration. Similar approach applies to 3D solid element. For rectilinear element, the method was equivalent to the incompatible element of Wilson et al. (1973) with bubble functions. The later was used for enhancing convergence in beam bending problem.

Taylor (1979) extended this method to shell element. Applying this approach to general quadrilateral elements, however, requires solving two (four for 3D) sets of four (eight for 3D) simultaneous equations.

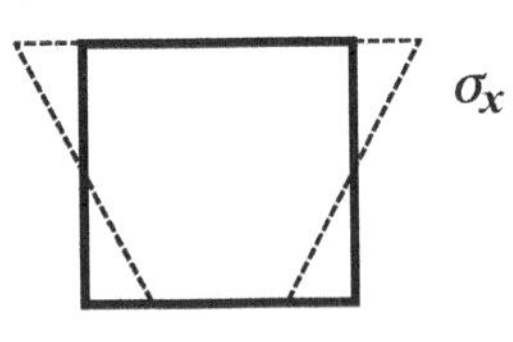

FIGURE 3.9 Linear distribution of stress at the element side.

Reduced integration for the bilinear interpolation uses a single quadrature point at the element center with $\xi = \eta = 0$. The evaluation of strain increment at the element center obtains the correct components for translation, stretch/compression, and shear, but zero hourglass components.

Perturbation hourglass control, an improved method, was developed in Flanagan and Belytschko (1981) for general four-node quadrilateral element and eight-node brick element with the one-point integration. The algorithm was developed with velocity-based formulation, which was suitable for large deformation and explicit finite element method. As mentioned before, the stiffness matrix is not formed in explicit finite element method. Instead, the increments of strain and stress are calculated from the increment of displacement in each time step. The procedure of taking care of hourglass modes was determined by the deformation pattern of the displacement increment. Belytschko and Tsay (1983) extended this method to R-M plate bending element. Belytschko et al. (1984) then further extended this method to a more general type of quadrilateral shell element for nonlinear transient dynamics, commonly known as the Belytschko–Tsay (B-T) shell element. We will discuss its fundamentals in the following sections.

3.5.2 Four-node Belytschko–Tsay Shell Element—Perturbation Hourglass Control

An element local system was suggested by Belytschko et al. (1984). For instance, $\boldsymbol{r}_{13} \times \boldsymbol{r}_{24}$ could be taken as the normal direction of the element and used for defining the unit vector $\boldsymbol{e}_3$ of the local system. Here and later, we denote $\boldsymbol{r}_{IJ}$ for the vector $\boldsymbol{x}_J - \boldsymbol{x}_I$, and do similarly for the other variables. The side or the diagonal could be used to define a unit vector $\tilde{\boldsymbol{e}}_1$. The unit vector $\boldsymbol{e}_2$ was defined by using $\boldsymbol{e}_3 \times \tilde{\boldsymbol{e}}_1$. Finally, an orthogonal system was set with $\boldsymbol{e}_1 = \boldsymbol{e}_2 \times \boldsymbol{e}_3$.

Alternatively, we can use mid-side points to define $\tilde{\boldsymbol{e}}_1 = \boldsymbol{r}_2 + \boldsymbol{r}_3 - \boldsymbol{r}_1 - \boldsymbol{r}_4$, $\tilde{\boldsymbol{e}}_2 = \boldsymbol{r}_3 + \boldsymbol{r}_4 - \boldsymbol{r}_1 - \boldsymbol{r}_2$, and $\tilde{\boldsymbol{e}}_3 = \tilde{\boldsymbol{e}}_1 \times \tilde{\boldsymbol{e}}_2$. Then, we obtain the normalized unit vectors. In this system, $\tilde{\boldsymbol{e}}_1$ and $\tilde{\boldsymbol{e}}_2$ lie in the tangent plane at the element center. The bilinear interpolation forms a woven pattern for the element.

The transformation mapping the element to the master element in the reference plane is $x_j = \sum x_j^N \Phi_N(\xi,\eta)$, using the same shape functions defined in (3.35). The associated 2D Jacobian is

$$
\begin{aligned}
\boldsymbol{J} &= \frac{\partial(x,y)}{\partial(\xi,\eta)} = \begin{bmatrix} x_\xi & x_\eta \\ y_\xi & y_\eta \end{bmatrix}, \\
|\boldsymbol{J}| &= \det(\boldsymbol{J}) = x_\xi y_\eta - x_\eta y_\xi = J_0 + J_1\xi + J_2\eta, \\
\boldsymbol{J}^{-1} &= \frac{\partial(\xi,\eta)}{\partial(x,y)} = \begin{bmatrix} \xi_x & \xi_y \\ \eta_x & \eta_y \end{bmatrix} = \frac{1}{|\boldsymbol{J}|}\begin{bmatrix} y_\eta & -x_\eta \\ -y_\xi & x_\xi \end{bmatrix}.
\end{aligned} \tag{3.53}
$$

We can verify at the center of element:

$$
\begin{aligned}
J_0 &= \det\left(\boldsymbol{J}_{(\xi,\eta)=(0,0)}\right) = |(x_\xi y_\eta - x_\eta y_\xi)|_{(\xi,\eta)=(0,0)}| \\
&= |\boldsymbol{r}_{13} \times \boldsymbol{r}_{24}|/8 = A/4.
\end{aligned} \tag{3.54}
$$

Here, A is the area of the element.

For convenience in formulating the finite element equations with derivatives of the bilinear shape functions, Belytschko and Bachrach (1986) introduced an alternate form of shape functions:

$$\Phi_N = \Delta_N + x b_{xN} + y b_{yN} + H\gamma_N, \tag{3.55}$$

where $H = \xi\eta$. Plugging the isoparametric transformation for x and y in (3.55) leads to

$$\begin{aligned}
\Delta_N &= \left(\Sigma_N - \Sigma_J x^J b_{xN} - \Sigma_J y^J b_{yN}\right)/4, \\
\gamma_N &= \left(\Gamma_N - \Gamma_J x^J b_{xN} - \Gamma_J y^J b_{yN}\right)/4, \\
b_{xN} &= \Phi_{N,x}|_{(0,0)}, \\
b_{yN} &= \Phi_{N,y}|_{(0,0)}.
\end{aligned} \tag{3.56}$$

It is easy to verify $H_{,x}|_{(0,0)} = H_{,y}|_{(0,0)} = 0$. In fact, we have

$$\begin{aligned}
\Phi_{N,x}|_{(0,0)} &= \left(\Lambda_N^1 \xi_x + \Lambda_N^2 \eta_x\right)|_{(0,0)}/4 = \{-y_{24}, y_{13}, y_{24}, -y_{13}\}/2A, \\
\Phi_{N,y}|_{(0,0)} &= \left(\Lambda_N^1 \xi_y + \Lambda_N^2 \eta_y\right)|_{(0,0)}/4 = \{x_{24}, -x_{13}, -x_{24}, x_{13}\}/2A.
\end{aligned} \tag{3.57}$$

Then

$$\begin{aligned}
\Phi_{N,x} &= b_{xN} + \gamma_N H_{,x}, \\
\Phi_{N,y} &= b_{yN} + \gamma_N H_{,y}.
\end{aligned} \tag{3.58}$$

Note that the original paper used different notations. $-\vartheta_y$ and ϑ_x of the reference correspond to our $\dot{\beta}_x$ and $\dot{\beta}_y$, respectively. r_{JI} of the reference corresponds to our r_{IJ}.

Using (3.58), we write the velocity strain based on (3.23) for R-M plate element:

$$\begin{aligned}
d_{xx} &= (\dot{u}_{xN} - z\dot{\beta}_{xN})\Phi_{N,x} = (\dot{u}_{xN} - z\dot{\beta}_{xN})(b_{xN} + H_{,x}\gamma_N), \\
d_{yy} &= (\dot{u}_{yN} - z\dot{\beta}_{yN})\Phi_{N,y} = (\dot{u}_{yN} - z\dot{\beta}_{yN})(b_{yN} + H_{,y}\gamma_N), \\
2d_{xy} &= (\dot{u}_{xN} - z\dot{\beta}_{xN})\Phi_{N,y} + (\dot{u}_{yN} - z\dot{\beta}_{yN})\Phi_{N,x} \\
&= (\dot{u}_{xN} - z\dot{\beta}_{xN})(b_{yN} + H_{,y}\gamma_N) + (\dot{u}_{yN} - z\dot{\beta}_{yN})(b_{xN} + H_{,x}\gamma_N), \\
2d_{xz} &= \dot{w}_N \Phi_{N,x} - \dot{\beta}_{xN}\Phi_N = \dot{w}_N(b_{xN} + H_{,x}\gamma_N) - \dot{\beta}_{xN}\Phi_N, \\
2d_{yz} &= \dot{w}_N \Phi_{N,y} - \dot{\beta}_{yN}\Phi_N = \dot{w}_N(b_{yN} + H_{,y}\gamma_N) - \dot{\beta}_{yN}\Phi_N.
\end{aligned} \tag{3.59}$$

As previously discussed, it would lead to shear locking if (3.59) is fully implemented. The reduced integration using one point at element center results in a reduced form of velocity strain:

$$\begin{aligned} d^0_{xx} &= (\dot{u}_{xN} - z\dot{\beta}_{xN})b_{xN}, \\ d^0_{yy} &= (\dot{u}_{yN} - z\dot{\beta}_{yN})b_{yN}, \\ 2d^0_{xy} &= (\dot{u}_{xN} - z\dot{\beta}_{xN})b_{yN} + (\dot{u}_{yN} - z\dot{\beta}_{yN})b_{xN}, \\ 2d^0_{xz} &= \dot{w}_N b_{xN} - \dot{\beta}_{xN}\Sigma_N/4, \\ 2d^0_{yz} &= \dot{w}_N b_{yN} - \dot{\beta}_{yN}\Sigma_N/4. \end{aligned} \tag{3.60}$$

This contributes to the constant part of the strain rate (3.59). For the nonconstant part of the velocity strain (3.59), it is clear that the vector γ_N plays a special role. In fact, Flanagan and Belytschko (1981) used this vector to detect the hourglass modes. Belytschko et al. (1984) extended the method to the four-node quadrilateral shell element and introduced generalized hourglass strain rate, defined by

$$\begin{aligned} \dot{q}^B_\alpha &= \gamma_N \dot{\beta}_{\alpha N}, \\ \dot{q}^B_3 &= \gamma_N \dot{w}_N, \\ \dot{q}^M_\alpha &= \gamma_N \dot{u}_{\alpha N}. \end{aligned} \tag{3.61}$$

The superscript B and M indicated bending- and membrane-related contributions. We denote by Q^B_α, Q^B_3, and Q^M_α for the generalized hourglass stress conjugated to the generalized hourglass strain rate. Work rate contributed from hourglass nodal force on the hourglass nodal velocity, and from the generalized hourglass stress on the generalized hourglass strain rate, are equivalent. Therefore, we obtain

$$\begin{aligned} f^{HG}_{iN}\dot{u}_{iN} + m^{HG}_{\alpha N}\dot{\beta}_{\alpha N} &= Q^M_\alpha \dot{q}^M_\alpha + Q^B_3 \dot{q}^B_3 + Q^B_\alpha \dot{q}^B_\alpha \\ &= \left(Q^M_\alpha \dot{u}_{\alpha N} + Q^B_3 \dot{w}_N + Q^B_\alpha \dot{\beta}_{\alpha N}\right)\gamma_N. \end{aligned}$$

This suggests

$$\begin{aligned} f^{HG}_{\alpha N} &= \gamma_N Q^M_\alpha, \\ f^{HG}_{3N} &= \gamma_N Q^B_3, \\ m^{HG}_{\alpha N} &= \gamma_N Q^B_\alpha. \end{aligned} \tag{3.62}$$

A simple "diagonal" form of the incremental relation for the generalized hourglass stress and strain was proposed in Belytschko et al. (1984),

$$\begin{aligned} \dot{Q}^B_\alpha &= C_1 \dot{q}^B_\alpha, & C_1 &= r_\theta E\zeta^3 A b_{\alpha N} b_{\alpha N}/192, \\ \dot{Q}^B_3 &= C_2 \dot{q}^B_3, & C_2 &= r_w \kappa G\zeta^3 b_{\alpha N} b_{\alpha N}/12, \\ \dot{Q}^M_\alpha &= C_3 \dot{q}^M_\alpha, & C_3 &= r_m E\zeta A b_{\alpha N} b_{\alpha N}/8. \end{aligned} \tag{3.63}$$

Based on numerical experiences, a range of [0.01, 0.05] for the scalar factors r_j in (3.63) was also recommended. Due to the choice of small values for these scalars, this method was named as perturbation hourglass control.

Take into account of the contribution of the hourglass mode $\boldsymbol{d}^H = \boldsymbol{d} - \boldsymbol{d}^0$ to the strain energy in elastic environment. The energy integral contains terms of $\int H_{,x} H_{,x} \mathrm{d}V$, etc. If the element is a rectangle, we can obtain

$$\begin{aligned}
\int H_{,x} H_{,x} \mathrm{d}V &= \int (\eta \xi_x + \xi \eta_x)^2 \mathrm{d}V = \int \zeta |J| (y_{,\eta} \eta - y_{,\xi} \xi)^2 / |J|^2 \mathrm{d}\xi \mathrm{d}\eta \\
&= \zeta \iint |J|^{-1} \left(y_N \left(\Lambda_N^2 + \Pi_N \xi \right) \eta - y_N \left(\Lambda_N^1 + \Pi_N \eta \right) \xi \right)^2 \mathrm{d}\xi \mathrm{d}\eta / 16 \\
&\cong \frac{\zeta}{12A} \left(\left(y_N \Lambda_N^2 \right)^2 + \left(y_N \Lambda_N^1 \right)^2 \right) = \frac{\zeta A}{3} b_{xN} b_{xN}, \\
\int H_{,y} H_{,y} \mathrm{d}V &= \frac{\zeta A}{3} b_{yN} b_{yN}.
\end{aligned}$$

These terms form the major factors in (3.63).

In (3.62), total quantity of the generalized hourglass stress $\boldsymbol{Q}$ is used, which is considered as the accumulated action by integration during the time history with $\dot{\boldsymbol{Q}} = \boldsymbol{C}\dot{\boldsymbol{q}}$ in (3.63). This method is called stiffness method. Since (3.63) is based on elasticity model, the method is also named as elastic hourglass control in some literatures, with the counterpart based on plasticity model named as plastic hourglass control. If using $\boldsymbol{Q} = \boldsymbol{C}_v \dot{\boldsymbol{q}}$, we can design a viscosity method, and a mixed method with $\boldsymbol{Q} = \alpha_1 \int \boldsymbol{C}\dot{\boldsymbol{q}} \mathrm{d}t + \alpha_2 \boldsymbol{C}_v \dot{\boldsymbol{q}}$. In view of a spring-damper model, with the deformation as the hourglass mode, these methods represent the resistance force from stiffness and damper, respectively. In the sense that the hourglass mode is not a wanted motion mode, the stiffness method will retain a limited oscillation but cannot stop. The viscosity method will finally damp the motion with an unrecoverable deformation, and the mixed method can finally eliminate the motion.

For illustration, an initial velocity $v_z = 2$ of hourglass mode is assigned to the four nodes of a rectangular element, shown in Figure 3.10a. In this example, the element is 10×10 with thickness $\zeta = 1.2$, $E = 200$, $\rho = 8 \times 10^{-6}$, $\nu = 0$. We use the commercial software LS-DYNA V971 to perform the computation. Without hourglass control, strain and stress are not found in element calculation. The nodes move like free particles without any resistance, depicted in Figure 3.10b. The nodal displacement and the energy calculated by different hourglass control methods are shown in Figures 3.11–3.13. The behavior is just like the spring and damper. Note that in these cases the usual deformation energy is always zero due to the one-point reduced integration. The hourglass energy plays the role of strain energy, but not the natural deformation energy. In this study, the mixed method uses half strength of stiffness method and half strength of viscous method.

There are few more technical features implemented in B-T element:

1. The corotational system was used for element stress calculation, so that the transformation for stress components could be saved. As discussed in

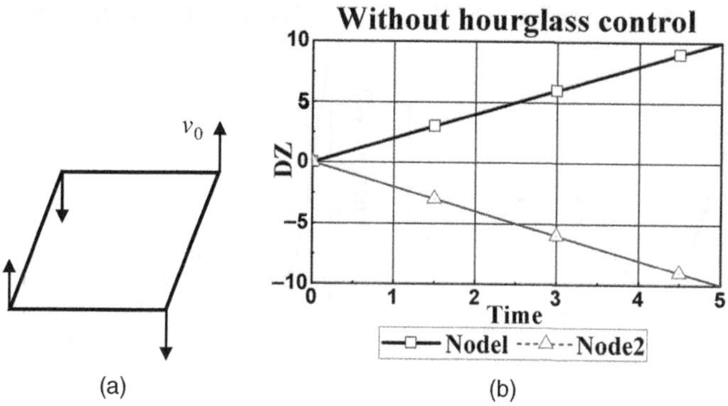

FIGURE 3.10 Response of an element with initial out-of-plane velocity of hourglass shape: (a) initial nodal velocity in an hourglass shape; (b) displacement without hourglass control.

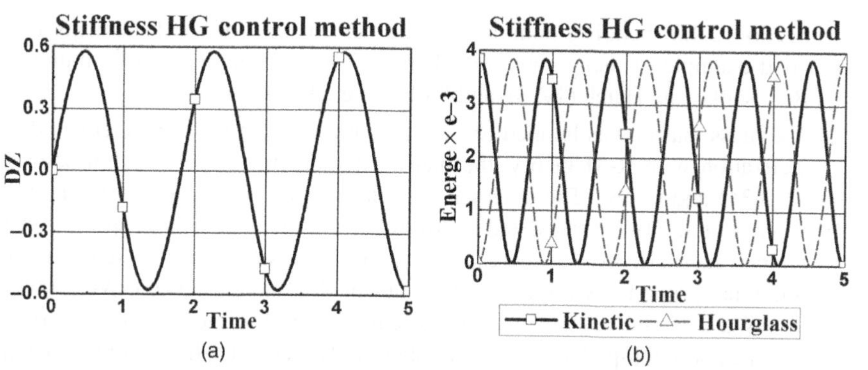

FIGURE 3.11 Stiffness method of hourglass control: (a) nodal displacement; (b) energy calculation.

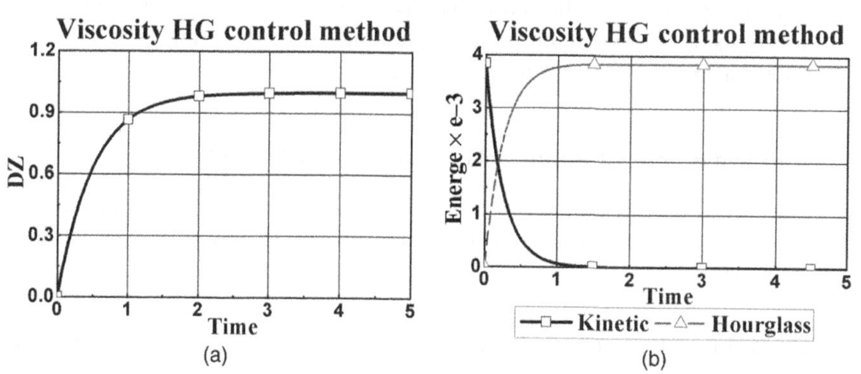

FIGURE 3.12 Viscous method of hourglass control: (a) nodal displacement; (b) energy calculation.

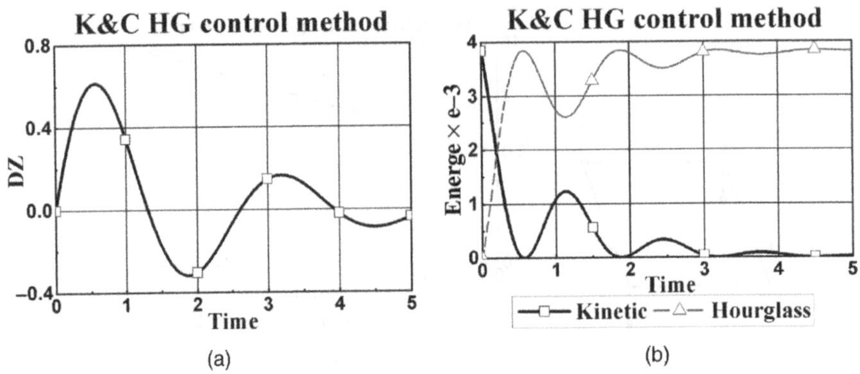

FIGURE 3.13 Mixed (viscoelastic) method of hourglass control: (a) nodal displacement; (b) energy calculation.

Section 2.4.3, bending dominated large deformation of shell elements is a good example of using corotational stress. More discussion can be found in Belytschko and Hsieh (1973), and Stolarski et al. (1995).

2. The mass matrix was lumped for fast execution. Using the finite element discretization with R-M theory, the matrix of moment of inertia is introduced in (3.34) and (3.38). The diagonal lumping was also adopted for the inertia. Hughes et al. (1978) suggested a term related to element area be added to the lumped inertia matrix to increase the stable time step, which is required by rotational property. See the recent studies of Wu and Gu (2003), Wu (2006), also Wu and Qiu (2009) for more discussions on the theory and numerical scheme about lumped mass of the explicit finite element. The examples shown in these publications were applications of B-T element.
3. The shear correction factor served as a penalty parameter for enforcing the Kirchhoff condition due to the nature of perturbed hourglass control. A reduced value proposed in Mindle and Belytschko (1983) was adopted. This approach also reduced the maximum frequency and increased the stable time step. See Babuška and Li (1992) for more discussion.

3.5.3 Improvement of Belytschko–Tsay Shell Element

The quadrilateral B-T element has been the main shell element in many applications using explicit finite element method since its inception. However, its shortcomings are also observed:

1. It failed to solve some of the test problems such as twisted beam and hemispherical shell due to warping.
2. It used artificial coefficients with empirical values, which lacked strong theoretical support.

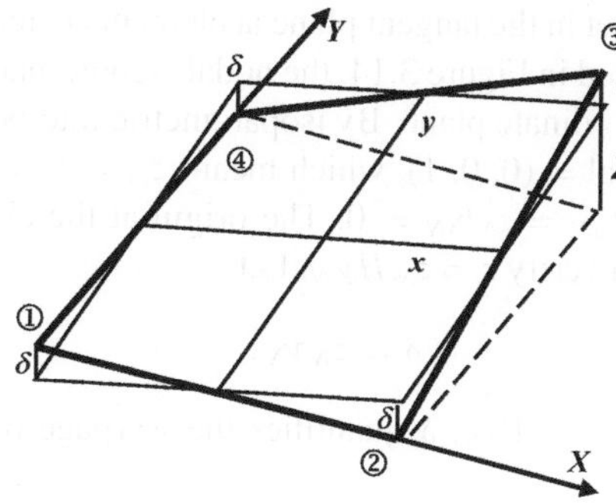

FIGURE 3.14 Description and measure of warping.

3. It did not pass the Kirchhoff-type patch test.
4. Its formulation was not variationally consistent.

The artificial coefficients for hourglass control are usually small. Most software has set their default values for general applications. When these parameters approach zero, however, the ability to control hourglass is lost. It is different from the usual perturbation method expecting a convergence with the small parameter approaching zero. On the other hand, locking will be evident if the parameters are set too large. The theoretical assessment remains to be challenging.

In this section, we discuss the improvement in handling warping.

The B-T element is formulated based on a flat configuration. Often, the element has warped geometry when the four nodes are not coplanar. Usually warping occurs when meshing the curved surfaces, as well as when bending along the diagonal during large deformation. Assume that we have a nearly squared element shown in Figure 3.14, with side length a and $z_1 = z_2 = z_4 = 0$ but $z_3 = z$. Consider a rigid rotation about the fixed left side with angular velocity ω. The corresponding velocity field is $\dot{u}_x = [0,\ 0,\ \omega z,\ 0]$, $\dot{u}_y = [0,\ 0,\ 0,\ 0]$, and $\dot{w} = [0,\ -\omega a,\ -\omega a,\ 0]$. In this case, $\sum \Gamma_N x^N = \sum \Gamma_N y^N = 0$ and $\gamma_N = \Gamma_N$. Now, the generalized hourglass strain rate $\dot{q}_x = \sum \gamma_N \dot{u}_x^N = \omega z \neq 0$. Then, the hourglass stress rate $\dot{Q}_x \neq 0$ and its contribution to the nodal force $F_x \neq 0$ due to the hourglass control procedure. This type of unexpected artificial force causes some uncertainty in the numerical analysis. To improve the handling of warping issue within the framework of B-T element, Belytschko et al. (1984) proposed to modify the hourglass strain rate of the membrane part as

$$\begin{bmatrix} \dot{q}_x^M \\ \dot{q}_y^M \end{bmatrix} = \gamma_N \begin{bmatrix} \dot{u}_x^N \\ \dot{u}_y^N \end{bmatrix} - 4\delta \begin{bmatrix} \dot{\beta}_x \\ \dot{\beta}_y \end{bmatrix}, \quad \begin{bmatrix} \dot{\beta}_x \\ \dot{\beta}_y \end{bmatrix} = \begin{bmatrix} \sum \dot{\beta}_x^N/4 \\ \sum \dot{\beta}_x^N/4 \end{bmatrix}. \tag{3.64}$$

Here, the nodal angular velocity was used to represent the action of rigid rotation. δ represented the quantity of warpage with an assumption that $z_N = \Gamma_N \delta$.

If the local system is set in the tangent plane at element center $\xi = \eta = 0$, using the mid-side points as depicted in Figure 3.14, the nodal z-coordinates z_N are the distance from the nodes to the coordinate plane. By isoparametric interpolation, $z = z_N \Phi_N$. At element center the normal $= (0, 0, 1)$, which means $z_{,x} = z_{,y} = 0$ at element center. This is equivalent to $z_N b_{xN} = z_N b_{yN} = 0$. The origin at the element center implies $z_N \Sigma_N = 0$. Then, we can verify $z = z_N H \gamma_N$. Let

$$\delta = z_N \gamma_N. \tag{3.65}$$

Hence, $z = H\delta$ and $z_N = \Gamma_N \delta$. δ quantifies the warpage of the element. $\delta = 0$ means a flat element.

In the example discussed above, the rigid rotation will not generate hourglass velocity with this approach. Note that (3.64) does not completely solve the issue in general applications. For some of the benchmark examples with curved surfaces, such as twisted beam, B-T element does not provide correct answer.

The issues related to warped element have motivated many investigations. Belytschko et al. (1992) (B-W-C) studied the quadrilateral geometry in 3D framework for the bilinear R-M plate element with possible warping. An element local system was defined with nodal locations at top and bottom surfaces, similar to the concept of Hughes and Liu (1981). The normal direction was assumed as variable of the element.

In B-W-C element, the Jacobian and subsequently the derivatives of shape functions contained terms with a factor of δ. The perturbation hourglass control of B-T element still applied to this situation with little change for coefficient C_1 of (3.63). This element improved from B-T element so that the twisted beam problem was solved correctly.

In practical applications of large deformation, bending along the element diagonal may happen in certain small area such as folding lines. This kind of deformation produces warping and initiates hourglass mode. Wu et al. (2006) and (2007) reported that mesh refinement could help reduce the zone affected by hourglass control method, and improve the robustness of computation.

3.5.4 About Convergence of Element using Reduced Integration

The B-T element using reduced integration and perturbation hourglass control does not pass patch test, although successful experiences provide engineers good confidence. Whereas mathematical assessment for B-T element remains a challenging task, some results regarding reduced integration are available. Jacquotte and Oden (1984) employed a projection method in postprocessing the finite element solution of Laplace equation, using reduced integration but without hourglass control, and proved that the postprocessed results can converge with optimal rates like the other standard method. It is named a posteriori control. See Jacquotte and Oden (1986) for more discussions.

It is shown that when mesh is refined, B-T element can converge and the results from fine mesh are close to that of the other elements that do not use perturbation hourglass control, cf. Wu et al. (2004).

3.6 PHYSICAL HOURGLASS CONTROL—BELYTSCHKO–LEVIATHAN (QPH) ELEMENT

In a series of improvement to B-T element, Belytschko and Leviathan (1994a, 1994b) developed a new technique, named physical hourglass control. The element adopted some techniques developed in that period of time. It used one-point quadrature but no parameters for hourglass control. The element is named Belytschko–Leviathan (B-L) element or QPH element in some literatures. We briefly describe the fundamentals of B-L element here, following Belytschko and Leviathan (1994a, 1994b).

3.6.1 Constant and Nonconstant Contributions

Similar to the method of Belytschko et al. (1992), Belytschko and Leviathan (1994a, 1994b) separated the constant and nonconstant contributions to velocity strain. The warped element geometry was mapped to the reference element. Based on the deformation modes, the velocity strain (rate of deformation) was decomposed into three parts: membrane, bending, and transverse shear. With five degrees of freedom per node, $(u_x, u_y, w, \beta_x, \beta_y)$, the velocity strain tensor was expressed in matrix form:

$$\begin{bmatrix} d_{xx} \\ d_{yy} \\ 2d_{xy} \\ 2d_{xz} \\ 2d_{yz} \end{bmatrix} = \sum_{N=1}^{4} (\mathbf{B}_N)_{5\times 5} \begin{bmatrix} \dot{u}_x^N \\ \dot{u}_y^N \\ \dot{w}^N \\ \dot{\beta}_x^N \\ \dot{\beta}_y^N \end{bmatrix}, \tag{3.66}$$

$$\boldsymbol{B}_N = \boldsymbol{B}_N^m + \boldsymbol{B}_N^b + \boldsymbol{B}_N^s. \tag{3.67}$$

Each of these $\boldsymbol{B}$ matrices consisted of a constant part corresponding to the one-point integration at element center and a nonconstant part corresponding to the contribution by the hourglass modes:

$$\begin{aligned} \boldsymbol{B}_N^m &= \left(\boldsymbol{B}_N^m\right)^0 + \left(\boldsymbol{B}_N^m\right)^H, \\ \boldsymbol{B}_N^b &= \left(\boldsymbol{B}_N^b\right)^0 + \left(\boldsymbol{B}_N^b\right)^H, \\ \boldsymbol{B}_N^s &= \left(\boldsymbol{B}_N^s\right)^0 + \left(\boldsymbol{B}_N^s\right)^H. \end{aligned} \tag{3.68}$$

The membrane-related matrix was proposed as

$$\left(\boldsymbol{B}_N^m\right)^0 = \begin{bmatrix} b_{xN} & 0 & 0 & 0 & 0 \\ 0 & b_{yN} & 0 & 0 & 0 \\ b_{yN} & b_{xN} & 0 & 0 & 0 \\ 0 & 0 & 0 & 0 & 0 \\ 0 & 0 & 0 & 0 & 0 \end{bmatrix}, \tag{3.69a}$$

$$\left(\boldsymbol{B}_N^m\right)^H = \begin{bmatrix} H_{,x}\gamma_N & 0 & 0 & H_{,x}Z_\gamma\Sigma_N/4 & 0 \\ 0 & H_{,y}\gamma_N & 0 & 0 & H_{,y}Z_\gamma\Sigma_N/4 \\ H_{,y}\gamma_N & H_{,x}\gamma_N & 0 & H_{,y}Z_\gamma\Sigma_N/4 & H_{,x}Z_\gamma\Sigma_N/4 \\ 0 & 0 & 0 & 0 & 0 \\ 0 & 0 & 0 & 0 & 0 \end{bmatrix}. \tag{3.69b}$$

Here, Z_γ is δ of (3.65). b_{xN} and b_{yN} are defined in (3.56). The terms included in $(\boldsymbol{B}_N^m)^0$ were from the usual calculations. The last two columns of $(\boldsymbol{B}_N^m)^H$ were added to what was resulted from the formal calculation as a correction for improving B-T element with warping. These terms coupled the membrane action with rotation of the normal. They vanish when the element is really flat. Note that they are different from (3.64), which is the formula used by Belytschko et al. (1984).

The bending-related matrix was proposed to be

$$\left(\boldsymbol{B}_N^b\right)^0 = \frac{\zeta}{2}\begin{bmatrix} b_{xN}^c & 0 & 0 & -b_{xN} & 0 \\ 0 & b_{yN}^c 0 & 0 & 0 & -b_{yN} \\ b_{yN}^c & b_{xN}^c & 0 & -b_{yN} & -b_{xN} \\ 0 & 0 & 0 & 0 & 0 \\ 0 & 0 & 0 & 0 & 0 \end{bmatrix}, \tag{3.70a}$$

$$\left(\boldsymbol{B}_N^b\right)^H = \frac{\zeta}{2}\begin{bmatrix} 0 & 0 & 0 & -H_{,x}\gamma_N & 0 \\ 0 & 0 & 0 & 0 & -H_{,y}\gamma_N \\ 0 & 0 & 0 & -H_{,y}\gamma_N & -H_{,x}\gamma_N \\ 0 & 0 & 0 & 0 & 0 \\ 0 & 0 & 0 & 0 & 0 \end{bmatrix}. \tag{3.70b}$$

Here, we denote

$$\begin{aligned} \left\{b_{xN}^c\right\} &= 2z_\gamma\{-x_{13}, x_{24}, x_{13}, -x_{24}\}/A^2, \\ \left\{b_{xN}^c\right\} &= 2z_\gamma\{-y_{13}, y_{24}, y_{13}, -y_{24}\}/A^2. \end{aligned} \tag{3.70c}$$

Note that the previous discussion about rotations of normal applies to this situation too. The first two columns in $(\boldsymbol{B}_N^b)^0$ are additional to the matrix from the usual calculation, with the coupling of membrane action and bending. For a warped element, nodal translation induces bending effect. These terms can improve the element performance.

3.6.2 Projection of Shear Strain

To avoid shear locking, B-L element adopted the projection method for transverse shear used by Wempner et al. (1982), Dvorkin and Bathe (1984), and Bathe and

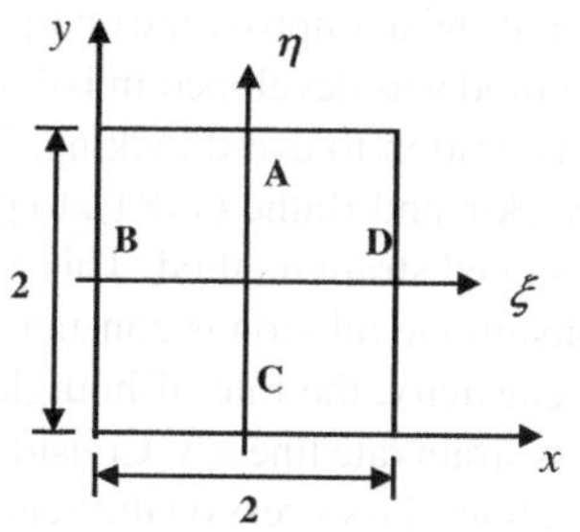

FIGURE 3.15 Mid-side points of B-D element. (Reproduced from Bathe and Dvorkin, 1985. Copyright (1985), with permission from Wiley-Blackwell.)

Dvorkin (1985). This method is quite different from the elements so far discussed here. The shear strain rates were interpolated using the values at mid-side points, illustrated in Figure 3.15:

$$\begin{aligned} d_{\xi\varsigma} &= \frac{1+\eta}{2} d_{\xi\varsigma}^{A} + \frac{1-\eta}{2} d_{\xi\varsigma}^{C}, \\ d_{\eta\varsigma} &= \frac{1+\xi}{2} d_{\eta\varsigma}^{D} + \frac{1-\xi}{2} d_{\eta\varsigma}^{B}. \end{aligned} \tag{3.71}$$

The shear strain rates at the mid-side points were calculated in the usual way. To express the shear strain of the physical domain in terms of the reference parameters, a tensor transform for the covariant components was adopted in Belytschko and Leviathan (1994a), due to the curvilinear system involved in the warped element. The results were also decomposed into the constant part and nonconstant part, with the detailed manipulation omitted here.

3.6.3 Physical Hourglass Control by One-point Integration

The strain rate of a general velocity field characterized by their nodal values can now be expressed by using these $\boldsymbol{B}$ matrices in terms of constant part and nonconstant part:

$$d_{ij} = d_{ij}^{0} + d_{ij}^{H}. \tag{3.72}$$

Each part has contributions from actions of membrane, bending, and shear.

The next step is to calculate the stress by using the incremental constitutive law. Using the one-point integration at element center to obtain the contribution from the constant part of the $\boldsymbol{B}$ matrix was a straightforward exercise, the same as what was done with the other element using one-point integration. The key development in B-L element was the treatment of the nonconstant part. Its contribution disappeared under the one-point integration. The perturbation hourglass control method of B-T element did not need detailed information but some artificial parameters. To avoid the artificial coefficients used by B-T element and to improve the one-point reduced integration, the physical hourglass control method was developed in B-L element. On the other hand, B-T element used reduced integration to avoid locking. B-L element used projected shear strain, also used by Dvorkin and Bathe (1984), to prevent the shear locking. It was also considered as an assumed strain method. This portion used full integration.

When the incremental constitutive relation is constant through the linear element, such as elasticity and super elasticity, the rate of hourglass stress can be calculated from the generalized hourglass strain rate linearly. Consider the fact that the rate of virtual work contributed from the hourglass stress on the generalized hourglass strain rate is equivalent to that contributed from the nodal hourglass force on the nodal velocity:

$$\begin{aligned}\delta W^H &= \int_\Omega (\sigma_{xx}^H \delta d_{xx}^H + \sigma_{yy}^H \delta d_{yy}^H + 2\sigma_{xy}^H \delta d_{xy}^H + 2\sigma_{xz}^H \delta d_{xz}^H + 2\sigma_{yz}^H \delta d_{yz}^H) d\Omega \\ &= f_{xN}^H \delta v_{xN} + f_{yN}^H \delta v_{yN} + f_{zN}^H \delta v_{zN} + m_{xN}^H \delta \omega_{xN} + m_{yN}^H \delta \omega_{yN}.\end{aligned}$$

The hourglass stress rate takes only the contribution of nonconstant part into account. In the rate form, we have with matrix notations

$$\dot{\boldsymbol{\sigma}}^H = \boldsymbol{E}\boldsymbol{d}^H = \boldsymbol{E}\boldsymbol{B}^H \boldsymbol{v}. \tag{3.73}$$

where $\boldsymbol{E}$ represents the constitutive relations, such as elasticity tensor and $\boldsymbol{d}$ is for the velocity strain tensor. The nodal hourglass forces are then calculated in the rate form. The integration is accomplished almost in a closed form:

$$\begin{aligned}(\boldsymbol{f}^H)^T \delta \boldsymbol{v} &= \int (\dot{\boldsymbol{\sigma}}^H)^T \delta \boldsymbol{d}^H \mathrm{d}\Omega = \int \boldsymbol{v}^T (\boldsymbol{B}^H)^T \boldsymbol{E} \boldsymbol{B}^H \delta \boldsymbol{v} \mathrm{d}\Omega \\ (\boldsymbol{f}^H) &= \int (\boldsymbol{B}^H)^T \boldsymbol{E} \boldsymbol{B}^H \boldsymbol{v} \mathrm{d}\Omega.\end{aligned} \tag{3.74}$$

The element formulated in this section performed better than B-T element for several benchmark examples. However, it did not solve the twisted beam problem correctly yet. In fact, the element had five degrees of freedom per node, the same as the B-T element. The drilling degree of freedom did not participate in the element formulation. If the neighboring elements joining at a node had the same or nearly the same tangents at this node, this formulation resulted in a singular or ill-conditioned global system (six degrees of freedom per node).

3.6.4 Drill Projection

Rankin and Nour-Omid (1988) proposed a method to extract the pure deformation from the displacement field, that is, to separate the deformation from the rigid body motion. An orthogonal decomposition for a general field of nodal velocity was proposed in the matrix form:

$$\begin{aligned} \boldsymbol{v} &= \boldsymbol{P}\boldsymbol{v} + \boldsymbol{Q}\boldsymbol{v}, \\ (\boldsymbol{P}\boldsymbol{v}) &\bullet (\boldsymbol{Q}\boldsymbol{v}) = 0. \end{aligned} \tag{3.75}$$

Let $\boldsymbol{R} = \{\boldsymbol{r}_1, \ldots, \boldsymbol{r}_K\}$ represent the K basic rigid body modes and $\sum \alpha_j \boldsymbol{r}_j = \boldsymbol{R}\boldsymbol{\alpha} = \boldsymbol{Q}\boldsymbol{v}$ represent any possible rigid body mode contained in the velocity field. The orthogonality requires $\boldsymbol{Q}\boldsymbol{v} \bullet (\boldsymbol{v} - \boldsymbol{Q}\boldsymbol{v}) = 0$. This means $\boldsymbol{\alpha}^T \boldsymbol{R}^T(\boldsymbol{v} - \boldsymbol{R}\boldsymbol{\alpha}) = 0$. We then obtain the projection $\bar{\boldsymbol{v}} = \boldsymbol{P}\boldsymbol{v}$,

$$\begin{aligned} \boldsymbol{\alpha} &= (\boldsymbol{R}^T\boldsymbol{R})^{-1}\boldsymbol{R}^T\boldsymbol{v}, \\ \boldsymbol{Q}\boldsymbol{v} &= \boldsymbol{R}(\boldsymbol{R}^T\boldsymbol{R})^{-1}\boldsymbol{R}^T\boldsymbol{v}, \\ \boldsymbol{P}\boldsymbol{v} &= \boldsymbol{v} - \boldsymbol{R}(\boldsymbol{R}^T\boldsymbol{R})^{-1}\boldsymbol{R}^T\boldsymbol{v} = \bar{\boldsymbol{v}}. \end{aligned} \tag{3.76}$$

With the equivalence of internal strain energy rate by the internal nodal force with the nodal velocity and the corresponding projected force with the projected velocity, we have

$$\boldsymbol{f}^T\boldsymbol{v} = \bar{\boldsymbol{f}}^T\bar{\boldsymbol{v}}. \tag{3.77}$$

Substituting the last equation of (3.76) in (3.77), we obtain

$$\boldsymbol{f} = \boldsymbol{P}^T\bar{\boldsymbol{f}}. \tag{3.78}$$

A rigid body motion does not generate strain energy. Then, we have $\bar{\boldsymbol{f}}^T(\boldsymbol{Q}\boldsymbol{v}) = \bar{\boldsymbol{f}}^T(\boldsymbol{v} - \bar{\boldsymbol{v}}) = 0$. Using (3.77), we obtain $\bar{\boldsymbol{f}} = \boldsymbol{f}$, that is, in this case there is no difference between the force and the projected force.

Rankin and Nour-Omid (1988) constructed a matrix $\boldsymbol{R}$ by using nodal coordinates to represent three rigid body rotations at each node. Belytschko and Leviathan (1994b) extended the concept to the drilling degree of freedom with the consideration of invariance of rigid body rotations. The nodal velocity contributed from the drilling effect was considered as a rotation with no translation. Matrix $\boldsymbol{R}$ was defined by using the unit normal vectors at the four nodes individually with no interactions from the other nodes.

For any of these projection methods, with $\boldsymbol{R}$ defined and $(\boldsymbol{R}^T\boldsymbol{R})^{-1}$ solved, the implementation of the projection method was straightforward. The procedure consisted of the following main steps:

1. Take the same steps as the usual procedure, up to nodal velocity $\boldsymbol{v}$.
2. Project velocity $\bar{\boldsymbol{v}} = \boldsymbol{P}\boldsymbol{v}$.
3. Calculate element nodal force and moment $\bar{\boldsymbol{f}}$, using $\bar{\boldsymbol{v}}$ in place of $\boldsymbol{v}$.
4. Project the nodal force and moment $\boldsymbol{f} = \boldsymbol{P}^T\bar{\boldsymbol{f}}$.
5. Go back to the usual procedure.

3.6.5 Improvement of B-L (QPH) Element

It is worth noting that with physical hourglass control method the stress calculation can be exact if the material is linear or the incremental constitutive relation is linear, such as elasticity or hyper elasticity. But this goal cannot be achieved for nonlinear material such as plasticity. Due to the complexity of nonlinear material behavior, the information at one quadrature point cannot simply represent the whole element. To improve the application for elastoplastic problems which will be discussed in Chapter 7, Zeng and Combescure (1998) proposed a plastic hourglass control method. The idea was to use the one-dimensional (1D) tangent modulus E_t to calculate the contribution from nonconstant (hourglass) part of strain rate to the stress rate:

$$\dot{\boldsymbol{\sigma}}^H = \boldsymbol{E}\boldsymbol{D}^H E_t / E. \tag{3.79}$$

The main procedure for this part was designed as following:

1. Use the stress from the one-point integration (the constant part) at quadrature points through the thickness, ζ_j, to evaluate tangent modulus $E_t(\zeta_j)$.
2. Find $\lambda^b = \min_j E_t(\zeta_j)/E$ and $\lambda^m = \text{mean}\, E_t(\zeta_j)/E$.
3. Update the generalized hourglass rate of deformation: replace $\dot{\boldsymbol{q}}^b$ and $\dot{\boldsymbol{q}}^m$ with $\lambda^b \dot{\boldsymbol{q}}^b$ and $\lambda^m \dot{\boldsymbol{q}}^m$, respectively. Here $\dot{\boldsymbol{q}}^b$ and $\dot{\boldsymbol{q}}^m$ are for the velocity contributions to the nonconstant part of the bending and membrane terms, respectively.
4. Use the updated hourglass rate of deformation to calculate the stress rate in the usual way.

Besides, Zeng and Combescure (1998) adopted the drill projection of Belytschko and Leviathan (1994b), but without the nodal projection for rigid body rotation by Rankin and Nour-Omid (1988). On the other hand, the nonconstant contribution of membrane part was revised from (3.69b)

$$\left(\boldsymbol{B}_N^m\right)^H = \begin{bmatrix} H_{,x}\gamma_N & 0 & Z_\gamma b_{xN} H_{,x} & 0 & 0 \\ 0 & H_{,y}\gamma_N & Z_\gamma b_{yN} H_{,y} & 0 & 0 \\ H_{,y}\gamma_N & H_{,x}\gamma_N & Z_\gamma (b_{xN} H_{,y} + b_{yN} H_{,x}) & 0 & 0 \\ 0 & 0 & 0 & 0 & 0 \\ 0 & 0 & 0 & 0 & 0 \end{bmatrix}. \tag{3.80}$$

3.7 SHEAR PROJECTION METHOD—BATHE–DVORKIN ELEMENT

3.7.1 Projection of Transverse Shear Strain

The method to prevent the shear locking developed in Dvorkin and Bathe (1984) and Bathe and Dvorkin (1985) was adopted by Belytschko and Leviathan (1994a, 1994b). It used an approach very different from the hourglass control related technologies so far discussed. In fact, it was a nonconfirming scheme used for transverse shear strain to solve the shear locking issue. In another word, an assumed strain method was involved in the element. The approach was first implemented in software NSC/NASTRAN in 1970s, see MacNeal (1982). The plate bending element is named Bathe–Dvorkin (B-D) element, also as fully integrated element in engineering applications because it does not use reduced integration. The element was developed for general nonlinear applications and initially implemented in the implicit static framework, then extended to the explicit software.

In this approach, the transverse shear strains were not directly differentiated from the displacement interpolations, but assumed, presented in the reference plane as

$$\begin{aligned} \gamma_{\xi\zeta} &= \frac{1+\eta}{2}\gamma_{\xi\zeta}^{A} + \frac{1-\eta}{2}\gamma_{\xi\zeta}^{C}, \\ \gamma_{\eta\zeta} &= \frac{1+\xi}{2}\gamma_{\eta\zeta}^{D} + \frac{1-\xi}{2}\gamma_{\eta\zeta}^{B}. \end{aligned} \tag{3.81}$$

Here, A, B, C, and D are the mid-points of the four sides, as shown in Figure 3.15. The shear strain at these points is calculated in a usual way. The component $\gamma_{\xi\zeta}$ is constant at the top and bottom sides and linear in the η-direction. It is constant in ξ-direction with given η inside the element. $\gamma_{\eta\zeta}$ does that in the other direction.

Note that at these mid-points, for instance, $\xi = 0$, $\eta = \pm 1$ at A and C, respectively,

$$\gamma_{\xi\zeta}|_{A,C} = \sum_{N=1}^{4} \left(w_N \left(\Lambda_N^1 \pm \Gamma_N\right) - \beta_{\xi N}\left(\Sigma_N \pm \Lambda_N^2\right)\right)/4. \tag{3.82}$$

For a pure bending about η-axis with $w_N = \beta_{\eta N} = 0$ and $\beta_{\xi N} = \beta_0 \Lambda_N^1$, we have $\gamma_{\xi\zeta}|_{A(C)} = 0$. Subsequently $\gamma_{\xi\zeta} = 0$ in the whole element. In this way, the shear locking is expected to be controlled.

To analyze the element behavior mathematically, Bathe and Brezzi (1985) introduced a projection operator in describing B-D element based on the reference system:

$$\begin{aligned} \gamma_{\xi\zeta} &= w_{,\xi} - \pi\beta_{\xi}, \\ \gamma_{\eta\zeta} &= w_{,\eta} - \pi\beta_{\eta}. \end{aligned} \tag{3.83}$$

Here, the projection operator $\boldsymbol{\pi}$ is defined below:

$$\begin{aligned} \pi\beta_{\xi} &= \sum \beta_{\xi}^{N}\left(\Sigma_N + \Lambda_N^2\eta\right)/4, \\ \pi\beta_{\eta} &= \sum \beta_{\eta}^{N}\left(\Sigma_N + \Lambda_N^1\xi\right)/4. \end{aligned} \tag{3.84}$$

The projection of the rotations uses the similar terms of the derivatives of w. In fact, (3.83) is coincident with (3.82). $\pi\beta_{\xi}$ is linear in η and independent of ξ. It is constant at the top and bottom sides and equals the mean value of the two nodes at the end. Hence, it is continuous when crossing the top and bottom boundaries of the element. Meanwhile, it is highly possibly discontinuous when crossing the left and right boundaries of the element. $\pi\beta_\eta$ is similar.

When transforming to the element coordinate system in the physical domain, the transverse shear strain components were treated as covariant tensor components. The related shear stress components were treated as contravariant tensor components; see Dvorkin and Bathe (1984), Bathe and Dvorkin (1985) and the references cited there for more discussions.

For the in-plane strain components related to membrane and bending, B-D element used 2×2 integration. Therefore, there was no issue of rank deficiency and no need for hourglass control. With the shear strains defined in (3.83) and (3.84), we can verify that rigid body rotation will not yield strain. For instance, consider a rectangular element in the reference system under the rigid body rotation about η-axis ($\xi = 0$), $\beta_\xi \equiv \beta$, $\beta_\eta \equiv 0$, and $w = \beta\xi$. The nodal values and shear strain components are

$$\begin{aligned} \beta_{\xi N} &= \beta, \\ \beta_{\eta N} &= 0, \\ w_N &= \beta\Lambda_N^1, \end{aligned} \tag{3.85}$$

$$\begin{aligned} \gamma_{\xi\zeta} &= \sum_{N=1}^{4} \left(\beta\Lambda_N^1 \left(\Lambda_N^1 + \Gamma_N\eta\right) - \beta\left(\Sigma_N + \Lambda_N^2\eta\right)\right)/4 = 0, \\ \gamma_{\eta\zeta} &= \sum_{N=1}^{4} \left(\beta\Lambda_N^1 \left(\Lambda_N^2 + \Gamma_N\xi\right)\right)/4 = 0. \end{aligned} \tag{3.86}$$

The B-D element has been used for linear and nonlinear static problems as well as frequency analysis. Its application to nonlinear transient dynamics has also been recommended by some researchers in recent years.

3.7.2 Convergence of B-D Element

For the static linear bending problem, Bathe and Brezzi (1985) proved the following theorem,

Assessment 3.5 *For a rectangular meshing, when the mesh is refined, the deflection and the rotation can converge not uniformly with optimal rates in H^1-norm, and uniformly with a rate that is half order lower,*

$$\begin{aligned} &||\boldsymbol{\beta}^s - \boldsymbol{\beta}||_1 + ||w^s - w||_1 + \zeta||\gamma^s - \gamma||_0 \\ &\quad \le C\min(h(||\boldsymbol{\beta}||_3 + \zeta||\boldsymbol{\gamma}||_1 + ||\boldsymbol{\gamma}||_0), h^{1/2}(||\boldsymbol{\beta}||_{5/2} + \zeta||\boldsymbol{\gamma}||_1 + ||\boldsymbol{\gamma}||_0)). \end{aligned} \tag{3.87}$$

Here, the constant C is independent of thickness, $\boldsymbol{\beta}$ and $\boldsymbol{\gamma}$ are the two-component vectors of rotation and transverse shear strain, respectively. $\zeta||\boldsymbol{\gamma}||_1 + ||\boldsymbol{\gamma}||_0$ is bounded uniformly with respect to thickness ζ. But $||\boldsymbol{\beta}||_3$ in the first estimate of the right-hand side is found to be thickness dependent due to the boundary layer shown in Arnold and Falk (1989b, 1990). Therefore, the first estimate is not thickness independent. $||\boldsymbol{\beta}||_{5/2}$ in the second estimate is bounded uniformly, hence the uniform convergence has been proven but with the rate half an order lower than the optimal one. For more discussion on the convergence theory of B-D element, see Brezzi et al. (1989), Durán and Liberman (1992), and Zhang and Zhang (1994). Wu (2005) extended the results to the L^2-estimate.

Assessment 3.6 *The deflection and the rotation can converge with optimal rates in L^2-norm,*

$$||\boldsymbol{\beta}^h - \boldsymbol{\beta}||_0 + ||w^h - w||_0 \le C\zeta^{-1}h\min(\zeta, h)(||\boldsymbol{\beta}||_3 + \zeta||\boldsymbol{\gamma}||_1 + ||\boldsymbol{\gamma}||_0). \quad (3.88)$$

Numerical examples presented by Lyly et al. (1993), Suri et al. (1995), Zhang and Zhang (1994), and Chapelle and Bathe (2003) showed results with the same optimal convergence rate for a quite large range of thickness.

Wu (2005) extended the results to linear transient dynamics by explicit finite element, and proved the following.

Assessment 3.7 *If certain conditions are satisfied by the initial conditions and boundary conditions, for a given thickness, the deflection, rotation, and their velocities can converge with optimal rates in both H^1-norm and L^2-norm,*

$$\begin{aligned} &||\text{err\_}\boldsymbol{\beta}||_0 + ||\text{err\_}w||_0 + ||\text{err\_}\dot{\boldsymbol{\beta}}||_0 + ||\text{err\_}\dot{w}||_0 \le Ch^2, \\ &||\text{err\_}\boldsymbol{\beta}||_1 + ||\text{err\_}w||_1 + ||\text{err\_}\dot{\boldsymbol{\beta}}||_1 + ||\text{err\_}\dot{w}||_1 + \zeta||\text{err\_}\gamma||_0 \le Ch. \end{aligned} \quad (3.89)$$

The results were not proven to be thickness independent either. On the other hand, optimal rates were achievable in the best scenario, but not always, depending on the loading conditions. Numerical examples showed that in some loading case, the B-D element converged with optimal rates for a range of thickness, similar to the static study mentioned above, cf. Wu (2005). But this is not true in some of the other loading cases, for example, excited with initial velocity. When the initial condition does not give zero initial velocity for the shear strain, then even if zero initial value is promised, the shear strain has a chance to grow in the dynamic process. The convergence deteriorates for the velocity-related items when the thickness is reduced a little bit, a typical situation of shear locking. It leads to no convergence to the K-L solution. As a matter of fact, the dynamic situation is more complicated than the static one.

3.8 ASSESSMENT OF FOUR-NODE R-M PLATE ELEMENT

3.8.1 Evaluations with Warped Mesh and Reduced Thickness

Here, we will use two examples to examine the performance of the four-node shell elements.

Example 3.1 Twisted beam B-T element is based on the assumption of a flat element, without accurate representation of warping. For some test problems with warping in major area, the element has failed to give correct answer. The twisted beam was defined in MacNeal and Harder (1985) as one of the test problems, depicted in Figure 3.16. The parameters are: length $= 12$, width $= 1.1$, thickness $= 0.32$, Young's modulus $E = 29.0 \times 10^{-6}$, and Poisson ratio $\nu = 0.22$. The beam is twisted 90 degrees. It is clamped at the left end and subjected to a lateral unit load tangent to the beam surface at the right end. To investigate the performance of B-T element, we use four meshes with 3×1, 6×1, 12×2, and 24×4 elements, respectively. When defining the load for the meshes with three elements and six elements, we apply the load at the node on the top edge. When defining the load for the meshes with 12×2 elements and 24×4 elements, we apply the load at the mid-point, that is, on the central line. The dynamic response of the beam is a type of vibration. The magnitude of the vibration is twice of the displacement of the beam at the equilibrium state under the same load. As stated in MacNeal and Harder (1985), the theoretical solution of the equilibrium problem is 0.005424. The dynamic solution expects a magnitude of vibration equal to 0.010848. We use the commercial software LS-DYNA V971 to perform the study with stiffness hourglass control method. Note that the implementation may vary in different software. B-T element does not give a solution close to the expected one, as shown in Figure 3.17.

This example has also been used to examine other elements to confirm the improvement from B-T element. Here, we study the B-D and B-L elements. As presented in Figure 3.18, the results of both elements appear to converge. The results of B-D

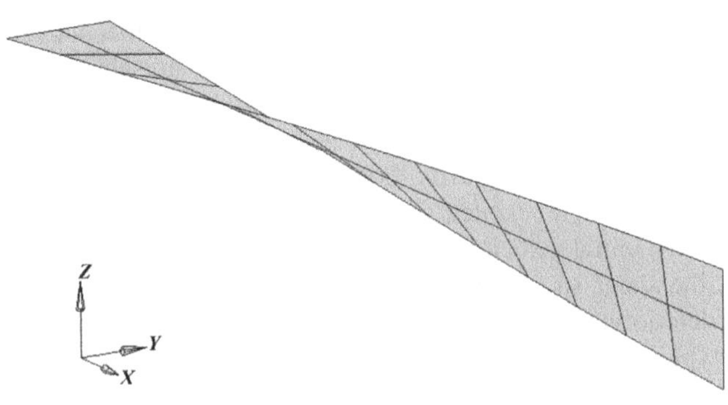

FIGURE 3.16 Geometry of the twisted beam.

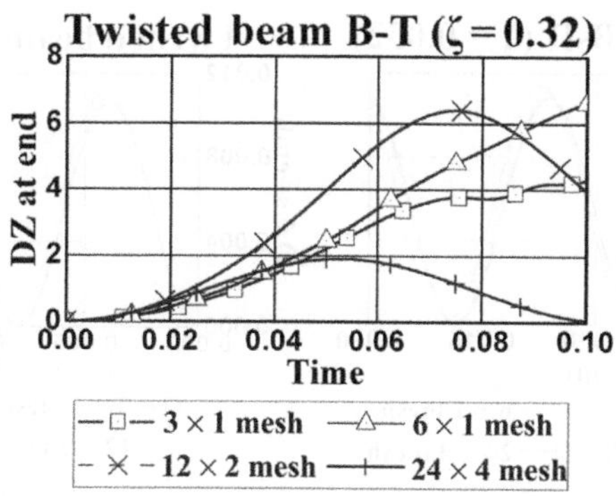

FIGURE 3.17 Solution of the twisted beam problem by B-T element.

element have small difference from coarse mesh to fine mesh. The results of B-L element show some difference from coarse mesh to fine mesh. The results from meshes with 12 × 2 elements and 24 × 4 elements are very close.

We further investigate the shear locking issue with reduced thickness using this example. The load scaling and mass scaling discussed in Sections 3.2.2 and 3.2.3 are applied. When we reduce the thickness by a factor of δ, we reduce the load by a factor of δ^3 and the mass density by a factor of δ^2. We consider two cases with $\delta = 0.1$ and 0.01 for thickness equals to 0.032 and 0.0032, respectively. The time history of displacement calculated by using B-D and B-L elements with the set of meshes is presented in Figure 3.19. As shown in Figure 3.19a, the results of B-D element with thickness equal to 0.032 have similar behavior of the results with the original thickness. But the results shown in Figure 3.19c with thickness equals to 0.0032 are not clearly converging yet, although they are close to each other. On the other hand, the results of B-L element with the reduced thickness, shown in Figures 3.19b and 3.19d demonstrate good convergence behavior, better than that in the case with the

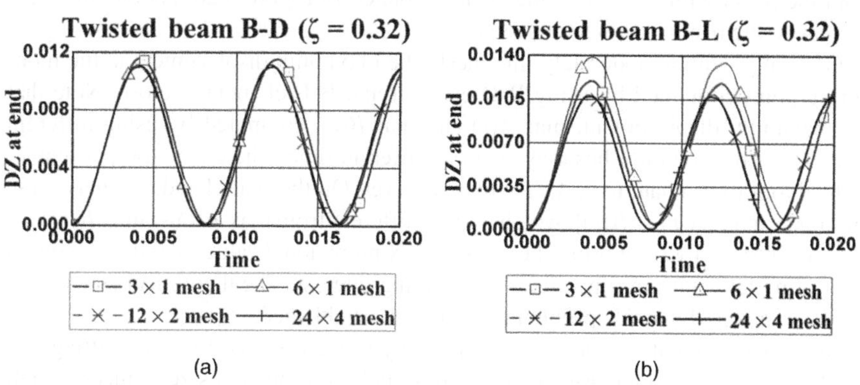

FIGURE 3.18 Solution of the twisted beam problem by other four-node elements: (a) B-D element; (b) B-L element.

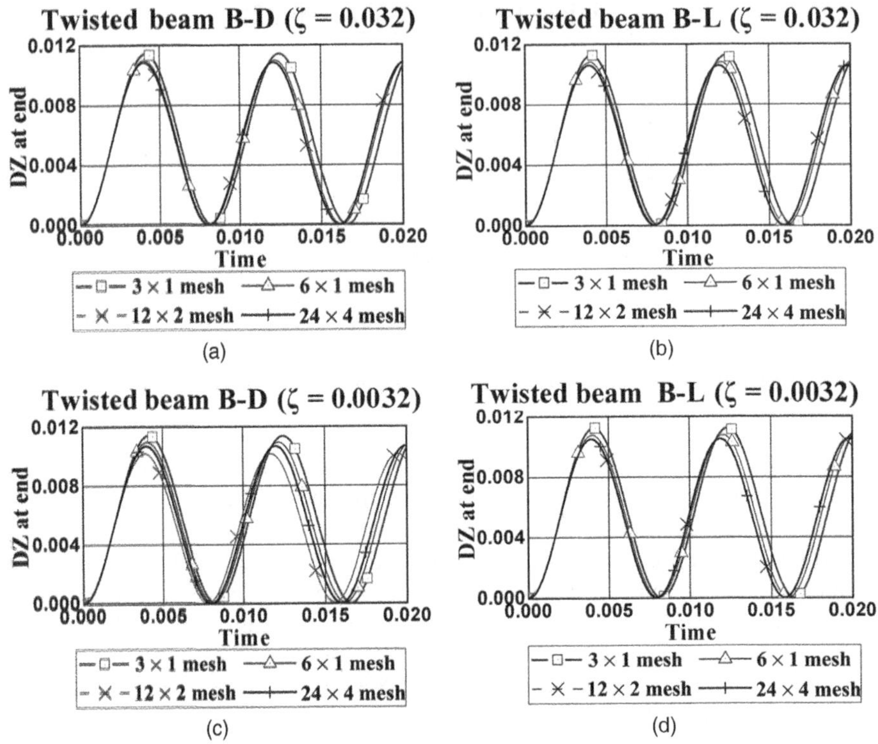

FIGURE 3.19 Solution of the twisted beam problem with reduced thickness: (a) B-D element with 1/10 thickness; (b) B-L element with 1/10 thickness; (c) B-D element with 1/100 thickness; (d) B-L element with 1/100 thickness.

original thickness. The results of these two elements for the three cases of thickness from the mesh with 24 $\times$ 4 elements are depicted in Figure 3.20. As the thickness decreases, the results seem to converge.

Generally, with a moderately fine mesh, the maximum displacement at the monitoring point computed by using B-D element and B-L element is close. Note that the maximum displacement, $\max A(t)$ and $\max B(t)$, computed by using different elements or different meshes may occur at different time. There is no proper mathematical tool to evaluate $\max A(t) - \max B(t)$ yet. On the other hand, the maximum difference $\max\{A(t) - B(t)\}$, which is quite different from $\max A(t) - \max B(t)$, can be characterized by L^{∞}-norm. The error in maximum nodal values and the maximum error of nodal values in time are of engineering interest. The theoretical assessment for the transient dynamics problems is yet to be established.

In this experiment, when mesh is uniformly refined by one-to-four splitting, the number of elements increases to four times, and the time step size reduces to half due to the reduced element size. Hence, eight times of computing time is expected. On the other hand, when the thickness reduces with a factor of δ, the mass density

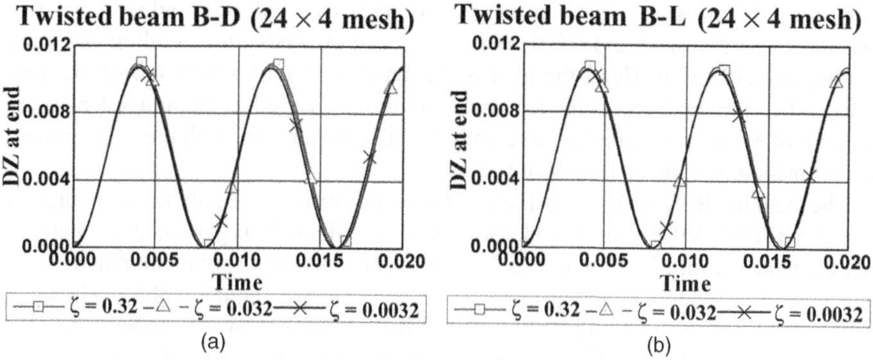

FIGURE 3.20 Solution of the twisted beam problem with various thickness, using 24 × 4 mesh: (a) B-D element; (b) B-L element.

reduces by a factor of δ^2 and the time step size reduces by a factor of δ. Therefore, a refined mesh and a reduced thickness roughly cost $8/\delta$ times of computing time.

Example 3.2 Hyperbolic paraboloid This example was proposed by Chapelle and Bathe (2003) to study the performance of R-M elements with reduced thickness. The saddle-shaped surface is doubly curved with negative Gaussian curvature. Since complete information of the physical parameters was not provided in the reference book, a model depicted in Figure 3.21 is created independently as following. The surface is defined by equation $z = y^2 - x^2$, for its mid-surface, in the domain $[-1, 1] \otimes [-1, 1]$. The material parameters are: Young's modulus $E = 210$, Poisson ratio $\nu = 0.3$, mass density $\rho = 7.85 \times 10^{-6}$, and thickness $\zeta = 0.1$. We use the weight for the distributed load, with gravity $g = 9.85 \times 10^{-3}$ in the negative z-direction. The hyperbolic paraboloid is clamped at one side, $x = -1$. We monitor the displacement at the center point of the opposite side, $x = 1$, $y = 0$, and $z = -1$. A set of four meshes with 10 × 10, 20 × 20, 40 × 40, and 80 × 80 elements are employed for

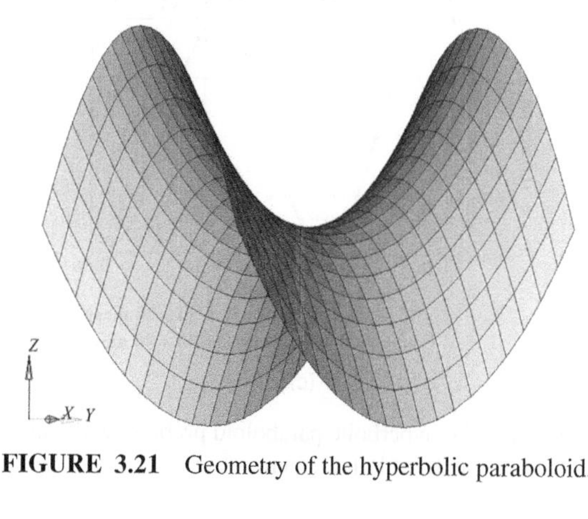

FIGURE 3.21 Geometry of the hyperbolic paraboloid.

this study. To investigate the element performance with reduced thickness, we also use thickness $\zeta = 0.01$ and 0.001. For the reduced thickness, load scaling and mass scaling are also used. Here, the load is the material weight, which is $\rho g \zeta$ per unit area. When the thickness is reduced by a factor of δ, we reduce the mass density by a factor of δ^2, as in the previous example. In this way, the load with the same gravity is automatically reduced by a factor of δ^3.

We examine B-T, B-D, and B-L elements for this example. Again the computation uses LS-DYNA V971. For B-T element, we use stiffness hourglass control method. We examine the time history of displacement at the monitoring point computed by using these three types of elements. For the case with thickness equals to 0.1, as shown in Figure 3.22, the results by B-T element and B-L element show the tendency to converge, but not yet for the B-D element. For the fine mesh with 80 × 80 elements, the results of B-T element and B-L element are close, but the result of B-D element is a little bit different. With the reduced thickness of 0.01, difference in results of B-T element from these meshes is relatively small. Tendency of convergence is demonstrated, as shown in Figure 3.23a. Small difference in results of B-D element from

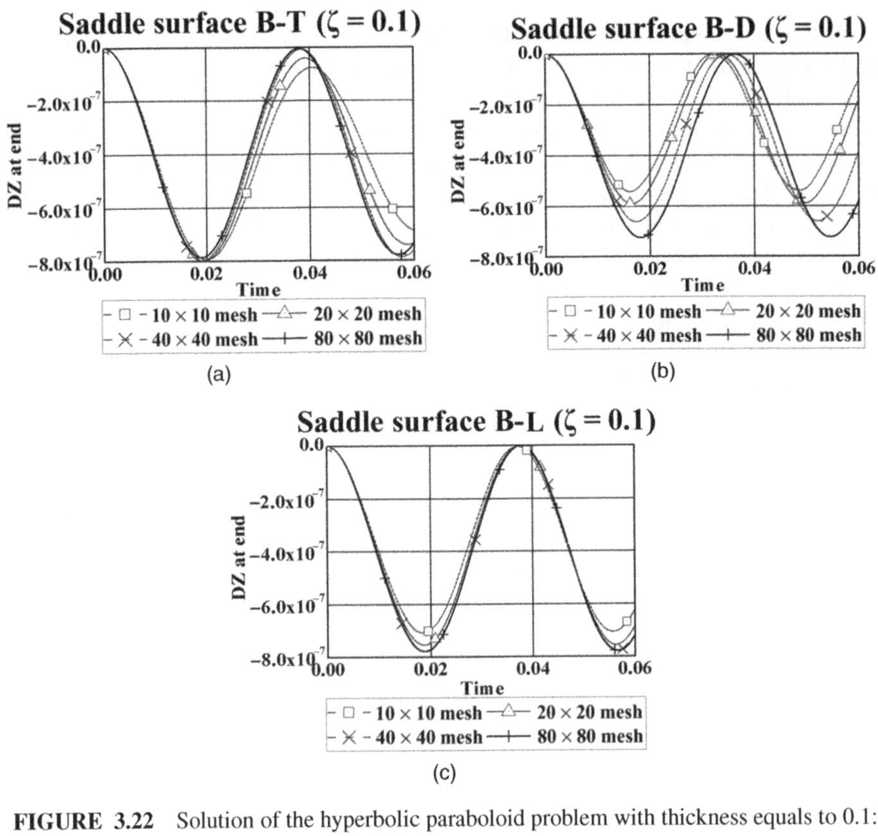

FIGURE 3.22 Solution of the hyperbolic paraboloid problem with thickness equals to 0.1: (a) B-T element; (b) B-D element; (c) B-L element.

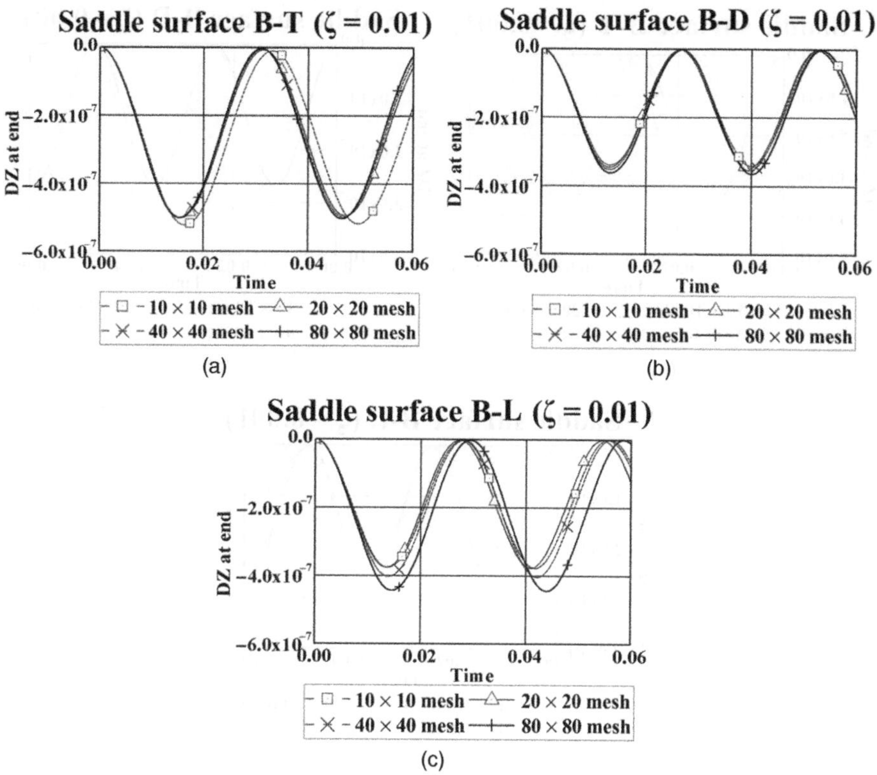

FIGURE 3.23 Solution of the hyperbolic paraboloid problem with thickness equals to 0.01: (a) B-T element; (b) B-D element; (c) B-L element.

meshes with 40 × 40 elements and 80 × 80 elements is still visible as shown in Figure 3.23b. On the other hand, as shown in Figure 3.23c, the results of B-L element do not have clear tendency of convergence yet. It is observed that there is sharp difference in the calculated maximum displacement at the monitoring point by using the three types of elements.

For the further reduced thickness of 0.001, the results of all three elements shown in Figure 3.24 present tendency to converge. The results of B-D element and B-L element with the 40 × 40 mesh are quite close, but the result of B-T element is different.

Note that when using B-T element, the results vary following the variation of coefficient of the hourglass control method. For all cases, the data presented here are results calculated with a fixed coefficient 0.02 (in LS-DYNA's environment) of stiffness method.

3.8.2 About the Locking-free Low Order Four-node R-M Plate Element

The R-M plate element without shear locking is not only the engineers' demand but also the mathematicians' goal. It has been one of the most active research fields

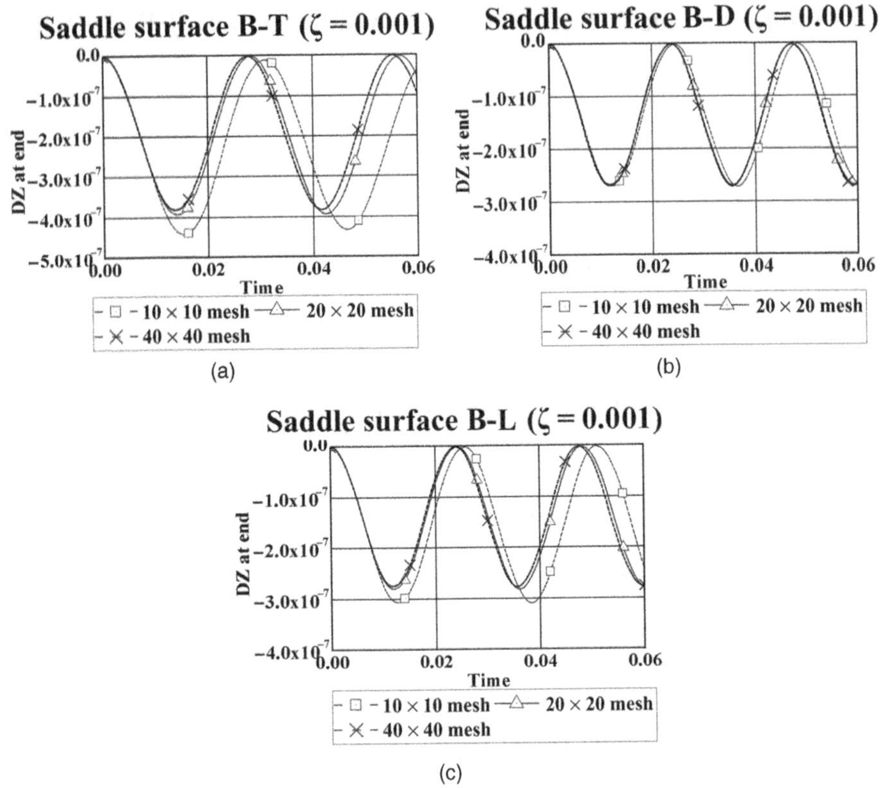

FIGURE 3.24 Solution of the hyperbolic paraboloid problem with thickness equals to 0.001: (a) B-T element; (b) B-D element; (c) B-L element.

of numerical mathematics. In the past decades, the mathematicians have applied various approaches and theories to develop R-M plate elements. Several of these elements have been rigorously proved to converge with optimal rates uniformly with respect to the thickness. On the other hand, some of the elements have been found with degradation of convergence rate when thickness becomes small, and therefore assessed to be not locking-free.

Arnold and Flak (1989a) successfully developed a low order element and proved its convergence with optimal rates uniformly with respect to thickness. This element has been recognized as the first locking-free R-M element computable in the primitive variables (β_α, w). Brezzi et al. (1989) used mixed method and developed several families of elements. Later Brezzi et al. (1991) derived error estimates for all variables of these elements, uniformly with respect to thickness. Durán and Liberman (1992) also used mixed method to study the R-M plate element. For their element, the deflection was modeled by bilinear interpolation. The rotations were modeled by bilinear interpolation plus the span of edge bubble functions. The transverse shear strains were defined in the vector form of $(a + by, c + dx)$, that is, the rotation of the lowest order Raviar–Thomas space. This element was proved to converge uniformly with respect to the thickness, hence locking-free.

For four-node elements, most of the theories were originally developed based on the rectangular meshes. Arnold et al. (2002) argued that when extending the finite element method from rectangular element to general quadrilateral element, there could be a loss of accuracy. This is due to the nonconstant Jacobian, which leads to approximation of the integrations. In general, only elements of rectangle or parallelogram shape have constant Jacobian. Durán–Liberman element was shown by Arnold et al. (2002) to have no loss of convergence rate with shape-regular meshes (for its definition, see Ciarlet, 2000).

If mesh refinement for the quadrilaterals is splitting one element into four by connecting the mid-points of opposite edges, no loss of convergence rate is found for some elements, under certain conditions.

It is worth noting that none of the four-node shell elements implemented in the commercial explicit finite element software has been proved locking-free, particularly for the transient dynamics. The elements proved locking-free mostly have complex formulation. Their implementation for practical applications remains to be a challenge.

CHAPTER 4

THREE-NODE SHELL ELEMENT (REISSNER–MINDLIN PLATE THEORY)

Many developments discussed in Chapter 3 are dealt with warping of curved surface for the four-node quadrilateral elements. This concern has led us to revisit the triangular element. The three-node triangular element with linear interpolation is always flat and has no warping. On the other hand, as discussed in Section 3.8.2, the locking-free element with rectangular mesh could lose convergence rate when extended to general quadrilateral mesh due to the nonconstant Jacobian. The three-node triangular element always has constant Jacobian. There is no need of a special mesh once an element is proven to be locking-free. Triangular mesh is also more flexible in modeling complex geometry.

The triangular element has been in demand as a candidate for solving the bending problems. It has been, however, another long journey to find a better answer. Obviously, with low order interpolation, shear locking is still a problem with the three-node triangular element based on R-M theory. The four-node quadrilateral element has a natural structure with tensor product, which, unfortunately does not exist in the triangular element. Some techniques developed for quadrilateral elements are not directly applicable to the triangular elements. It is found that in many applications, people prefer quadrilateral mesh to the triangular mesh due to the concern about the stiff C^0 element. Some researchers have even warned against the use of triangular mesh except in areas where the geometry is difficult to be meshed by quads. This exception is still conditional, for example, with limit on the total number of triangles to be less than a few percentages of the whole model.

Introduction to the Explicit Finite Element Method for Nonlinear Transient Dynamics, First Edition.
Shen R. Wu and Lei Gu.

In this chapter, we discuss some of the developments in triangular elements, including the projection method and the discrete Kirchhoff theory.

4.1 FUNDAMENTALS OF A THREE-NODE C^0 ELEMENT

4.1.1 Transformation and Jacobian

The isoparametric element is considered for the three-node triangles. The linear interpolation uses the following shape functions:

$$\begin{aligned} \varphi_1 &= 1 - \xi - \eta, \\ \varphi_2 &= \xi, \\ \varphi_3 &= \eta. \end{aligned} \tag{4.1}$$

A general triangle in O-XY plane is mapped to the master isosceles right-angled triangle depicted in Figure 4.1 by

$$\begin{aligned} x &= \sum x_J \varphi_J(\xi, \eta), \\ y &= \sum y_J \varphi_J(\xi, \eta). \end{aligned} \tag{4.2}$$

The Jacobian is

$$\begin{aligned} \boldsymbol{J} &= \frac{\partial(x, y)}{\partial(\xi, \eta)} = \begin{bmatrix} x_\xi & x_\eta \\ y_\xi & y_\eta \end{bmatrix} = \begin{bmatrix} x_2 - x_1 & x_3 - x_1 \\ y_2 - y_1 & y_3 - y_1 \end{bmatrix}, \\ D &= \det(\boldsymbol{J}) = (x_2 - x_1)(y_3 - y_1) - (x_3 - x_1)(y_2 - y_1) = 2A. \end{aligned} \tag{4.3}$$

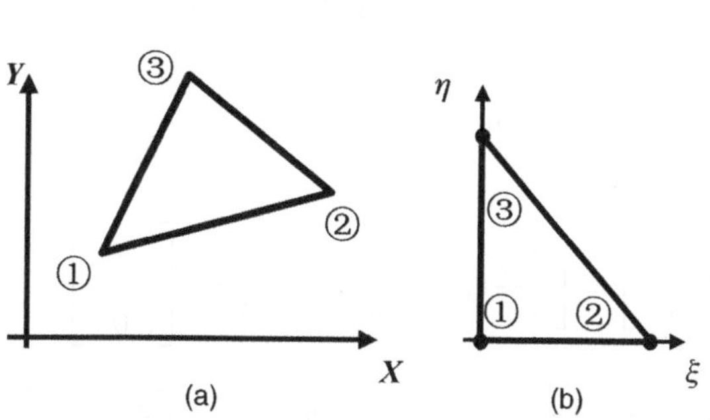

FIGURE 4.1 Three-node triangular element: (a) in the physical domain; (b) in the reference plane.

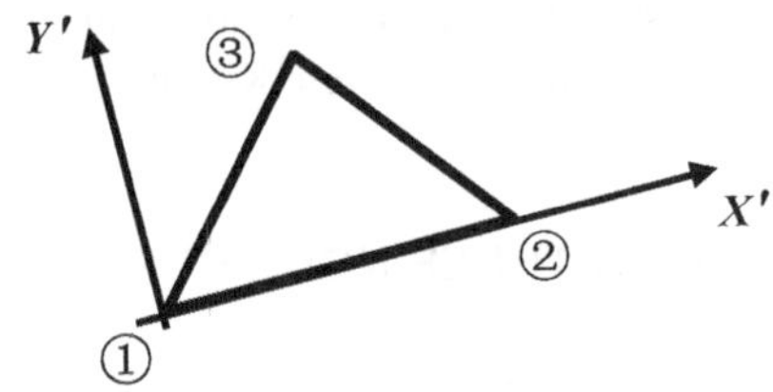

FIGURE 4.2 Local system of the triangular element.

Here, A is the area of the triangle. The inverse of Jacobian is

$$\boldsymbol{J}^{-1} = \frac{\partial(\xi, \eta)}{\partial(x, y)} = \begin{bmatrix} \xi_x & \xi_y \\ \eta_x & \eta_y \end{bmatrix} = \frac{1}{D}\begin{bmatrix} y_\eta & -x_\eta \\ -y_\xi & x_\xi \end{bmatrix}$$

$$= \frac{1}{D}\begin{bmatrix} y_3 - y_1 & x_1 - x_3 \\ y_1 - y_2 & x_2 - x_1 \end{bmatrix}. \tag{4.4}$$

All the transformations are linear with constant Jacobians.

We consider a simple way to define the local system proposed by Belytschko et al. (1984), depicted in Figure 4.2. In this situation, we have $x_1 = y_1 = y_2 = 0$. The Jacobian and its inverse are simplified as

$$\overline{\boldsymbol{J}} = \begin{bmatrix} x_2 & x_3 \\ 0 & y_3 \end{bmatrix},$$
$$D = x_2 y_3, \tag{4.5}$$
$$\overline{\boldsymbol{J}}^{-1} = \frac{1}{x_2 y_3}\begin{bmatrix} y_3 & -x_3 \\ 0 & x_2 \end{bmatrix}.$$

The derivatives of the shape functions are

$$\left[\frac{\partial \varphi_J}{\partial x}\right] = \left[\frac{\partial \varphi_J}{\partial \xi}\xi_x + \frac{\partial \varphi_J}{\partial \eta}\eta_x\right] = \frac{1}{x_2 y_3}\begin{bmatrix} -y_3 \\ y_3 \\ 0 \end{bmatrix},$$
$$\left[\frac{\partial \varphi_J}{\partial y}\right] = \left[\frac{\partial \varphi_J}{\partial \xi}\xi_y + \frac{\partial \varphi_J}{\partial \eta}\eta_y\right] = \frac{1}{x_2 y_3}\begin{bmatrix} x_3 - x_2 \\ -x_3 \\ x_2 \end{bmatrix}. \tag{4.6}$$

4.1.2 Numerical Quadrature for In-plane Integration

Consider integration over the master element. We have the following for the monomials:

$$\begin{aligned}
&\iint 1\mathrm{d}\xi\mathrm{d}\eta = 1/2,\\
&\iint \xi\mathrm{d}\xi\mathrm{d}\eta = \iint \eta\mathrm{d}\xi\mathrm{d}\eta = 1/6,\\
&\iint \xi^2\mathrm{d}\xi\mathrm{d}\eta = \iint \eta^2\mathrm{d}\xi\mathrm{d}\eta = 1/12,\\
&\iint \xi\eta\mathrm{d}\xi\mathrm{d}\eta = 1/24.
\end{aligned} \tag{4.7}$$

When devicing a numerical integration, we use

$$\iint f(\xi,\eta)\mathrm{d}\xi\mathrm{d}\eta \approx \sum w_j f(\xi_j,\eta_j). \tag{4.8}$$

We notice the difference from that of the quadrilateral element. The latter utilizes the structure of tensor product, which does not exist in the triangular element. The traditional Gauss quadrature used for quadrilateral element does not apply to the triangular element. Therefore, we need a different rule: a tri-symmetric method. That means if we use one-point quadrature, the integration point should be the centroid, which has coordinates $(1/3, 1/3)$ in the master element. In higher order, a set of three points with identical weights is a basic structure for the integration rule.

We can verify:

1. Quadrature with one point at element center $(1/3, 1/3)$ and weight $w = 1/2$ can make exact integration up to linear terms;
2. Quadrature with three points at mid-side points $(1/2, 0)$, $(1/2, 1/2)$, and $(0, 1/2)$, and weight $w_j = 1/6$ can make exact integration up to quadratic terms.

4.1.3 Shear Locking with C^0 Triangular Element

Similar to the situation discussed in Chapter 3, the three-node triangular bending element based on R-M theory has a shear locking issue.

For simplicity, consider an example with the master element. Assume a pure bending mode with nodal deflection and rotations as

$$\begin{aligned}
&w_J = \beta_{yJ} = 0, \quad J = 1,2,3,\\
&\beta_{x1} = \beta_{x3} = -\alpha, \quad \beta_{x2} = \alpha.
\end{aligned} \tag{4.9}$$

Using interpolation with the shape functions (4.1), we have

$$\begin{aligned}
&w \equiv 0,\\
&\beta_x = -\alpha(1-x-y) + \alpha x - \alpha y = \alpha(-1+2x),\\
&\gamma_{xz}^h = w_{,x} - \beta_x = \alpha(1-2x) \neq 0.
\end{aligned} \tag{4.10}$$

The linear interpolation results in a nonzero transverse shear strain. As discussed in Chapter 3, the bending stiffness is proportional to ζ^3 (ζ is the thickness) while the shear stiffness is proportional to ζ. When the thickness becomes small, the nonzero transverse shear strain energy will dominate the bending strain energy. This results in shear locking.

One-point reduced integration has been used for the quadrilateral shell element to avoid shear locking, in the unique situation that the shear strain is zero at the integration point. For the C^0 triangular element, three-point quadrature is needed to fully integrate the shear strain energy for elasticity. From (4.1), however, we realize that the one-point reduced quadrature at (1/3, 1/3) does not eliminate the shear strain.

In the history of explicit software development, efficient C^0 quadrilateral element, such as B-T element with good accuracy, has been available earlier than the counterpart of C^0 triangular element. The C^0 triangular element used to be considered too stiff and therefore was recommended to be avoided or to be applied sparingly.

4.2 DECOMPOSITION METHOD FOR C^0 TRIANGULAR ELEMENT WITH ONE-POINT INTEGRATION

4.2.1 A C^0 Element with Decomposition of Deflection

Among many efforts towards obtaining a reliable and efficient C^0 triangular element, Belytschko et al. (1984) proposed a decomposition method for the plate bending element. The deflection and rotations of the normal were decomposed to study the contributions to bending energy and shear energy. We describe the fundamentals in this section. Symbolically, let

$$\begin{aligned} w &= w^b + w^s, \\ \beta_\alpha &= \beta_\alpha^b + \beta_\alpha^s. \end{aligned} \tag{4.11}$$

Correspondingly, the strain and the curvature are decomposed into two parts, which are indicated by superscripts b and s for bending and transverse shear, respectively,

$$\begin{aligned} \gamma_\alpha &= \gamma_\alpha^b + \gamma_\alpha^s = w_{,\alpha} - \beta_\alpha, \\ \kappa_{\alpha\beta} &= \kappa_{\alpha\beta}^b + \kappa_{\alpha\beta}^s = -(\beta_{\alpha,\beta} + \beta_{\beta,\alpha}). \end{aligned} \tag{4.12}$$

Note that the quoted paper used different definition for normal rotations. Here, β_x and β_y correspond to $-\theta_x$ and $-\theta_y$ in the reference paper.

For simplicity, we consider the local system shown in Figure 4.2 and the Jacobian of (4.5). Then, (4.12) is rewritten in vector form as

$$
\begin{aligned}
(\boldsymbol{\beta}, w)^T &= (\beta_{1x}, \beta_{1y}, \beta_{2x}, \beta_{2y}, \beta_{3x}, \beta_{3y}, w_1, w_2, w_3), \\
\kappa &= \boldsymbol{B}^b \begin{bmatrix} \boldsymbol{\beta} \\ w \end{bmatrix}, \quad \gamma = \boldsymbol{B}^s \begin{bmatrix} \boldsymbol{\beta} \\ w \end{bmatrix}, \\
(\boldsymbol{B}^b)_{3\times 9} &= \left(\boldsymbol{B}_r^b, \boldsymbol{O}_{3\times 3}\right), \quad (\boldsymbol{B}^s)_{2\times 9} = \left(\boldsymbol{B}_r^s, \boldsymbol{B}_d^s\right).
\end{aligned}
\tag{4.13}
$$

Here, the differential operators are also decomposed. The subscripts r and d are for the contributions from rotations and deflection respectively. Using (4.6), we have

$$
\begin{aligned}
\left(\boldsymbol{B}_r^b\right)_{3\times 6} &= \frac{-1}{x_2 y_3} \begin{bmatrix} -y_3 & 0 & y_3 & 0 & 0 & 0 \\ 0 & x_3 - x_2 & 0 & -x_3 & 0 & x_2 \\ x_3 - x_2 & -y_3 & -x_3 & y_3 & x_2 & 0 \end{bmatrix}, \\
\left(\boldsymbol{B}_r^s\right)_{2\times 6} &= -\begin{bmatrix} \varphi_1 & 0 & \varphi_2 & 0 & \varphi_3 & 0 \\ 0 & \varphi_1 & 0 & \varphi_2 & 0 & \varphi_3 \end{bmatrix}, \\
\left(\boldsymbol{B}_d^s\right)_{2\times 3} &= \frac{1}{x_2 y_3} \begin{bmatrix} -y_3 & y_3 & 0 \\ x_3 - x_2 & -x_3 & x_2 \end{bmatrix}.
\end{aligned}
\tag{4.14}
$$

In many applications, bending is the main deformation mode and the bending strain energy is the main part of deformation energy. With the concept of decomposition, we assume that only the bending mode contributes to the bending energy and only the shear mode contributes to the shear strain energy. For linear elasticity, the strain energy is decomposed:

$$
\begin{aligned}
2U &= \int_A \left((\kappa^b)^T D_b \kappa^b + (\gamma^s)^T D_s \gamma^s\right) \mathrm{d}A \\
&= \int_A \left((\boldsymbol{\beta}^b, w^b)(\boldsymbol{B}^b)^T \boldsymbol{D}_b \boldsymbol{B}^b \begin{bmatrix} \boldsymbol{\beta}^b \\ w^b \end{bmatrix} + (\boldsymbol{\beta}^s, w^s)(\boldsymbol{B}^s)^T \boldsymbol{D}_s \boldsymbol{B}^s \begin{bmatrix} \boldsymbol{\beta}^s \\ w^s \end{bmatrix}\right) \mathrm{d}A,
\end{aligned}
\tag{4.15a}
$$

$$
\begin{aligned}
\boldsymbol{D}_b &= \frac{E\zeta^3}{12(1-\nu^2)} \begin{bmatrix} 1 & \nu & 0 \\ \nu & 1 & 0 \\ 0 & 0 & 1-\nu/2 \end{bmatrix}, \\
\boldsymbol{D}_s &= \kappa\mu\zeta \boldsymbol{I}_2, \quad \kappa \sim 5/6.
\end{aligned}
\tag{4.15b}
$$

As the curvature only involves the normal rotations, we consider the decomposition of deflection, which is devised to avoid shear locking. A portion of the deflection, denoted by w^k, is proposed to construct equivalent Kirchhoff configuration so that the curvature is the same as what is defined in K-L theory:

$$\begin{aligned}\kappa_x &= -w^k_{,x^2} = -\beta_{x,x},\\ \kappa_y &= -w^k_{,y^2} = -\beta_{y,y},\\ \kappa_{xy} &= -2w^k_{,xy} = -(\beta_{x,y} + \beta_{y,x}).\end{aligned} \tag{4.16}$$

The fact that $\beta_{\alpha,\beta}$ are constants in the C^0 triangular element with linear interpolation suggests a quadratic form for w^k:

$$\begin{aligned}w^k = &-((x^2 - x_2x - x_3(x_3 - x_2)y/y_3)\kappa_x\\ &+ (y^2 - y_3y)\kappa_y + (xy - x_3y)\kappa_{xy})/2 - \alpha_x x - \alpha_y y + \delta.\end{aligned} \tag{4.17}$$

The nodal values in vector form are

$$\left(w^k_J\right) = \begin{bmatrix} 1 & 0 & 0 \\ 1 & -x_2 & 0 \\ 1 & -x_3 & -y_3 \end{bmatrix} \begin{bmatrix} \delta \\ \alpha_x \\ \alpha_y \end{bmatrix}. \tag{4.18}$$

In fact δ represents a rigid body translation and is omitted in the following discussion.

Kirchhoff condition is enforced at the three nodes for rotations associated with w^k, for $J = 1, 2, 3$:

$$\begin{aligned}\left(\gamma^k_{xz}\right)_J &= \left(w^k_{,x} - \beta^k_x\right)_J = 0,\\ \left(\gamma^k_{yz}\right)_J &= \left(w^k_{,y} - \beta^k_y\right)_J = 0.\end{aligned} \tag{4.19}$$

From (4.16) and (4.17), we obtain the nodal values of $\boldsymbol{\beta}^k_\alpha$:

$$\begin{aligned}\left[\left(\beta^k_x\right)_J\right] &= \frac{1}{2}\begin{bmatrix} x_2 & 0 & 0 \\ -x_2 & 0 & 0 \\ -2x_3 + x_2 & 0 & -y_3 \end{bmatrix} \begin{bmatrix} \kappa_x \\ \kappa_y \\ \kappa_{xy} \end{bmatrix} - \alpha_x \begin{bmatrix} 1 \\ 1 \\ 1 \end{bmatrix},\\ \left[\left(\beta^k_y\right)_J\right] &= \frac{1}{2}\begin{bmatrix} x_3(x_3 - x_2)/y_3 & y_3 & x_3 \\ x_3(x_3 - x_2)/y_3 & y_3 & x_3 - x_2 \\ x_3(x_3 - x_2)/y_3 & -y_3 & 0 \end{bmatrix} \begin{bmatrix} \kappa_x \\ \kappa_y \\ \kappa_{xy} \end{bmatrix} - \alpha_y \begin{bmatrix} 1 \\ 1 \\ 1 \end{bmatrix}.\end{aligned} \tag{4.20}$$

Rewriting in the matrix form, we have

$$\boldsymbol{\beta}^k = \boldsymbol{S\kappa} + \boldsymbol{R}^T\boldsymbol{\alpha}. \tag{4.21}$$

Here, S is the matrix of coefficients in (4.20). Other notations are defined by

$$\begin{aligned}
\boldsymbol{\alpha}^{\mathrm{T}} &= [\,\alpha_x \quad \alpha_y\,], \\
\boldsymbol{R} &= \begin{bmatrix} 1 & 0 & 1 & 0 & 1 & 0 \\ 0 & 1 & 0 & 1 & 0 & 1 \end{bmatrix}, \\
\boldsymbol{\kappa}^T &= [\,\kappa_x \quad \kappa_y \quad \kappa_{xy}\,], \\
[\boldsymbol{\beta}^k]^T &= [\,\beta_{x1}^k \quad \beta_{y1}^k \quad \beta_{x2}^k \quad \beta_{y2}^k \quad \beta_{x3}^k \quad \beta_{y3}^k\,].
\end{aligned} \tag{4.22}$$

Recall (4.15), using interpolation and (4.14), we have

$$\begin{aligned}
\boldsymbol{\kappa} &= \boldsymbol{B}_r^b \boldsymbol{\beta}, \\
\boldsymbol{\beta}^k &= \boldsymbol{A}\boldsymbol{\beta} + \boldsymbol{R}^T \boldsymbol{\alpha}, \quad \boldsymbol{A} = \boldsymbol{S}\boldsymbol{B}_r^b.
\end{aligned} \tag{4.23}$$

Assuming no decomposition for $\boldsymbol{\beta}$, we let $\boldsymbol{\beta}^k = \boldsymbol{\beta}$. Solve (4.23) by a left multiplication with $\boldsymbol{R}$

$$\boldsymbol{\alpha} = \boldsymbol{R}(\boldsymbol{I}_6 - \boldsymbol{A})\boldsymbol{\beta}/3. \tag{4.24}$$

Here, we use the relation $\boldsymbol{R}\boldsymbol{R}^T = 3\boldsymbol{I}_2$. The decomposition is then completed with removing the rigid body mode δ:

$$\begin{aligned}
(w_J^k) = \boldsymbol{w}^k &= -\begin{bmatrix} 0 & 0 & 0 \\ 0 & -x_2 & 0 \\ 0 & -x_3 & -y_3 \end{bmatrix} \begin{bmatrix} \delta \\ \alpha_x \\ \alpha_y \end{bmatrix} = \boldsymbol{X}(\boldsymbol{I}_6 - \boldsymbol{A})\boldsymbol{\beta}/3, \\
\boldsymbol{X} &= \begin{bmatrix} 0 & 0 & 0 & 0 & 0 & 0 \\ x_2 & 0 & x_2 & 0 & x_2 & 0 \\ x_3 & y_3 & x_3 & y_3 & x_3 & y_3 \end{bmatrix}, \\
\boldsymbol{w}^s &= \boldsymbol{w} - \boldsymbol{w}^k, \\
\boldsymbol{\beta}^b &= \boldsymbol{\beta}, \quad \boldsymbol{\beta}^s = 0.
\end{aligned} \tag{4.25}$$

According to (4.15), rotations are wholly responsible for the bending strain energy, whereas w^s alone is responsible for the shear strain energy:

$$2U = \int_A \left(\boldsymbol{\beta}^T (\boldsymbol{B}^b)^T \boldsymbol{D}_b \boldsymbol{B}^b \boldsymbol{\beta} + (\boldsymbol{w}^s)^T (\boldsymbol{B}^s)^T \boldsymbol{D}_s \boldsymbol{B}^s \boldsymbol{w}\right) \mathrm{d}A. \tag{4.26}$$

The implementation for linear application is ready. We may use the power (work rate) to replace the strain energy with incremental formulation when extended to nonlinear problems.

4.2.2 A C^0 Element with Decomposition of Rotations

Alternatively, Kennedy et al. (1986) proposed the decomposition for the rotations:

$$\beta_\alpha = \beta_\alpha^b + \beta_\alpha^s. \tag{4.27}$$

Consider the element configuration to be the same as described in Section 4.2.1. Let $\boldsymbol{\beta}^b$ play the role of Kirchhoff structure with

$$\begin{aligned} w_{,x} - \beta_x^b &= 0, \\ w_{,y} - \beta_y^b &= 0. \end{aligned} \tag{4.28}$$

Note that the quoted paper used different definition for the normal rotations. Here β_x and β_y correspond to $-\theta_y$ and θ_x in the reference paper, respectively. Using the operator matrix described in (4.14) with linear interpolation for deflection, $\boldsymbol{\beta}^b = \boldsymbol{B}_d^s w$, we have

$$\begin{aligned} \beta_x^b &= \frac{y_3(w_2 - w_1)}{x_2 y_3} = \frac{w_2 - w_1}{x_2}, \\ \beta_y^b &= \frac{(x_3 - x_2)w_1 - x_3 w_2 + x_2 w_3}{x_2 y_3} = \frac{(w_3 - w_1)x_2 - (w_2 - w_1)x_3}{x_2 y_3}. \end{aligned} \tag{4.29}$$

These are constants and considered properties of the C^0 triangular element. These relations hold true for the corresponding velocity terms. The physical meaning is implied in a rigid body rotation mode (out of plane). In fact, the constant rotation at all nodes matches rigid body rotation with the same rotating motion. Hence, the rest contents of the normal rotations $\boldsymbol{\beta}^s = \boldsymbol{\beta} - \boldsymbol{\beta}^b$ represent the deformation.

As usual, using $\boldsymbol{B}_r^b$ of (4.14), the rates of membrane strain are

$$\begin{bmatrix} d_x \\ d_y \\ 2d_{xy} \end{bmatrix} = \begin{bmatrix} v_{x,x} \\ v_{y,y} \\ v_{x,y} + v_{x,y} \end{bmatrix}$$

$$= \frac{1}{x_2 y_3} \begin{bmatrix} -y_3 & 0 & y_3 & 0 & 0 & 0 \\ 0 & x_3 - x_2 & 0 & -x_3 & 0 & x_2 \\ x_3 - x_2 & -y_3 & -x_3 & y_3 & x_2 & 0 \end{bmatrix} \begin{bmatrix} V_{x1} \\ V_{y1} \\ V_{x2} \\ V_{y2} \\ V_{x3} \\ V_{y3} \end{bmatrix}. \tag{4.30a}$$

The rates of curvature are calculated from $\dot{\beta}^s$, which is the deformation part of the normal rotation:

$$\begin{bmatrix} \dot{\kappa}_x \\ \dot{\kappa}_y \\ 2\dot{\kappa}_{xy} \end{bmatrix} = \begin{bmatrix} \dot{\beta}^s_{x,x} \\ \dot{\beta}^s_{y,y} \\ \dot{\beta}^s_{x,y} + \dot{\beta}^s_{y,x} \end{bmatrix}$$

$$= \frac{1}{x_2 y_3} \begin{bmatrix} -y_3 & 0 & y_3 & 0 & 0 & 0 \\ 0 & x_3 - x_2 & 0 & -x_3 & 0 & x_2 \\ x_3 - x_2 & -y_3 & -x_3 & y_3 & x_2 & 0 \end{bmatrix} \begin{bmatrix} \dot{\beta}^s_{x1} \\ \dot{\beta}^s_{y1} \\ \dot{\beta}^s_{x2} \\ \dot{\beta}^s_{y2} \\ \dot{\beta}^s_{x3} \\ \dot{\beta}^s_{y3} \end{bmatrix}. \quad (4.30b)$$

The rates of transverse shear strain were suggested in the reference paper:

$$\begin{bmatrix} 2d_{xz} \\ 2d_{yz} \end{bmatrix} = \begin{bmatrix} -\dot{\beta}^s_x \\ -\dot{\beta}^s_y \end{bmatrix} = \boldsymbol{B}^s_\beta \dot{\boldsymbol{\beta}}^s,$$

$$\boldsymbol{B}^s_\beta = \frac{1}{6x_2 y_3} \times \begin{bmatrix} -y_3(2x_2 + x_3) & -(y_3)^2 & y_3(x_3 - 3x_2) & (y_3)^2 & -x_2 y_3 & 0 \\ (x_3)^2 - (x_2)^2 & y_3(x_3 - 2x_2) & x_3(2x_2 - x_3) & -y_3(x_2 + x_3) & x_2(x_2 - 2x_3) & -3x_2 y_3 \end{bmatrix}. \quad (4.30c)$$

The rest of the element formulation and software implementation is straightforward.

4.3 DISCRETE KIRCHHOFF TRIANGULAR ELEMENT

Apart from improving the C^0 element, another method is using the discrete Kirchhoff theory whose early development dates back to the late 1960s, cf. Wempner et al. (1968), Stricklin et al. (1969), and Dhatt (1969, 1970). The element had been "forgotten" for about 10 years, but regained substantial development after being reported by Batoz et al. (1980) as still the most efficient and reliable one among the 9-dof triangular plate bending elements. The nonlinear applications of the Discrete Kirchhoff Triangular (DKT) element can be found in, for example, Bathe et al. (1983) and Wenzel and Schoop (2004). The discrete Kirchhoff theory has also been extended to quadrilateral shell element and axisymmetric shell element. Li et al. (2001), and Wu et al. (2005) reported the application of DKT element, which is implemented in explicit software for application in the nonlinear transient dynamics. In fact, it is the combination of the DKT plate bending element and a constant strain two-directional (2D) element for membrane stress.

The main idea is using different interpolations for variables and enforcing the element with Kirchhoff conditions at discrete locations to handle the shear locking issue.

The usual three-node linear interpolation is employed for the in-plane motion u_x and u_y, with $2 \times 3 = 6$ unknowns:

$$\begin{aligned} \varphi_1 &= 1 - \xi - \eta, \\ \varphi_2 &= \xi, \\ \varphi_3 &= \eta. \end{aligned} \tag{4.31}$$

The quadratic interpolation with additional three mid-side nodes is used for the normal rotations β_x and β_y, $2 \times 6 = 12$ unknowns:

$$\begin{aligned} \beta_x &= \sum_{J=1}^{6} \beta_{xJ} \ \ \psi_J(\xi, \eta), \\ \beta_y &= \sum_{J=1}^{6} \beta_{yJ} \ \ \psi_J(\xi, \eta). \end{aligned} \tag{4.32}$$

In the master (parametric) element, shown in Figure 4.3, the shape functions are

$$\begin{aligned} \psi_1 &= (1 - \xi - \eta)(1 - 2\xi - 2\eta), \\ \psi_2 &= \xi(2\xi - 1), \\ \psi_3 &= \eta(2\eta - 1), \\ \psi_4 &= 4\xi\eta, \\ \psi_5 &= 4\eta(1 - \xi - \eta), \\ \psi_6 &= 4\xi(1 - \xi - \eta). \end{aligned} \tag{4.33}$$

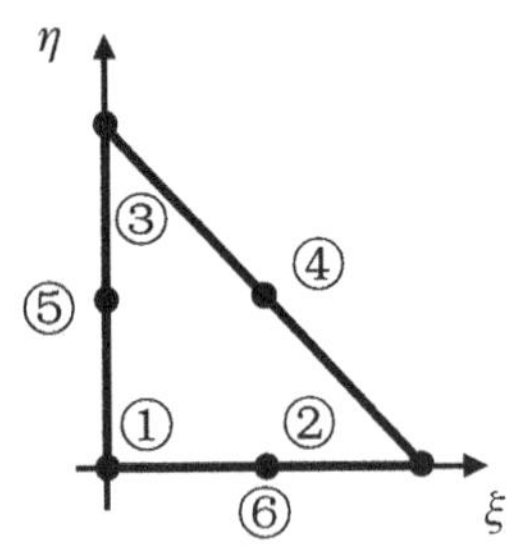

FIGURE 4.3 Triangular element with three mid-points.

For deflection, the element uses two-node Hermite cubic interpolation along each of the three sides. The detailed formula for the element interior is not critical. For example, using parameter s from 0 to 1 along side 1 with node 1 and node 2, we have

$$w|_{\text{Side 1}} = w_1\chi_1 + w_2\chi_2 + w_{1S}\chi_1^1 + w_{2S}\chi_2^1. \tag{4.34}$$

Here, w_{1S} and w_{2S} represent the s-directional derivatives of deflection at the end nodes. The shape functions are

$$\begin{aligned} \chi_1 &= 1 - 3s^2 + 2s^3, \\ \chi_2 &= 3s^2 - 2s^3, \\ \chi_1^1 &= (s - 2s^2 + s^3)L, \\ \chi_2^1 &= (-s^2 + s^3)L. \end{aligned} \tag{4.35}$$

The unknowns introduced are the nodal values and the s-directional derivatives at both ends of each side, total $2 \times 3 + 3 = 9$. So far, a total of 21 unknowns have been introduced for bending and shear.

Now, we impose the discrete Kirchhoff conditions:

(K1) $\gamma_{xz} = \gamma_{yz} = 0$ at the three corner nodes ($2 \times 3 = 6$ constraints).

(K2) The transverse shear strain $\gamma_{zt} = 0$ at three mid-side nodes (3 constraints).

(K3) The rotation component β_n is linear along three sides (3 constraints).

Here, t and n indicate the directions tangential and normal to the sides, shown in Figure 4.4. Note that the rates of β_x and β_y correspond to $-\omega_2$ and ω_I in Wu et al. (2005), respectively. The rates of β_n and β_t correspond to $-\omega_t$ and ω_n, respectively. Conditions (K1) and (K2) are some remedies to the shear locking corresponding to the case discussed in Section 4.1.3. Condition (K3) is for the C^0-continuity of rotation around element boundaries. In fact, γ_{xz} and γ_{yz} form a 2D vector. Condition (K1) is equivalent to the fact that any component of the shear strain is zero at the corners.

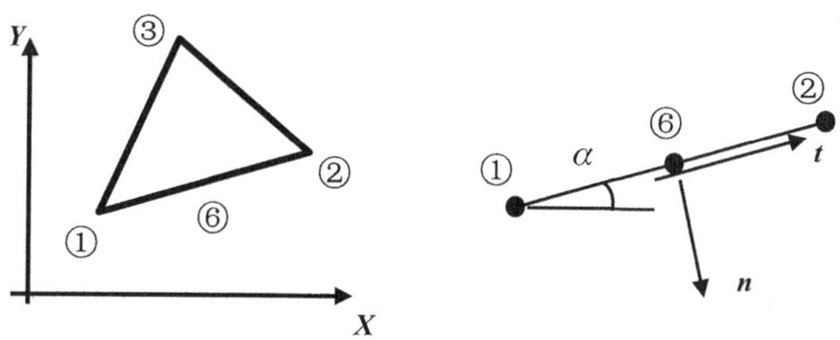

FIGURE 4.4 A side with a mid-point.

Quadratic and cubic interpolations are introduced here for the rotation and deflection, respectively. Other types of discrete Kirchhoff conditions are also possible.

We examine side 3 with nodes 1, 2, and 6 for illustration. Condition (K1) yields

$$\begin{aligned}\gamma_{zt}\Big|_{s=0} &= \left(\frac{\partial w}{L_3 \partial s} - \beta_t\right)\Big|_{s=0} = w_{1S} - \beta_{1t} = 0,\\ \gamma_{zt}|_{s=1} &= w_{2S} - \beta_{2t} = 0.\end{aligned} \tag{4.36}$$

Condition (K2) requires

$$\gamma_{zt}|_{s=0.5} = \frac{3}{2}\frac{w_2 - w_1}{L_3} - \frac{1}{4}(w_{1S} + w_{2S}) - \beta_{6t} = 0. \tag{4.37}$$

Condition (K3) requires

$$\beta_{6n} = (\beta_{1n} + \beta_{2n})/2. \tag{4.38}$$

Express the rotation in a format of 2D vector with (4.36)–(4.38)

$$\begin{aligned}\boldsymbol{\beta}_6 &= \beta_{6t}\boldsymbol{e}_n - \beta_{6n}\boldsymbol{e}_t = \beta_{6y}\boldsymbol{e}_x - \beta_{6x}\boldsymbol{e}_y\\ &= \frac{\boldsymbol{\beta}_1 + \boldsymbol{\beta}_2}{2} + \left(\frac{3}{2}\frac{w_2 - w_1}{L_3} - \frac{3}{4}(\beta_{1t} + \beta_{2t})\right)\boldsymbol{e}_n.\end{aligned} \tag{4.39}$$

With $\boldsymbol{e}_x = \boldsymbol{e}_t \cos\alpha_3 + \boldsymbol{e}_n \sin\alpha_3$ and $\boldsymbol{e}_y = \boldsymbol{e}_t \sin\alpha_3 - \boldsymbol{e}_n \cos\alpha_3$, for $\alpha = 1, 2$, we have $\beta_{\alpha t} = \beta_{\alpha x}\cos\alpha_3 + \beta_{\alpha y}\sin\alpha_3$. With $\boldsymbol{e}_n = \boldsymbol{e}_x \sin\alpha_3 - \boldsymbol{e}_y \cos\alpha_3$, we find

$$\begin{aligned}\beta_{6x} &= 0.5(\beta_{1x} + \beta_{2x}) + (1.5(w_2 - w_1)\cos\alpha_3/L_3 - 0.75((\beta_{1x} + \beta_{2x})\cos\alpha_3\\ &\quad + (\beta_{1y} + \beta_{2y})\sin\alpha_3))\cos\alpha_3,\\ \beta_{6y} &= 0.5(\beta_{1y} + \beta_{2y}) + (1.5(w_2 - w_1)\sin\alpha_3/L_3 - 0.75((\beta_{1x} + \beta_{2x})\cos\alpha_3\\ &\quad + (\beta_{1y} + \beta_{2y})\sin\alpha_3))\sin\alpha_3.\end{aligned} \tag{4.40}$$

$\boldsymbol{\beta}_4$ and $\boldsymbol{\beta}_5$ can be obtained by permutation. Thus the derivatives of deflection at corner nodes and the rotation at the mid-side nodes can all be eliminated internally. The system is condensed to three degrees of freedom per node (β_x, β_y, w) for bending-shear, denoted by

$$[\theta_k] = \left(\beta_x^1, \beta_x^2, \beta_x^3, \beta_y^1, \beta_y^2, \beta_y^3, w^1, w^2, w^3\right). \tag{4.41}$$

Utilize the fact that $\psi_1 + (\psi_5 + \psi_6)/2 = \varphi_1$, etc., denote

$$[\Phi_J] = (\varphi_1, \varphi_2, \varphi_3, \psi_4, \psi_5, \psi_6). \tag{4.42}$$

We then obtain the condensed form:

$$\beta_x = \sum_{J=1}^{6} \beta_x^J \psi_J = \sum_{K=1}^{9}\sum_{J=1}^{6} \theta_k (H_x)_{KJ} \Phi_J,$$
$$\beta_y = \sum_{J=1}^{6} \beta_y^J \psi_J = \sum_{K=1}^{9}\sum_{J=1}^{6} \theta_k (H_y)_{KJ} \Phi_J. \tag{4.43}$$

The transformation matrices are derived:

$$(H_x)_{9\times 6} = \begin{bmatrix} \boldsymbol{I}_3 & \begin{bmatrix} 0 & CC_2 & CC_3 \\ CC_1 & 0 & CC_3 \\ CC_1 & CC_2 & 0 \end{bmatrix} \\ \boldsymbol{O}_3 & \begin{bmatrix} 0 & SC_2 & SC_3 \\ SC_1 & 0 & SC_3 \\ SC_1 & SC_2 & 0 \end{bmatrix} \\ \boldsymbol{O}_3 & \begin{bmatrix} 0 & CL_2 & -CL_3 \\ -CL_1 & 0 & CL_3 \\ CL_1 & -CL_2 & 0 \end{bmatrix} \end{bmatrix}, \tag{4.44a}$$

$$(H_y)_{9\times 6} = \begin{bmatrix} \boldsymbol{O}_3 & \begin{bmatrix} 0 & SC_2 & SC_3 \\ SC_1 & 0 & SC_3 \\ SC_1 & SC_2 & 0 \end{bmatrix} \\ \boldsymbol{I}_3 & \begin{bmatrix} 0 & SS_2 & SS_3 \\ SS_1 & 0 & SS_3 \\ SS_1 & SS_2 & 0 \end{bmatrix} \\ \boldsymbol{O}_3 & \begin{bmatrix} 0 & SL_2 & -SL_3 \\ -SL_1 & 0 & SL_3 \\ SL_1 & -SL_2 & 0 \end{bmatrix} \end{bmatrix}. \tag{4.44b}$$

The following notations are used in (4.44): $SL_j = 1.5 \sin \alpha_j / L_j$, $CL_j = 1.5 \cos \alpha_j / L_j$, $SS_j = -0.75 \sin^2 \alpha_j$, $CC_j = -0.75 \cos^2 \alpha_j$, and $SC_j = -0.75 \sin \alpha_j \cos \alpha_j$.

The rest of the element formulation is rather straightforward. Generally, three-point in-plane quadrature is needed due to inclusion of quadratic shape functions. The application of a two-point quadrature was reported in Li et al. (2001), which provided close results.

It is worth noting that DKT element has more complex formulation than C^0 elements has, which is discussed in previous sections. On the other hand, higher order interpolation used in DKT element requires smaller stable time step. Therefore, it generally uses more computing time than C^0 element does.

4.4 ASSESSMENT OF THREE-NODE R-M PLATE ELEMENT

4.4.1 Evaluations with Warped Mesh and Reduced Thickness

To study the performance of triangular elements, we use the same examples discussed in Section 3.8.1 for quadrilateral elements. We examine C^0 element discussed in Section 4.2.2 and DKT element discussed in Section 4.3. We use the commercial software LS-DYNA V971 to perform the study.

Example 4.1 Twisted beam The problem and parameters were defined by MacNeal and Harder (1985) as one of the test problems, cf. discussion in Example 3.1. A set of four triangular meshes is generated by one-to-two splitting of the quadrilateral meshes, depicted in Figure 4.5. The mesh is graded with a consistent orientation for the triangulation. In this way the triangular mesh has the same number of nodes and double number of elements of the corresponding quadrilateral mesh.

The time history of the displacement at the end point for the original thickness is shown in Figure 4.6. Obviously, the result of C^0 element, shown in Figure 4.6a, is questionable with a tendency to diverge. The maximum displacement at the end point calculated from a fine mesh is away from the reference value, as well as the solution of B-D element or B-L element, which are presented in Example 3.1. The results of DKT element shown in Figure 4.6b seem to converge with small difference

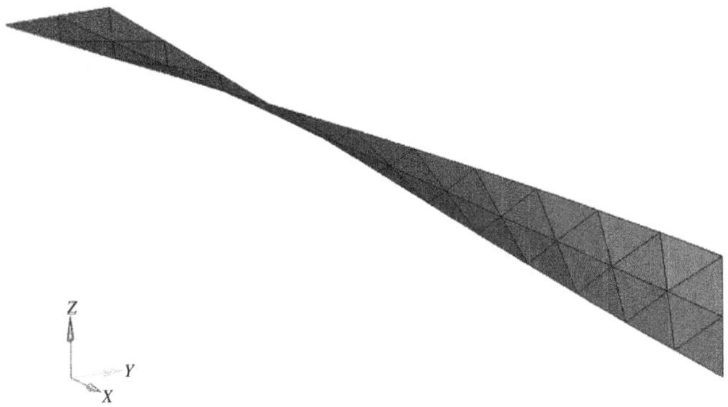

FIGURE 4.5 A triangular mesh of the twisted beam.

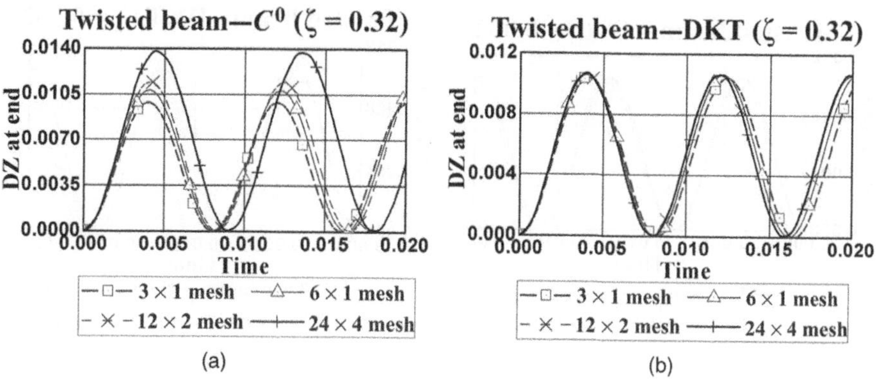

FIGURE 4.6 Solution of the twisted beam problem by the triangular elements: (a) C^0 element; (b) DKT element.

from coarse mesh to fine mesh. The results of DKT element are close to those of B-D element and B-L element. With the reduced thickness 0.032 and 0.0032, load scaling and mass scaling are applied in the same way as discussed in Example 3.1. As shown in Figure 4.7, both results of C^0 element and DKT element converge. The results of C^0 element from the meshes with $40 \times 40 \times 2$ elements and $80 \times 80 \times 2$ elements are close to each other, shown in Figure 4.7a and c, better than what is observed in the case with the original thickness. The results of these two elements are close, and also close to the results of the quadrilateral B-D and B-L elements.

Example 4.2 Hyperbolic paraboloid This example was proposed by Chapelle and Bathe (2003) to study the performance of R-M elements with reduced thickness. It was also used in Wu et al. (2005) to study the performance of DKT element for transient dynamics problem. The problem and parameters are defined in Example 3.2. A set of four triangular meshes is generated by one-to-two splitting of quadrilateral meshes, graded with an alternate orientation to form a cross pattern, and is depicted in Figure 4.8. The triangular mesh has the same number of nodes and double number of elements of the corresponding quadrilateral mesh. Reduction of thickness by 10 times and 100 times with the load scaling and mass scaling is also included in the study.

For thickness equal to 0.1, the results of C^0 element depicted in Figure 4.9 show that convergence has not been achieved with the mesh refinement. But it has a better chance to converge for the case of reduced thickness of 0.01, with fine mesh's result close to DKT's results. The results of DKT element seem to converge, as shown in Figure 4.10. Even with the reduced thickness, the difference in results of DKT element from fine meshes is small. It is also observed that there is a certain type of difference in the results of quadrilateral elements and triangular elements.

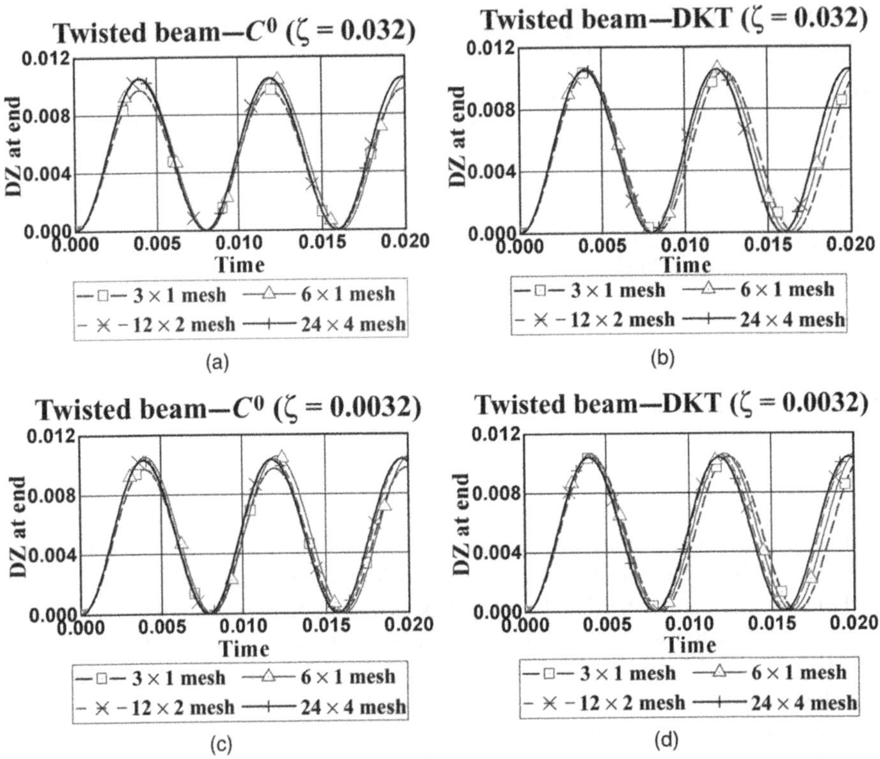

FIGURE 4.7 Solution of the twisted beam problem with reduced thickness by the triangular elements: (a) C^0 element for the reduced thickness 0.032; (b) DKT element for the reduced thickness 0.032; (c) C^0 element for the reduced thickness 0.0032; (d) DKT element for the reduced thickness 0.0032.

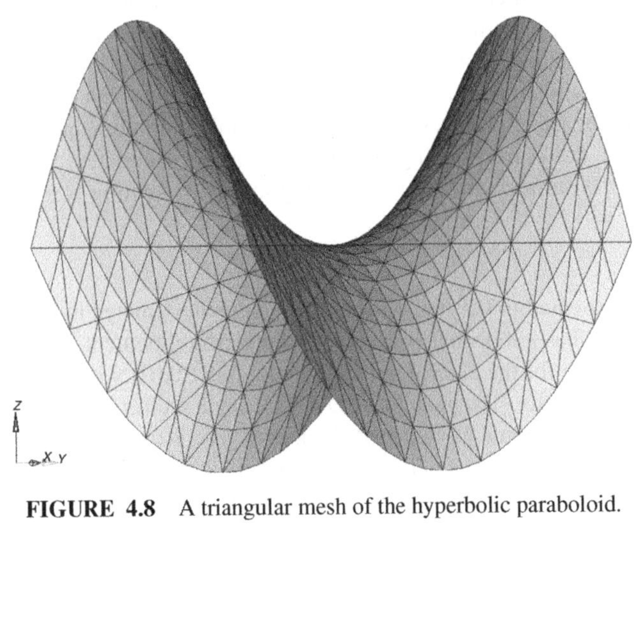

FIGURE 4.8 A triangular mesh of the hyperbolic paraboloid.

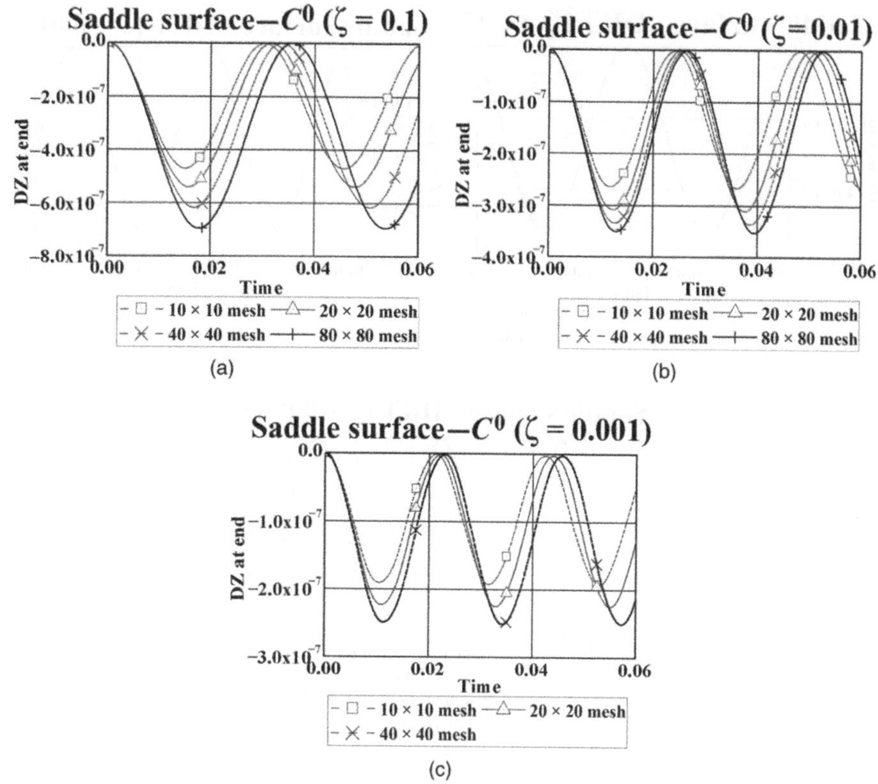

FIGURE 4.9 Solution of the hyperbolic paraboloid problem by the C^0 triangular elements: (a) C^0 element for the original thickness 0.1; (b) C^0 element for the reduced thickness 0.01; (c) C^0 element for the reduced thickness 0.001.

4.4.2 About the Locking-free Low Order Three-node R-M Plate Element

All the developments discussed here have focused on the subject of removing shear locking. Unfortunately, like the quadrilateral elements discussed in Chapter 3, these elements have not been proven to be locking-free even for static linear applications. The first recognized locking-free R-M element was developed by Arnold and Falk (1989). It was a three-node element using nonconforming method. The rotations used linear interpolation plus one cubic bubble function $\lambda_1\lambda_2\lambda_3$ (expressed in barycentric coordinates), which was condensed inside the element. The deflection used nonconforming linear interpolation with element continuity at mid-points of the three sides only. The transverse shear strains were approximated by piecewise constants and were discontinuous cross the elements.

Durán and Liberman (1992) developed a locking-free triangular element using low order interpolation with the same concept of the rival four-node element

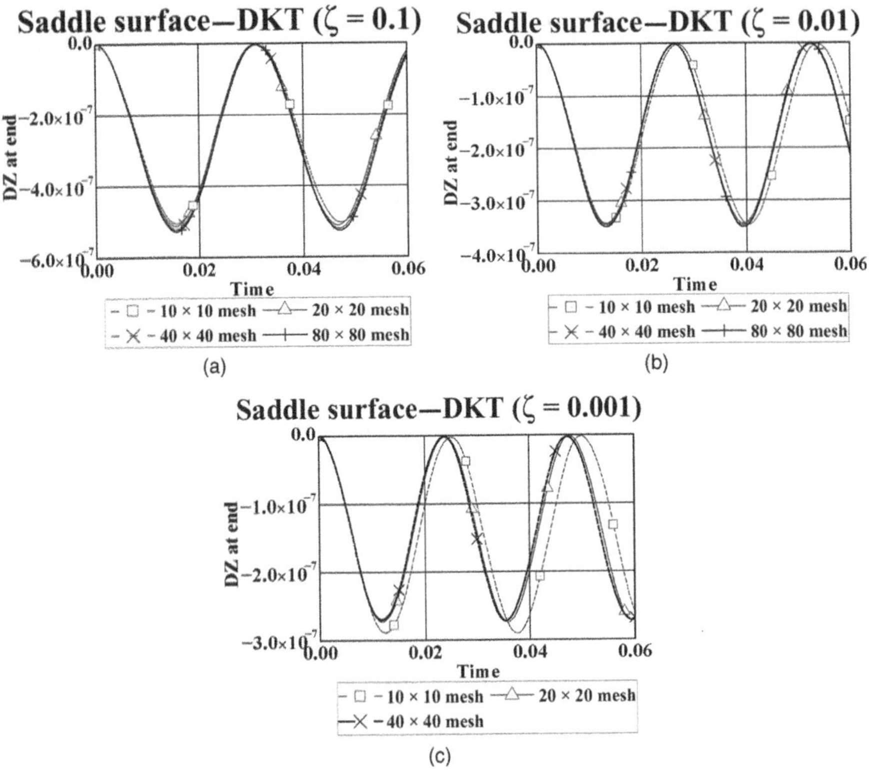

FIGURE 4.10 Solution of the hyperbolic paraboloid problem by the DKT triangular elements: (a) DKT element for the original thickness 0.1; (b) DKT element for the reduced thickness 0.01; (c) DKT element for the reduced thickness 0.001.

described in Section 3.8.2. The rotations used linear interpolation plus the span of edge bubble functions. The transverse shear strains used the rotation of the lowest order Raviart–Thomas space.

Locking-free elements with quadratic or higher order interpolation for rotations or deflection have also been developed; see Falk and Tu (2000) and Arnold et al. (2002) for commentaries. These elements used bubble functions, or nonconforming approaches, etc. Their development has never been a simple task. The implementation of these elements for general applications of nonlinear transient dynamic problems deserves further investigations.

CHAPTER 5

EIGHT-NODE SOLID ELEMENT

The eight-node hexahedron element, called solid element or brick element, is one of the commonly used elements for three-dimensional (3D) applications. The trilinear eight-node hexahedron element belongs to the class of first-order approximation and uses trilinear shape functions in the form of tensor product. Due to the complexity of the 3D geometry, other types of solid elements, for example, six-node wedge element and four-node tetrahedron element are also used in many applications.

Solid elements can be used to analyze the complex loading cases. In particular, when the structure has large variation in thickness or even stepped shape, solid element is a better choice than shell element. If the dimensions of the material body in three directions are in the same order of magnitude, or if the deformation in the thickness direction is important, such as fogging process, solid element might be a better choice.

5.1 TRILINEAR INTERPOLATION FOR THE EIGHT-NODE HEXAHEDRON ELEMENT

For the element shown in Figure 5.1, the shape function corresponding to the eight nodes are defined below with the reference coordinate system:

$$\begin{aligned}\Phi_N &= (1+\xi^N\xi)(1+\eta^N\eta)(1+\zeta^N\zeta)/8 \\ &= \left(\Sigma_N + \Lambda_N^1\xi + \Lambda_N^2\eta + \Lambda_N^3\zeta + \Gamma_N^1\eta\zeta + \Gamma_N^2\zeta\xi + \Gamma_N^3\xi\eta + \Gamma_N^4\xi\eta\zeta\right)/8,\end{aligned} \tag{5.1}$$

Introduction to the Explicit Finite Element Method for Nonlinear Transient Dynamics, First Edition. Shen R. Wu and Lei Gu.

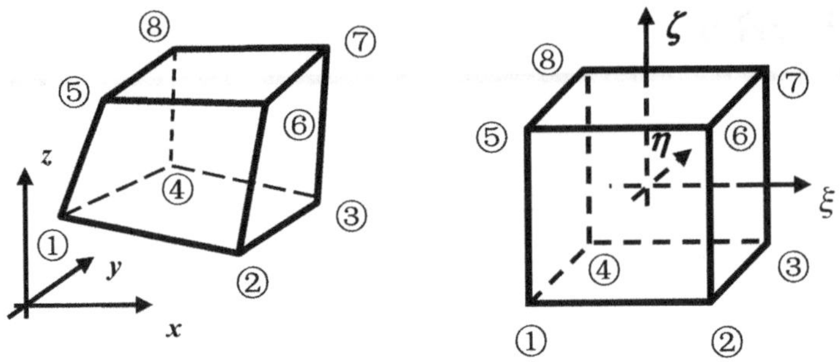

FIGURE 5.1 Configuration of brick element.

where we denote the nodal values by

$$\begin{aligned}
&\Sigma_N = 1, \\
&\Lambda_N^1 = \xi^N, \quad \Lambda_N^2 = \eta^N, \quad \Lambda_N^3 = \zeta^N, \\
&\Gamma_N^1 = \eta^N \zeta^N, \quad \Gamma_N^2 = \zeta^N \xi^N, \quad \Gamma_N^3 = \xi^N \eta^N, \quad \Gamma_N^4 = \xi^N \eta^N \zeta^N.
\end{aligned} \tag{5.2}$$

An isoparametric mapping serves as the transformation of coordinates from the physical domain Ω to the reference domain $\hat{\Omega}$, applying the same formula as what used for interpolation of displacement, velocity, and acceleration. For $i = 1, 2, 3$, with summation on the repeated index N, we have

$$\begin{aligned}
x_i &= x_i^N \Phi_N(\xi, \eta, \zeta), \\
u_i &= u_i^N \Phi_N(\xi, \eta, \zeta), \\
v_i &= v_i^N \Phi_N(\xi, \eta, \zeta), \\
a_i &= a_i^N \Phi_N(\xi, \eta, \zeta).
\end{aligned} \tag{5.3}$$

Representation of a function by interpolation is determined by its nodal values. The union of the shape functions contributes to the basis of interpolation for the finite element space. When considering (ξ, η, ζ) as parameters, $\Phi_N(\xi, \eta, \zeta)$ is a linear combination of Σ_N, ${\Lambda^1}_N$, ${\Lambda^2}_N$, ${\Lambda^3}_N$, ${\Gamma^1}_N$, ${\Gamma^2}_N$, ${\Gamma^3}_N$, and ${\Gamma^4}_N$. These eight vectors of eight-dimensional (8D) space, defined in (5.2), are determined by the nodal coordinates of the cubic element in the reference space which is listed in Table 5.1. It is obvious that the eight vectors are orthogonal to each other, and hence form a basis of the 8D linear vector space. They represent essentially the eight independent deformation modes. As an example, the x-displacement represented by the eight base vectors is depicted in Figure 5.2. Σ_N is the rigid body translation mode. ${\Lambda^1}_N$ is the tension/compression mode. ${\Lambda^2}_N$ and ${\Lambda^3}_N$ are the shear modes in xy-plane and xz-plane, respectively. ${\Gamma^1}_N$, ${\Gamma^2}_N$, ${\Gamma^3}_N$, and ${\Gamma^4}_N$ are the four hourglass modes. The existence of four hourglass modes in one displacement component is just one of the

TABLE 5.1 Coefficients of the Shape Functions for the Eight-Node Element

Node	ξ	η	ζ	Σ_I	Λ_{1I}	Λ_{2I}	Λ_{3I}	Γ_{1I}	Γ_{2I}	Γ_{3I}	Γ_{4I}
1	−1	−1	−1	1	−1	−1	−1	1	1	1	−1
2	1	−1	−1	1	1	−1	−1	1	−1	−1	1
3	1	1	−1	1	1	1	−1	−1	−1	1	−1
4	−1	1	−1	1	−1	1	−1	−1	1	−1	1
5	−1	−1	1	1	−1	−1	1	−1	−1	1	1
6	1	−1	1	1	1	−1	1	−1	1	−1	−1
7	1	1	1	1	1	1	1	1	1	1	1
8	−1	1	1	1	−1	1	1	1	−1	−1	−1

Source: Reprinted from Flanagan and Belytschko, 1981. Copyright (1981), with permission from Wiley-Blackwell.

features of the 3D applications, which are more complex than the two-dimensional (2D) applications.

Similar to the four-node quadrilateral shell element, we write the shape functions in the alternative way, cf. Belytschko and Bachrach (1986):

$$\begin{aligned}\Phi_N &= \Delta_N + b_{1N}x + b_{2N}y + b_{3N}z + C_{1N}\eta\zeta + C_{2N}\zeta\xi + C_{3N}\xi\eta + C_{4N}\xi\eta\zeta \\ &= \Delta_N + b_{iN}x_i + C_{\alpha N}h_\alpha,\end{aligned} \tag{5.4}$$

where b_{iN} and $C_{\alpha N}$ are constants. Here and later in this chapter, the summation on repeated lowercase Latin indices has range from 1 to 3; the summation on the Greek indices has range from 1 to 4, and the summation on repeated capital Latin indices has range from 1 to 8 for the nodal values. We also denote for simplicity:

$$h_1 = \eta\zeta, \quad h_2 = \zeta\xi, \quad h_3 = \xi\eta, \quad h_4 = \xi\eta\xi. \tag{5.5}$$

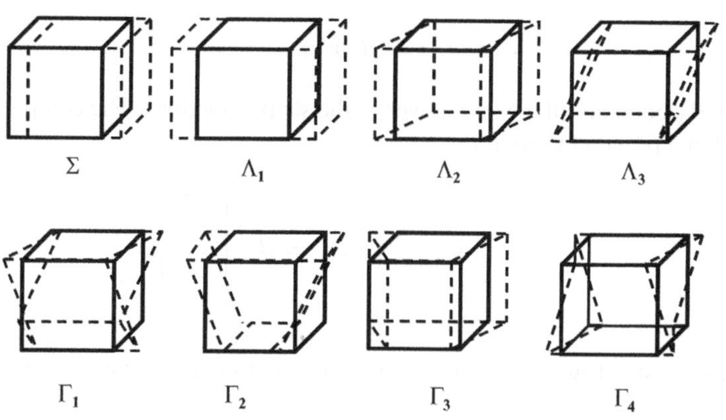

FIGURE 5.2 *x*-displacement modes of the brick element. (Reprinted from Flanagan and Belytschko, 1981. Copyright (1981), with permission from Wiley-Blackwell.)

This format is convenient for calculation involving first spatial derivatives:

$$\begin{aligned}\frac{\partial \Phi_N}{\partial x_i} &= b_{iN} + C_{\alpha N}\frac{\partial h_\alpha}{\partial x_i} \\ &= b_{iN} + C_{1N}\left(\zeta\frac{\partial \eta}{\partial x_i} + \eta\frac{\partial \zeta}{\partial x_i}\right) + C_{2N}\left(\xi\frac{\partial \zeta}{\partial x_i} + \zeta\frac{\partial \xi}{\partial x_i}\right) \\ &\quad + C_{3N}\left(\eta\frac{\partial \xi}{\partial x_i} + \xi\frac{\partial \eta}{\partial x_i}\right) + C_{4N}\left(\eta\zeta\frac{\partial \xi}{\partial x_i} + \zeta\xi\frac{\partial \eta}{\partial x_i} + \xi\eta\frac{\partial \zeta}{\partial x_i}\right). \end{aligned} \tag{5.6}$$

We can simply verify

$$b_{iN} = \partial \Phi_N/\partial x_i|_{\xi=\eta=\zeta=0}, \quad i = 1, 2, 3. \tag{5.7}$$

Furthermore, by using the isoparametric transformation $x_i = x_i^I \Phi_I(\zeta)$ with (5.1), the shape functions can be expressed as

$$\Phi_N = \Delta_N + b_{iN} x_i^I \left(\Sigma_I + \Lambda_I^j \xi_j + \Gamma_I^\alpha h_\alpha\right)/8 + C_{\alpha N} h_\alpha.$$

Comparing with (5.1), we obtain

$$\begin{aligned}\Delta_N &= \left(\Sigma_N - b_{iN} x_i^I \Sigma_I\right)/8, \\ C_{\alpha N} &= \left(\Gamma_N^\alpha - b_{iN} x_i^I \Gamma_I^\alpha\right)/8.\end{aligned} \tag{5.8}$$

Hence, the coefficients of (5.4) are all determined in (5.7) and (5.8). In addition, we obtain an equality $b_{iN} x_i^I \Lambda_I^j = \Lambda_N^j$.

Components of small strain rate $\dot{\varepsilon}_{ij} = (\partial v_i/\partial x_j + \partial v_j/\partial x_i)/2$ are then expressed with nodal velocities, in matrix form:

$$\dot{\boldsymbol{\varepsilon}} = \boldsymbol{B}_N \boldsymbol{v}^N. \tag{5.9}$$

Matrix $\boldsymbol{B}_N$ contains all the derivatives of the shape functions. The components are presented in a pseudo matrix form:

$$[\dot{\varepsilon}_x, \dot{\varepsilon}_y, \dot{\varepsilon}_z, \dot{\gamma}_{xy}, \dot{\gamma}_{yz}, \dot{\gamma}_{zx}]^T = \boldsymbol{B}_N \begin{bmatrix} v_x^N \\ v_y^N \\ v_z^N \end{bmatrix}. \tag{5.10}$$

$\boldsymbol{B}$ matrix consists of two parts, the constant part and the nonconstant part, referring to (5.6):

$$\boldsymbol{B}_N = \boldsymbol{B}_{cN} + \boldsymbol{B}_{nN}, \tag{5.11}$$

$$\boldsymbol{B}_{cN} = \begin{bmatrix} b_{1N} & 0 & 0 \\ 0 & b_{2N} & 0 \\ 0 & 0 & b_{3N} \\ b_{2N} & b_{1N} & 0 \\ 0 & b_{3N} & b_{2N} \\ b_{3N} & 0 & b_{1N} \end{bmatrix}, \tag{5.12}$$

$$\boldsymbol{B}_{nN} = \begin{bmatrix} C_{\alpha N}\partial h_\alpha/\partial x & 0 & 0 \\ 0 & C_{\alpha N}\partial h_\alpha/\partial y & 0 \\ 0 & 0 & C_{\alpha N}\partial h_\alpha/\partial z \\ C_{\alpha N}\partial h_\alpha/\partial y & C_{\alpha N}\partial h_\alpha/\partial x & 0 \\ 0 & C_{\alpha N}\partial h_\alpha/\partial z & C_{\alpha N}\partial h_\alpha/\partial y \\ C_{\alpha N}\partial h_\alpha/\partial z & 0 & C_{\alpha N}\partial h_\alpha/\partial x \end{bmatrix}. \tag{5.13}$$

To further elaborate the formulation related to the shape functions, we need to examine the transformation from physical space to the reference space, or Jacobian, defined by

$$\boldsymbol{J} = \frac{\partial(x, y, z)}{\partial(\xi, \eta, \zeta)} = \begin{bmatrix} x_\xi & x_\eta & x_\zeta \\ y_\xi & y_\eta & y_\zeta \\ z_\xi & z_\eta & z_\zeta \end{bmatrix}, \quad (\boldsymbol{J})_{ij} = \frac{\partial x_i}{\partial \xi_j}. \tag{5.14}$$

Denote the determinant of Jacobian matrix and its inverse by

$$D = |\boldsymbol{J}| = \epsilon_{ijk}\frac{\partial x}{\partial \xi_i}\frac{\partial y}{\partial \xi_j}\frac{\partial z}{\partial \xi_k}, \tag{5.15}$$

$$\boldsymbol{J}^{-1} = \frac{\partial(\xi, \eta, \zeta)}{\partial(x, y, z)} = \begin{bmatrix} \xi_x & \xi_y & \xi_z \\ \eta_x & \eta_y & \eta_z \\ \zeta_x & \zeta_y & \zeta_z \end{bmatrix}, \quad (\boldsymbol{J}^{-1})_{ij} = \frac{\partial \xi_i}{\partial x_j}. \tag{5.16}$$

The values of the Jacobian and its inverse at the element center are of particular interest. At $\xi = \eta = \zeta = 0$, we have

$$\begin{aligned} (\boldsymbol{J})_{ij}(0) &= \frac{\partial x_i}{\partial \xi_j}(0) = x_i^I \Lambda_I^j/8, \\ D_{ij} &= (\boldsymbol{J}^{-1})_{ij}(0) = \frac{\partial \xi_i}{\partial x_j}(0). \end{aligned} \tag{5.17}$$

5.2 LOCKING ISSUES OF THE EIGHT-NODE SOLID ELEMENT

Similar to the shear-locking issue of the Reissner–Mindlin plate element, the trilinear solid element also has shear-locking issue when the structure appears to be thin. On the other hand, volumetric locking may occur when the material is nearly incompressible (Poisson ratio $\nu \sim 0.5$). It means that, with certain pattern of deformation, the strain energy calculated from volume change is unrealistically large that the contribution from other deformation mode is underestimated.

To illustrate with a simple example of elasticity: assume that a rectangular solid element is aligned with the coordinate system, occupying the domain $[-L_x/2, L_x/2] \otimes [-L_y/2, L_y/2] \otimes [-L_z/2, L_z/2]$. The element experiences a deformation mode of Γ_2 as depicted in Figure 5.2,

$$u_x = \varepsilon_0 xz/L_z, \; u_y = u_z = 0. \tag{5.18}$$

The components of strain and stress are

$$\begin{aligned} \varepsilon_x &= \varepsilon_0 z/L_z, \quad \varepsilon_y = \varepsilon_z = 0, \\ \varepsilon_{xz} &= \varepsilon_0 x/2L_z, \quad \varepsilon_{xy} = \varepsilon_{yz} = 0, \end{aligned} \tag{5.19}$$

$$\begin{aligned} \sigma_x &= (\lambda + 2\mu)\varepsilon_0 z/L_z, \quad \sigma_y = \sigma_z = \lambda\varepsilon_0 z/L_z, \\ \tau_{xz} &= \mu\varepsilon_0 x/L_z, \quad \tau_{xy} = \tau_{yz} = 0, \end{aligned} \tag{5.20}$$

where λ and μ are the Lamé elasticity constants. The strain energy is

$$\begin{aligned} U &= \int_\Omega \sigma_{ij}\varepsilon_{ij} \mathrm{d}\Omega/2 = \varepsilon_0^2 \int_\Omega \left((\lambda + 2\mu)z^2/L_z^2 + \mu x^2/L_z^2\right) \mathrm{d}\Omega/2 \\ &= (\lambda + 2\mu)\varepsilon_0^2 L_x L_y L_z^3/24L_z^2 + \mu\varepsilon_0^2 L_x^3 L_y L_z/24L_z^2 \\ &= (\lambda + 2\mu)\varepsilon_0^2 V/24 + \mu\varepsilon_0^2 L_x^2 V/24L_z^2. \end{aligned} \tag{5.21}$$

Note that $\lambda + 2\mu = 2\mu(1 - \nu)/(1 - 2\nu)$ (calculated from (1.7)). When the material is nearly incompressible with $\nu \sim 0.5$, the first term on the right-hand side of (5.21) becomes dominant. This deformation does not change the volume, but is not represented properly. This is a type of volumetric locking. On the other hand, the deformation in fact represents a kind of pure bending. We find out that if L_z becomes small, then shear deformation dominates strain energy. This is similar to shear locking with the R-M plate element.

Several techniques have been developed to alleviate these troublesome locking issues. Naturally, we may consider the one-point reduced integration scheme. Similar to what has been discussed for the four-node shell elements, the contribution of the troubled terms in this example disappears when the increments of strain and stress are evaluated at the element center. Comparing to shell element, one-point integration is even more attractive. $2^3 = 8$ Gauss points may be needed for now for integrating the stress of the 3D solid element accurately, instead of having $2 \times 2 = 4$ points for the shell element. However, this will result in rank deficiency for stress calculation. Various stabilization techniques have been developed to solve the issue. We will discuss the basic ideas in the following sections.

5.3 ONE-POINT REDUCED INTEGRATION AND THE PERTURBED HOURGLASS CONTROL

To explore the nature of the hourglass modes, Flanagan and Belytschko (1981), and Belytschko (1983) examined the linear portion of the velocity using Taylor expansion about the center of the element:

$$v_i^{\mathrm{Lin}} = \bar{v}_i + \bar{v}_{i,j}(x_j - \bar{x}_j). \tag{5.22}$$

The bared quantities represented their values at the center of the element. By the trilinear interpolation, the center values $\bar{x}_j$ and $\bar{v}_i$ are equivalent to the means of their nodal values:

$$\begin{aligned} \bar{x}_j &= \Sigma_{I=1,8} x_j^I/8, \\ \bar{v}_i &= \Sigma_{I=1,8} v_i^I/8. \end{aligned} \tag{5.23}$$

Using (5.7), we have

$$\bar{v}_{i,j} = v_i^I \Phi_{i,j}|_0 = v_i^I b_{jI}. \tag{5.24}$$

Now the linear part of the velocity at node I is expressed as $v_{iI}^{\mathrm{Lin}} = \bar{v}_i \Sigma_I + \bar{v}_{i,j}(x_j^I - \bar{x}_j \Sigma_I)$ and the hourglass mode of the velocity is defined by the nodal values as

$$v_{iI}^{HG} = v_i^I - v_{iI}^{\mathrm{Lin}} = v_i^I - \bar{v}_i \Sigma_I - \bar{v}_{i,j}\left(x_j^I - \bar{x}_j \Sigma_I\right). \tag{5.25}$$

Apparently,

$$v_{iI}^{HG} \Sigma_I = 0. \tag{5.26}$$

Using orthogonality of the base vectors with $\Lambda_N^i = 8\partial\Phi_N/\partial\xi_i(0)$, we have

$$\begin{aligned} v_{iI}^{HG} \Lambda_I^n &= v_i^I \Lambda_I^n - \bar{v}_{i,j} x_j^I \Lambda_I^n \\ &= v_i^I \Lambda_I^n - v_i^N \frac{\partial \Phi_N}{\partial x_j(0)} 8 \frac{\partial x_j}{\partial \xi_n(0)} = v_i^I \Lambda_I^n - 8 v_i^N \frac{\partial \Phi_N}{\partial \xi_n(0)} = 0. \end{aligned} \tag{5.27}$$

Hence, as a vector of the 8D space, v_{iI}^{HG} is the linear combination of the four hourglass base vectors only, and can be expressed as

$$v_{iI}^{HG} = \dot{q}_{i\alpha} \Gamma_I^{\alpha}. \tag{5.28}$$

This is similar to (3.61). This also provides the meaning of hourglass mode of the velocity field. The above discussion applies to the B-T shell element. Applying the orthogonality of the base vectors and (5.24) and (5.25), we derive from (5.28),

$$\begin{aligned} 8\dot{q}_{i\alpha} &= v_{iI}^{HG}\Gamma_I^\alpha = v_i^I\Gamma_I^\alpha - \bar{v}_{i,j}x_j^I\Gamma_I^\alpha = v_i^I\Gamma_I^\alpha - v_i^N b_{jN}x_j^I\Gamma_I^\alpha \\ &= v_i^I\left(\Gamma_I^\alpha - b_{jI}x_j^J\Gamma_J^\alpha\right) = 8v_i^I C_{\alpha I}. \end{aligned}$$

The last equality is due to (5.8). Therefore,

$$\dot{q}_{i\alpha} = v_i^I C_{\alpha I}. \tag{5.29}$$

Note that $C_{\alpha I}$ are the coefficients used in the alternate form of the shape functions. As a matter of fact, (5.29) provides a way to detect the hourglass components from the nodal velocity.

We are ready to process the key steps of stabilization after these preparations. As discussed in the previous section, one-point integration scheme for stress evaluation eliminates the locking issue, with zero stress contributions from the terms that cause locking. From (5.6) and (5.13), it is clear that the components of strain rate at this quadrature point (the center of element) are the constant part of the rate of strain. If the deformation is just an hourglass mode, the one-point quadrature gives zero strain rate and zero strain energy increment, and leads to unstable situation.

The stabilization can be achieved by using a viscous damping method or a stiffness method, as discussed in Flanagan and Belytschko (1981) or their combination, similar to the case of B-T shell element discussed in Section 3.5.2. We can consider $\dot{q}_{i\alpha}$ as hourglass velocity and introduce the artificial damping force:

$$Q_{i\alpha} = C_c\dot{q}_{i\alpha}. \tag{5.30}$$

On the other hand, we can consider $\Delta q_{i\alpha} = \dot{q}_{i\alpha}\Delta t$ as the increment of hourglass deformation and introduce the artificial stiffness in the incremental form:

$$\begin{aligned} \Delta Q_{i\alpha} &= C_k\Delta q_{i\alpha}, \\ Q_{i\alpha} &= C_k\int \dot{Q}_{i\alpha}\mathrm{d}t = C_k\sum \Delta Q_{i\alpha}. \end{aligned} \tag{5.31}$$

The nodal force contributed by the hourglass control is determined by the work rate:

$$v_i^I f_{iI}^{HG} = Q_{i\alpha}\dot{q}_{i\alpha}. \tag{5.32}$$

From (5.29), we obtain

$$f_{iI}^{HG} = C_{\alpha I}Q_{i\alpha}. \tag{5.33}$$

Flanagan and Belytschko (1981) recommended using maximum frequency and stiffness of the element to determine coefficients C_c and C_k:

$$\begin{aligned} C_c &= 2\varepsilon M_8 \omega_{\max}, \\ C_k &= \kappa K_{\max}. \end{aligned} \tag{5.34}$$

Here, $M_8 = \rho V/8$ was the lumped nodal mass with V for the volume of the element. $K_{\max}$ denoted the maximum nodal stiffness of the element and $\omega_{\max} = \sqrt{K_{\max}/M_8}$ was the corresponding frequency. By investigating the linear eigen modes, these authors obtained the estimates for maximum stiffness:

$$(\lambda + 2\mu) B_{iI} B_{iI}/3V \leq K_{\max} \leq (\lambda + 2\mu) B_{iI} B_{iI}/V. \tag{5.35}$$

The parameters ε and κ in (5.34) are chosen for application. Their default values as small parameters are implemented in the software. Usually these parameters can work properly for certain classes of applications, but are not promised for an arbitrary case. This stabilization approach is named perturbation hourglass control due to the appearance of the small parameters.

5.4 ASSUMED STRAIN METHOD AND SELECTIVE/REDUCED INTEGRATION

The perturbation hourglass control needs artificial damping or stiffness, both with user-determined coefficients. The default values are set in the software after extensive studies with various examples, but the parameters are not able to work perfectly for all the applications that contain various meshes, loading conditions, and materials. The stabilization method without relying on additional control parameters has been in high demand. Among many efforts and the successful developments, we briefly discuss the assumed strain method in Belytschko and Bindeman (1993) here, and some other techniques in the next sections.

Note that locking is caused by the nonconstant terms of strain components in the example discussed in the previous section. Based on the concept of assumed strain developed in Simo and Hughes (1986), Belytschko and Bindeman (1993) proposed a stabilization scheme of assumed strain to prevent volumetric locking and shear locking; see Fish and Belytschko (1988) for more discussions about the assumed strain method. We adopt the shorthand notations, for example,

$$X_1^N = C_{1N} \partial h_1/\partial x; \quad Y_{23}^N = C_{2N} \partial h_2/\partial y + C_{3N} \partial h_3/\partial y. \tag{5.36}$$

Assumed rate of strain is defined, by modifying the nonconstant part of $\boldsymbol{B}$ matrix from (5.11) to (5.13):

$$\bar{\boldsymbol{B}}_N = \boldsymbol{B}_{cN} + \bar{\boldsymbol{B}}_{nN}, \tag{5.37}$$

$$\bar{\boldsymbol{B}}_{nN} = \begin{bmatrix} X^N_{1234} & -\bar{\nu}Y^N_3 - \nu Y^N_{24} & -\bar{\nu}Z^N_2 - \nu Z^N_{34} \\ -\bar{\nu}X^N_3 - \nu X^N_{14} & Y^N_{1234} & -\bar{\nu}Z^N_1 - \nu Z^N_{34} \\ -\bar{\nu}X^N_2 - \nu X^N_{14} & -\bar{\nu}Y^N_1 - \nu Y^N_{24} & Z^N_{1234} \\ Y^N_{12} & X^N_{12} & 0 \\ 0 & Z^N_{23} & Y^N_{23} \\ Z^N_{13} & 0 & X^N_{13} \end{bmatrix}. \tag{5.38}$$

Here, $\bar{\nu} = \nu/1 - \nu$. The assumed strain is applied to the nonconstant part only. In fact, with this formula, we have the contribution of the nonconstant part to the volume change as follows:

$$\begin{aligned} \dot{\varepsilon}_{kk} &= v^N_x \left(X^N_{1234} - \bar{\nu}X^N_{23} - 2\nu X^N_{14}\right) + v^N_y \left(Y^N_{1234} - \bar{\nu}Y^N_{13} - 2\nu Y^N_{24}\right) \\ &\quad + v^N_z \left(Z^N_{1234} - \bar{\nu}Z^N_{12} - 2\nu Z^N_{34}\right) \\ &= v^N_x \left((1-\bar{\nu})X^N_{23} + (1-2\nu)X^N_{14}\right) + v^N_y \left((1-\bar{\nu})Y^N_{13} + (1-2\nu)Y^N_{24}\right) \\ &\quad + v^N_z \left((1-\bar{\nu})Z^N_{12} + (1-2\nu)Z^N_{34}\right). \end{aligned} \tag{5.39}$$

Note that $1 - \bar{\nu} = (1-2\nu)/(1-\nu)$. When $\nu \to 0.5$, we have $\dot{\varepsilon}_{kk} \to 0$. The strain energy will not be dominated by the volumetric strain. On the other hand, for shear strain calculated by using the modified $\bar{\boldsymbol{B}}_{nN}$ defined in (5.38), the terms causing locking are dropped. For instance, in calculating $\dot{\varepsilon}_{xz}$ contributed by v_x of (5.18) discussed in the example of Section 5.4.2, the terms $v^N_x C_{2N}\partial(\zeta\xi)/\partial z$ and $v^N_x C_{4N}\partial(\xi\eta\zeta)/\partial z$ are dropped. Now, only Z^N_{13} has contribution. In this case, the nodal velocity v^N_x is proportional to Γ^2_N. With the aligned rectilinear element, we can verify that $\Gamma^2_N C_{1N} = \Gamma^2_N C_{3N} = 0$ and

$$v^N_x Z^N_{13} = \varepsilon_0 L_x \Gamma^2_N (C_{1N}\partial(\eta\zeta)/\partial z + C_{3N}\partial(\xi\eta)/\partial z) = 0.$$

After the stress is obtained from the constitutive laws, the nodal forces are calculated in the usual way:

$$\begin{aligned} f_{iI} &= \int \sigma_{ij}\Phi_{i,j} = \int \sigma_{ij}(b_{jI} + C_{\alpha I}\partial h_\alpha/\partial x_j) \\ &= \int \sigma_{ij}b_{jI} + \int \sigma_{ij}C_{\alpha I}\partial h_\alpha/\partial x_j. \end{aligned} \tag{5.40}$$

For evaluating the stress, usually the eight-point quadrature of standard $2 \times 2 \times 2$ Gauss integration is needed. A four-point quadrature was also suggested by Belytschko and Bindeman (1993), using half of the eight points, at reference coordinates $(-1/\sqrt{3}, -1/\sqrt{3}, -1/\sqrt{3})$, $(1/\sqrt{3}, 1/\sqrt{3}, -1/\sqrt{3})$, $(-1/\sqrt{3}, 1/\sqrt{3}, 1/\sqrt{3})$, and $(1/\sqrt{3}, -1/\sqrt{3}, 1/\sqrt{3})$, with double weight.

In a simple case of a rectilinear element, the element coordinate system is aligned with the reference coordinate system by the corotational stress system. Then $\partial x_i/\partial \xi_j = 0$ if $i \neq j$. For linear elasticity case, the stress can be determined directly from the assumed strain. All the strain components are calculated exactly with the information at the element center. The nodal force can then be integrated exactly by one-point quadrature. Belytschko and Bindeman (1993) then proposed a stabilization scheme for the general nonlinear applications. It used one-point integration for the nodal force associated with the constant part of $\partial \Phi_i/\partial x_j$, which was the first term in the right-hand side of (5.40). The stabilization part (the second term in the right-hand side of (5.40)) that associated with the nonconstant part took the same form as (5.33)

$$f_{iI}^{\text{Stab}} = C_{\alpha I} Q_{i\alpha}. \tag{5.41}$$

The rates of the generalized hourglass stress $Q_{i\alpha}$ were calculated by using the elastic relation between nodal force and nodal velocity, with the modified shear modulus below:

$$2\bar{\mu} = \sqrt{\Delta S_{ij}\Delta S_{ij}/\Delta e_{ij}\Delta e_{ij}}, \tag{5.42}$$

where ΔS_{ij} and Δe_{ij} are the deviatoric components of the stress and strain increments, respectively. For example, we have

$$\begin{aligned}
\dot{Q}_{11} &= \bar{\mu}((H_{22} + H_{33})\dot{q}_{11} + H_{12}\dot{q}_{22} + H_{13}\dot{q}_{33}), \\
\dot{Q}_{1\mathrm{j}} &= 2\bar{\mu}(C_1 H_{11}\dot{q}_{1\mathrm{j}} + C_3 H_{1\mathrm{j}}\dot{q}_{\mathrm{j}1}), \quad j = 2, 3, \\
\dot{Q}_{14} &= 2\bar{\mu}C_2 H_{11}\dot{q}_{14}.
\end{aligned} \tag{5.43}$$

Other components are obtained by permutation, where $\dot{q}_{i\alpha}$ is the detected hourglass velocity and the same as (5.29) discussed in the previous section. Coefficients H_{ij} and C_i are explicitly calculated, for example,

$$\begin{aligned}
H_{11} &= \int (\partial h_2/\partial x)^2 = \int (\partial h_3/\partial x)^2 \\
&= 3\int (\partial h_4/\partial x)^2 = \Lambda_J^2 y^J \Lambda_K^3 z^K / 3\Lambda_I^1 x^I,
\end{aligned} \tag{5.44a}$$

$$H_{12} = \int (\partial h_1/\partial y)(\partial h_2/\partial x) = \Lambda_K^3 z^K/3, \tag{5.44b}$$

$$C_1 = 1/1 - \nu, \quad C_2 = (1+\nu)/3, \quad C_3 = \bar{\nu}. \tag{5.45}$$

In summary, the nodal forces associated with the constant part of the shape function's derivatives use the reduced one-point integration. Other terms related to the assumed strain use exact integration with the approximation of an elastic model. This is the concept of selective reduced integration. For early developments and applications of selective reduced integration, we refer to Zienkiewicz et al. (1971), Kavanagh and Key (1972), Markus and Hughes (1978), and the references cited.

5.5 ASSUMED DEVIATORIC STRAIN

In Belytschko and Bindeman (1993), the assumed deviatoric strain was also proposed as an extension from Hughes (1980) in the view of the theory that volumetric strain calculated from the deviatoric components alone vanishes. Nagtegaal et al. (1974) developed, for finite element of plasticity, the concept of mean dilatation. The assumed deviatoric strain applied only to the nonconstant part was proposed with the following form:

$$\bar{\boldsymbol{B}}_{nN} = \begin{bmatrix} (2/3)X_{1234}^N & -Y_{1234}^N/3 & -Z_{1234}^N/3 \\ -X_{1234}^N/3 & (2/3)Y_{1234}^N & -Z_{1234}^N/3 \\ -X_{1234}^N/3 & -Y_{1234}^N/3 & (2/3)Z_{1234}^N \\ Y_{12}^N & X_{12}^N & 0 \\ 0 & Z_{23}^N & Y_{23}^N \\ Z_{13}^N & 0 & X_{13}^N \end{bmatrix}. \tag{5.46}$$

By this definition, the Poisson ratio did not appear in the formula, hence there was no issue of volumetric locking. The shear locking was treated in the same way as discussed in Section 5.4.

Stabilization procedure and nodal force calculation were the same as described in Section 5.4, except that coefficients C_i of (5.45) were replaced with

$$C_1 = 2/3, \quad C_2 = 2/9, \quad C_3 = -1/3. \tag{5.47}$$

5.6 AN ENHANCED ASSUMED STRAIN METHOD

Another assumed strain method was developed by Puso (2000). It also used one-point integration and a physically stabilized scheme, which is different from the approach of the element discussed in Section 5.4. The following brief description covers only the stabilization part, using the notations of Section 5.1.

The mean values of $\partial\Phi_N/\partial x_i$ are calculated as follows:

$$\bar{b}_{iN} = \frac{1}{V_e}\int_{\Omega_e} \frac{\partial\varphi_N}{\partial x_i} \mathrm{d}\Omega_e. \tag{5.48}$$

The coefficients of the shape functions in (5.8) are modified:

$$\begin{aligned}\bar{\Delta}_N &= \left(\Sigma_N - \bar{b}_{iN} x_i^I \Sigma_I\right)/8, \\ \bar{C}_{\alpha N} &= \left(\Gamma_N^\alpha - \bar{b}_{iN} x_i^I \Gamma_I^\alpha\right)/8.\end{aligned} \tag{5.49}$$

$\boldsymbol{B}_N$ matrix in (5.11)–(5.13) for strain calculation is now modified with both constant part and nonconstant part:

$$\bar{\boldsymbol{B}}_N = \bar{\boldsymbol{B}}_{cN} + \bar{\boldsymbol{B}}_{nN}. \tag{5.50}$$

Entities b_{iN} in $\boldsymbol{B}_{cN}$ are replaced by $\bar{b}_{iN}$. Coefficients $C_{\alpha N}$ in $\boldsymbol{B}_{nN}$ are replaced by $\bar{C}_{\alpha N}$.

Expressing by the six-component form (with the symmetry of stress and strain tensors); and writing with parameters in the reference system, the stabilization portion of strain rate is assumed:

$$\dot{\boldsymbol{\varepsilon}}_N = \hat{\boldsymbol{J}}_0^{-1} \tilde{\boldsymbol{B}}_s(\xi) \tilde{\dot{\boldsymbol{u}}}, \tag{5.51}$$

where $\tilde{\dot{\boldsymbol{u}}}$ is the nodal velocity in the reference system. $\hat{\boldsymbol{J}}_0^{-1}$ is assumed and replaces $\boldsymbol{J}^{-1}$ for improving mesh distortion insensitivity, and is defined below:

$$\hat{\boldsymbol{J}}_0^{-1} = \begin{bmatrix} \|\boldsymbol{j}_1\|^{-2} &&&&& \\ & \|\boldsymbol{j}_2\|^{-2} &&&& \\ && \|\boldsymbol{j}_3\|^{-2} &&& \\ &&& \|\boldsymbol{j}_1\|^{-1}\|\boldsymbol{j}_2\|^{-1} && \\ &&&& \|\boldsymbol{j}_2\|^{-1}\|\boldsymbol{j}_3\|^{-1} & \\ &&&&& \|\boldsymbol{j}_3\|^{-1}\|\boldsymbol{j}_1\|^{-1} \end{bmatrix}. \tag{5.52}$$

$\boldsymbol{j}_i = \partial \boldsymbol{x}/\partial \xi_i|_0$ is the ith column of $\boldsymbol{J}_0$. $\tilde{\boldsymbol{B}}_s(\boldsymbol{\xi})$ is defined below to alleviate the shear locking:

$$\hat{\boldsymbol{J}}_0^{-1} = \begin{bmatrix} \bar{C}_{\alpha N}\partial h_\alpha/\partial \xi & 0 & 0 \\ 0 & \bar{C}_{\alpha N}\partial h_\alpha/\partial \eta & 0 \\ 0 & 0 & \bar{C}_{\alpha N}\partial h_\alpha/\partial \zeta \\ \bar{C}_{1N}\zeta & \bar{C}_{2N}\zeta & 0 \\ 0 & \bar{C}_{2N}\xi & \bar{C}_{3N}\xi \\ \bar{C}_{1N}\eta & 0 & \bar{C}_{3N}\eta \end{bmatrix}. \tag{5.53}$$

In addition, an enhanced assumed strain rate is introduced to eliminate the Poisson-type locking in bending and volumetric locking as follows:

$$\dot{\boldsymbol{\varepsilon}}_{\text{ehance}} = \hat{\boldsymbol{J}}_0^{-1} \tilde{\boldsymbol{G}}(\boldsymbol{\xi}) \dot{\boldsymbol{\alpha}}. \tag{5.54}$$

Finally,

$$\dot{\boldsymbol{\varepsilon}} = \hat{\boldsymbol{J}}_0^{-1}(\tilde{\boldsymbol{B}}_s(\boldsymbol{\xi})\tilde{\dot{\boldsymbol{u}}} + \tilde{\boldsymbol{G}}(\boldsymbol{\xi})\dot{\boldsymbol{\alpha}}). \tag{5.55}$$

Here, $\dot{\boldsymbol{\alpha}}$ is determined based on the virtual power with the requirement that the corresponding internal nodal forces $f_\alpha = 0$.

5.7 TAYLOR EXPANSION OF ASSUMED STRAIN ABOUT THE ELEMENT CENTER

Note that in the stabilization algorithms of assumed strain discussed in Sections 5.4 and 5.5, an assumption implied that the element was a regular brick. The implementation was performed using the eight-point regular Gauss quadrature, or a four-point quadrature, or the one-point reduced quadrature. To apply the technique to the general situations, Taylor expansion of the assumed strain was used in Liu et al. (1985) and (1994). Schulz (1985) initiated the concept that used Taylor series on stress. The trilinear shape function contains products of the reference parameters of second and third degrees. Their derivatives contain the linear terms and monomials $\eta\zeta$, $\zeta\xi$, and $\xi\eta$ of second degree. Accordingly, $\boldsymbol{B}$ matrix used for calculating strain components is defined as the Taylor series:

$$\begin{aligned}\bar{\boldsymbol{B}}_N &= \boldsymbol{B}_N(0) + \boldsymbol{B}_{N,\xi}(0)\xi + \boldsymbol{B}_{N,\eta}(0)\eta + \boldsymbol{B}_{N,\zeta}(0)\zeta \\ &\quad + 2\boldsymbol{B}_{N,\eta\zeta}(0)\eta\zeta + 2\boldsymbol{B}_{N,\zeta\xi}(0)\zeta\xi + 2\boldsymbol{B}_{N,\xi\eta}(0)\xi\eta.\end{aligned} \tag{5.56}$$

All these submatrices are derived in terms of the reference coordinates. We need Taylor expansion of $\partial\Phi_N/\partial x_i$, which compose the $\boldsymbol{B}$ matrix. The linear terms can be directly derived from (5.6), for example,

$$\begin{aligned}b_{xN,\xi} &= \frac{\partial}{\partial\xi}\frac{\partial\Phi_N}{\partial x}(0) = \left(C_{2N}\frac{\partial\zeta}{\partial x} + C_{3N}\frac{\partial\eta}{\partial x}\right)(0) = C_{2N}D_{31} + C_{3N}D_{21}, \\ b_{xN,\eta} &= \frac{\partial}{\partial\eta}\frac{\partial\Phi_N}{\partial x}(0) = \left(C_{1N}\frac{\partial\zeta}{\partial x} + C_{3N}\frac{\partial\xi}{\partial x}\right)(0) = C_{3N}D_{11} + C_{1N}D_{31}, \\ b_{xN,\zeta} &= \frac{\partial}{\partial\zeta}\frac{\partial\Phi_N}{\partial x}(0) = \left(C_{1N}\frac{\partial\eta}{\partial x} + C_{2N}\frac{\partial\xi}{\partial x}\right)(0) = C_{1N}D_{21} + C_{2N}D_{11}.\end{aligned} \tag{5.57}$$

Here, D_{ij} are the elements of the inverse of Jacobian matrix at element center, defined in (5.17).

For the quadratic terms, we have for instance

$$\begin{aligned}b_{xN,\xi\eta} &= \frac{\partial^2}{\partial\xi\partial\eta}\frac{\partial\Phi_N}{\partial x}(0) = -\frac{1}{8}(C_{1N}D_{3i} + C_{3N}D_{1i})X_i^N\left(\Gamma_N^2 D_{31} + \Gamma_N^3 D_{21}\right) \\ &\quad - \frac{1}{8}(C_{2N}D_{3i} + C_{3N}D_{2i})X_i^N\left(\Gamma_N^3 D_{11} + \Gamma_N^1 D_{31}\right).\end{aligned} \tag{5.58}$$

The other terms can be derived by permutation; see Liu et al. (1985), also (1994) for detailed formulation of all terms.

With assumed deviatoric strain method discussed in Section 5.5, Liu et al. (1985) used assumed strain of the nonconstant part in the same way as (5.46), denoted by a superscript "dev." $\boldsymbol{B}$ matrix was then defined as

$$\begin{aligned}\bar{\boldsymbol{B}}_N &= \boldsymbol{B}_N(0) + \boldsymbol{B}^{\text{dev}}_{N,\xi}(0)\xi + \boldsymbol{B}^{\text{dev}}_{N,\eta}(0)\eta + \boldsymbol{B}^{\text{dev}}_{N,\zeta}(0)\zeta \\ &\quad + 2\boldsymbol{B}^{\text{dev}}_{N,\eta\zeta}(0)\eta\zeta + 2\boldsymbol{B}^{\text{dev}}_{N,\zeta\xi}(0)\zeta\xi + 2\boldsymbol{B}^{\text{dev}}_{N,\xi\eta}(0)\xi\eta. \end{aligned} \tag{5.59}$$

The rate of strain was calculated from $\dot{\boldsymbol{\varepsilon}} = \bar{\boldsymbol{B}}_N \boldsymbol{v}^N$. The rate of stress was calculated following the usual process with the material constitutive law, for example, $\dot{\boldsymbol{\sigma}} = \boldsymbol{E}_t\dot{\boldsymbol{\varepsilon}}$ (see discussion later in Part 3). The polynomial form was proposed for the rate of stress corresponding to the Taylor expansion, denoted by

$$\dot{\boldsymbol{\sigma}} = \dot{\boldsymbol{\sigma}} + \dot{\boldsymbol{\sigma}}^{\xi}\xi + \dot{\boldsymbol{\sigma}}^{\eta}\eta + \dot{\boldsymbol{\sigma}}^{\zeta}\zeta + 2\dot{\boldsymbol{\sigma}}^{\eta\zeta}\eta\zeta + 2\dot{\boldsymbol{\sigma}}^{\zeta\xi}\zeta\xi + 2\dot{\boldsymbol{\sigma}}^{\xi\eta}\xi\eta. \tag{5.60}$$

It was assumed that the components of the stress rate were obtained individually from the corresponding components of the rate of strain. It was further assumed that the same constitutive relation being used for calculating all these components (defined in (5.60)) of the rate of stress. For instance, $\dot{\boldsymbol{\sigma}}^{\xi} = \boldsymbol{E}_t\dot{\boldsymbol{\varepsilon}}^{\xi}$, $\dot{\boldsymbol{\sigma}}^{\xi\eta} = \boldsymbol{E}_t\dot{\boldsymbol{\varepsilon}}^{\xi\eta}$, etc; see Liu et al. (1984) for more discussions.

The assumed deviatoric strain solves volumetric locking. To eliminate shear locking and membrane locking, which may happen to the thin structure, Liu et al. (1994) used another form of assumed strain with Taylor expansion approach. The assumed tensile strain uses all the terms of (5.59) for the nonconstant part of $\boldsymbol{B}$ matrix. The assumed shear strain uses fewer terms as defined below:

$$\begin{aligned}\bar{\boldsymbol{B}}_{xyN} &= \boldsymbol{B}_{xyN}(0) + \boldsymbol{B}^{\text{dev}}_{xyN,\zeta}(0)\zeta, \\ \bar{\boldsymbol{B}}_{yzN} &= \boldsymbol{B}_{yzN}(0) + \boldsymbol{B}^{\text{dev}}_{yzN,\xi}(0)\xi, \\ \bar{\boldsymbol{B}}_{zxN} &= \boldsymbol{B}_{zxN}(0) + \boldsymbol{B}^{\text{dev}}_{zxN,\eta}(0)\eta. \end{aligned} \tag{5.61}$$

$\boldsymbol{B}$ matrix associated with the nonconstant part of the assumed strain then become

$$\bar{\boldsymbol{B}}_{nN} = \begin{bmatrix} (2/3)D_{11}g_{1N} & -D_{22}g_{2N}/3 & -D_{33}g_{3N}/3 \\ -D_{11}g_{1N}/3 & (2/3)D_{22}g_{2N} & -D_{33}g_{3N}/3 \\ -D_{11}g_{1N}/3 & -D_{22}g_{2N}/3 & (2/3)D_{33}g_{3N} \\ D_{22}C_{1N}\zeta & D_{11}C_{2N}\zeta & 0 \\ 0 & D_{33}C_{2N}\xi & D_{22}C_{3N}\xi \\ D_{33}C_{1N}\eta & 0 & D_{11}C_{3N}\eta \end{bmatrix}, \tag{5.62}$$

where

$$g_{1N} = C_{3N}\eta + C_{2N}\zeta + 2C_{4N}\eta\zeta,$$
$$g_{2N} = C_{1N}\zeta + C_{3N}\xi + 2C_{4N}\zeta\xi, \quad (5.63)$$
$$g_{3N} = C_{2N}\xi + C_{1N}\eta + 2C_{4N}\xi\eta.$$

Note that coefficients $C_{\alpha N}$ and D_{ij} are defined in (5.8) and (5.17), respectively. Using this approach, when viewing in the corotational system, the hourglass mode causing shear locking has no contribution to the strain energy, therefore, does not generate spurious mode.

For nonlinear applications like elastic–plastic deformation with complex loading and unloading, Liu et al. (1994) recommended four-point quadrature scheme to more accurately compute the rate of stress. The integration points were chosen at reference coordinates $(1/\sqrt{3}, 1/\sqrt{3}, 1/\sqrt{3})$, $(-1/\sqrt{3}, -1/\sqrt{3}, 1/\sqrt{3})$, $(-1/\sqrt{3}, 1/\sqrt{3}, -1/\sqrt{3})$, and $(1/\sqrt{3}, -1/\sqrt{3}, -1/\sqrt{3})$. This was essentially the same quadrature rule recommended by Belytschko and Bindeman (1993). The nodal forces were calculated by

$$f_{iN}^{\text{int}} = \sum_{k=1}^{4} \frac{V}{4} \bar{B}_{jN}(\xi_k)\sigma_{ij}(\xi_k). \quad (5.64)$$

From (5.62) to (5.63), the nodal forces consist of contributions from the constant and nonconstant parts of the assumed strain:

$$\begin{bmatrix} f_{xN}^{\text{int}} \\ f_{yN}^{\text{int}} \\ f_{zN}^{\text{int}} \end{bmatrix} = \sum_{k=1}^{4} \frac{V}{4} \begin{bmatrix} b_{1N}\sigma_{11} + b_{2N}\sigma_{12} + b_{3N}\sigma_{13} \\ b_{1N}\sigma_{12} + b_{2N}\sigma_{22} + b_{3N}\sigma_{23} \\ b_{1N}\sigma_{13} + b_{2N}\sigma_{23} + b_{3N}\sigma_{33} \end{bmatrix}$$
$$+ \sum_{k=1}^{4} \frac{V}{4} \begin{bmatrix} g_{1N}\sigma_{11}^{\text{dev}} + D_{22}C_{1N}\zeta\sigma_{12} + D_{33}C_{1N}\eta\sigma_{13} \\ g_{2N}\sigma_{22}^{\text{dev}} + D_{33}C_{2N}\zeta\sigma_{12} + D_{11}C_{2N}\xi\sigma_{23} \\ g_{3N}\sigma_{33}^{\text{dev}} + D_{11}C_{3N}\eta\sigma_{13} + D_{22}C_{3N}\xi\sigma_{23} \end{bmatrix}. \quad (5.65)$$

As mentioned in Section 5.4, for elastic applications, one-point quadrature scheme can provide an exact integration for the stress and the nodal force. Successful applications of this assumed strain method can be found in Liu et al. (1985), also (1994).

Note that in the Taylor series method of Schulz (1985), derivatives of stress with respect to the reference coordinates at the element center were introduced as additional variables being used in Taylor expansion. Schulz (1985) named the method as stress-derivative hourglass control. The methods developed in Liu et al. (1985) and (1994) did not use additional variables but used more coefficients, which were explicitly integrated for strain or rate of strain. When the element is nearly rectilinear (a brick) or parallelepiped, the Jacobian can be approximated as constant and computation is simplified.

Hu and Nagy (1997) used the one-point quadrature scheme of Liu et al. (1994) for brake squeal analysis, involving high frequency response of brake rotor under impact and friction load caused by braking action. The material was assumed to be elastic. Due to high frequency in the range of 1,000–16,000 Hz, plus contact and impact loading, the traditional method of eigenvalues had been unsuccessful. Due to the lack of eigen solver in the explicit finite element code, the frequencies were obtained by using Fast Fourier Transform after the deformation time history was obtained. The results presented for the first time, a good correlation to the test data.

In the direction of selective/reduced integration and according to Taylor expansion method which applied to assumed strain, a method of directional selective reduced integration was developed in Koh and Kikuchi (1987). The method chose the direction based on geometry characteristics under integration. In previously discussed methods, reduced integration was applied to certain terms of strain or rate of strain, but applied equally for the three dimensions. Koh and Kikuchi (1987) recommended reduced integration with respect to the larger dimension (in element local system). The strain was decomposed into deviatoric normal, shear, and volumetric components. Shear components could be set to zero to avoid shear locking; or volumetric components could be set to zero to avoid volumetric locking for the applications with incompressible materials.

5.8 EVALUATION OF EIGHT-NODE SOLID ELEMENT

We use the two examples investigated with the shell elements to evaluate the performance of solid elements, regarding the shear-locking issue. We use commercial software LS-DYNA V971 to perform the numerical experiment. So far we have only discussed the solid elements with the one-point integration scheme. We choose the element using stiffness hourglass control method in Flanagan and Belytschko (1981) discussed in Section 5.3; assumed strain method in Belytschko and Bindeman (1993) discussed in Section 5.4; and enhanced assumed strain method in Puso (2000) discussed in Section 5.6.

Example 5.1 Twisted beam We start with the quadrilateral meshes used for the shell elements to investigate the solid elements. Letting quadrilateral mesh represent the mid-surface, the thickness with four layers of solid elements is being built. The model has equal thickness of the shell element, depicted in Figure 5.3.

With thickness equal to 0.32, the time history of displacement at the tip calculated by these three types of elements is presented in Figure 5.4. Here, following the definition of LS-DYNA, IHQ = 4 indicates the stiffness hourglass control method; IHQ = 6 indicates the assumed strain method; and IHQ = 9 indicates the enhanced assumed strain method. All three elements show tendency to converge and the results with fine mesh are close to the calculated results of B-D and B-L shell elements shown in Figure 3.18. For the reduced thickness equal to 0.032, the result of the stiffness hourglass control method presented in Figure 5.5a has not converged yet. The result seems to be too stiff, but is improved with refined meshes. It is a type of locking phenomenon. On the other hand, the result of the assumed strain method has better performance, shown in Figure 5.5b. The enhanced assumed strain method performs

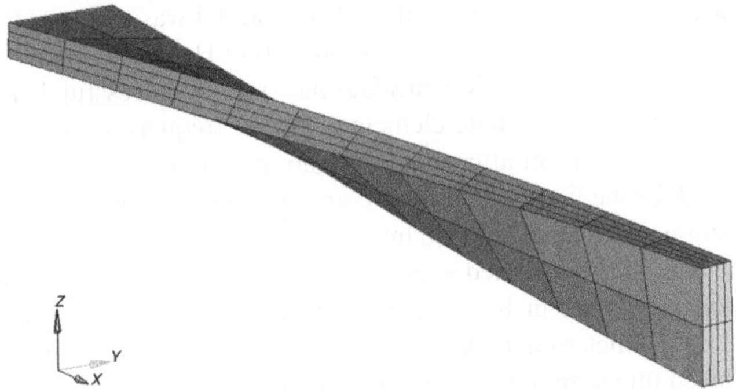

FIGURE 5.3 Solid element mesh of the twisted beam.

even better, which is shown in Figure 5.5c. The results of stiffness hourglass control method depend on the coefficient. We use 0.05 for all cases in this example.

Note that in this study, with 24×4 mesh, the length and width of the quadrilateral shell element are about 0.50 and 0.275, respectively and the same measurement is applied to solid element. For the original thickness, solid element is 0.08 in the depth (in thickness direction), which is smaller than the length and width. Hence, it needs

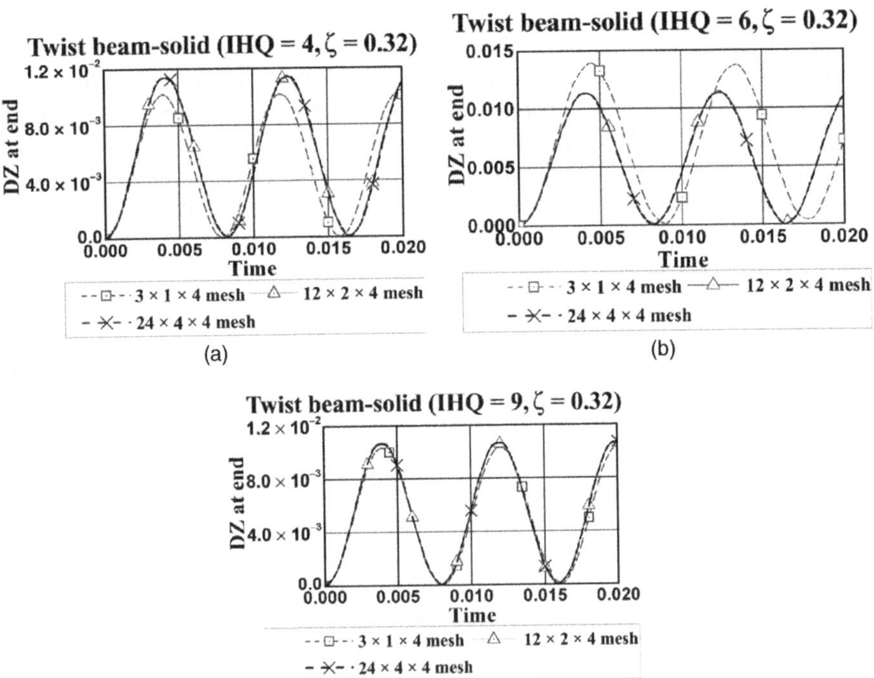

FIGURE 5.4 Solution of the twisted beam problem by solid element: (a) stiffness hourglass control method; (b) assumed strain method; (c) enhanced assumed strain method.

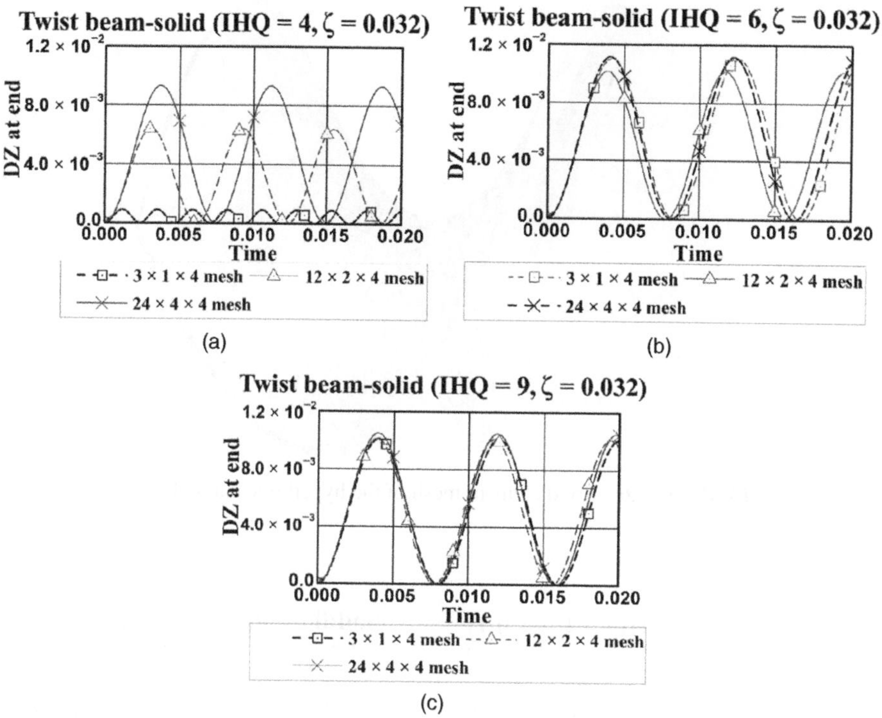

FIGURE 5.5 Solution of the twisted beam problem by solid element with reduced thickness: (a) stiffness hourglass control method; (b) assumed strain method; (c) enhanced assumed strain method.

smaller time step size than what is needed by the quadrilateral mesh. Since we keep using four layers of solid elements from coarse mesh to fine mesh, the time step size is dominated by the thickness and is the same in this example. When the thickness is reduced to one-tenth of the original thickness, the length and width remain to be the same, but the depth is reduced to one-tenth. Hence, the time step size is expected to be reduced by a factor of 0.1. Along with the required load scaling and mass scaling, the time step size will be reduced by another factor of 0.1. All together a time step size that is 100 times smaller than that of the original thickness is expected.

Example 5.2 Hyperbolic paraboloid We start with the set of four quadrilateral meshes used for the shell elements to investigate the solid elements. The solid meshes are created in the same way as the previous example, shown in Figure 5.6. In this example, we use a coefficient 0.001 for the stiffness hourglass control method.

With thickness equal to 0.1, the time history of displacement at the monitoring point calculated by using these three types of elements is presented in Figure 5.7. The

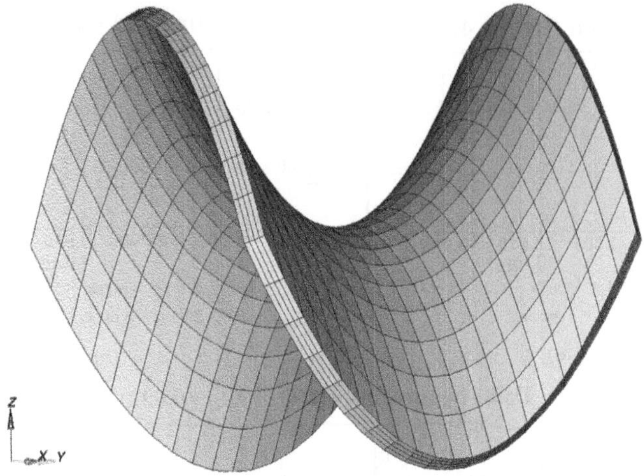

FIGURE 5.6 Solid element mesh of the hyperbolic paraboloid.

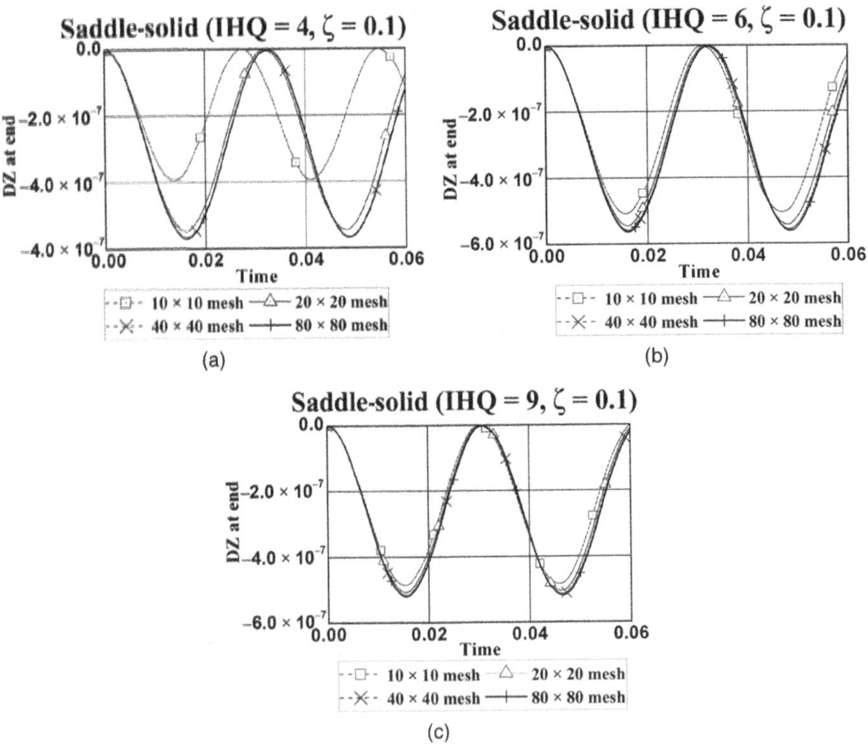

FIGURE 5.7 Solution of the hyperbolic paraboloid problem by solid element: (a) stiffness hourglass control method; (b) assumed strain method; (c) enhanced assumed strain method.

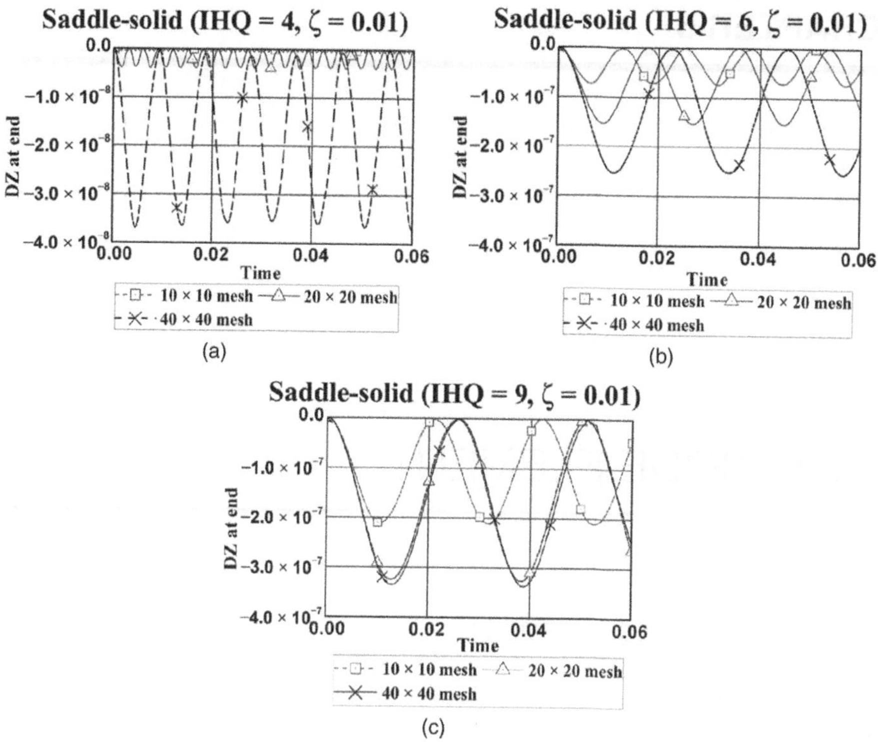

FIGURE 5.8 Solution of the hyperbolic paraboloid problem by solid element with reduced thickness: (a) stiffness hourglass control method; (b) assumed strain method; (c) enhanced assumed strain method.

results of all the elements have a tendency to converge. With the reduced thickness equal to 0.01, the result of the stiffness hourglass control method, depicted in Figure 5.8a, is far away from convergence. The result changes drastically when the mesh is refined. On the other hand, the result of the assumed strain method has not yet converged either, depicted in Figure 5.8b. The results seem to be too stiff for coarse meshes, and that is a typical shear-locking behavior. But the result of the enhanced assumed strain method seems to have converged, depicted in Figure 5.8c.

CHAPTER 6

TWO-NODE ELEMENT

6.1 TRUSS AND ROD ELEMENT

Both truss and rod are a two-node element, which transmits the axial load only. Each node has three translational degrees of freedom. In the element local system, shown in Figure 6.1, the displacement is approximated by linear interpolation with $\varphi_1(x) = (L - x)/L$ and $\varphi_2(x) = x/L$. The velocity strain is hence constant in the element, characterized by their axial components (or the projections) $\dot{u} = \boldsymbol{e} \bullet \dot{\boldsymbol{u}}$:

$$\begin{aligned} \dot{u} &= \dot{u}^\alpha \varphi_\alpha(x), \\ \dot{\varepsilon} &= \dot{u}^\alpha \varphi'_\alpha(x) = \left(\dot{u}_x^2 - \dot{u}_x^1\right)/L. \end{aligned} \tag{6.1}$$

Here, $\boldsymbol{e}$ is the unit vector of the local system. Subsequently, the stress rate is also a constant $\dot{\sigma} = \dot{\sigma}(\dot{\varepsilon})$, so is the stress σ no matter if the element is modeling a linear or nonlinear material. The calculation of internal nodal force involves, cf. (2.19):

$$F_x^\alpha = -\int \sigma \varphi'_\alpha(x) \mathrm{d}\Omega = -A\sigma \int \varphi'_\alpha(x) \mathrm{d}x = \pm A\sigma. \tag{6.2}$$

When the effect of Poisson's ratio is taken into account for linear elasticity, the cross-section area is $A = A_0(1 - 2\nu\varepsilon)$. Here, A_0 is the original cross-section area. If the material is considered incompressible, for example, plastic, then we use $\dot{A} = A(1 - \dot{\varepsilon})$.

Introduction to the Explicit Finite Element Method for Nonlinear Transient Dynamics, First Edition.
Shen R. Wu and Lei Gu.

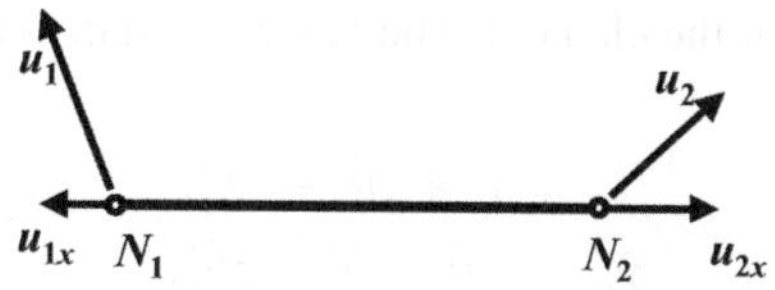

FIGURE 6.1 Truss element.

When the truss element connects to other part of the structure, the connection works like a hinge, which transmits axial force only. In particular, when it connects to a shell element, the bending moment is not transmitted from the shell to the truss.

6.2 TIMOSHENKO BEAM ELEMENT

In large deformation analysis, the beam element can be used for simplified modeling of connection between two structural parts and not just for a real beam type of structure. The beam is a two-node element. It can transmit axial force and shear forces in two transverse directions; also bending moments, and torque at its two nodes. This means three translational and three rotational degrees of freedom. Similar to R-M plate discussed in Chapter 3, we are interested in Timoshenko beam illustrated in Figure 6.2. The velocity components of a generic point at (x, y, z) in the cross section are represented by the velocity at the central line ($y = z = 0$) and rotational velocity of the centerline and the cross section:

$$\begin{aligned} \dot{U}_x &= \dot{u}_x + z\dot{\theta}_y - y\dot{\theta}_z, \\ \dot{U}_y &= \dot{u}_y - z\dot{\theta}_x, \\ \dot{U}_z &= \dot{u}_z + y\dot{\theta}_x. \end{aligned} \tag{6.3}$$

All these variables are functions of t and x only, and are interpolated with linear shape functions, which are the same as those used for truss element:

$$\begin{aligned} \varphi_1 &= (1-\xi)/2, \quad \varphi_{1,x} = -1/L, \\ \varphi_2 &= (1+\xi)/2, \quad \varphi_{2,x} = 1/L. \end{aligned} \tag{6.4}$$

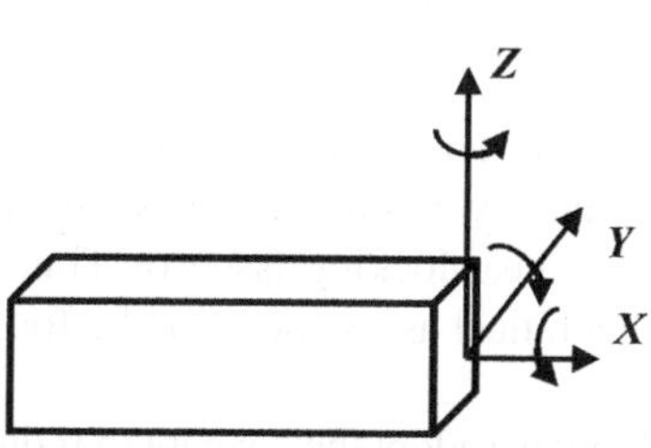

FIGURE 6.2 Beam element.

Here, L is the length of the element. The velocity strain components are

$$\begin{aligned} \dot{\varepsilon}_x &= \dot{u}'_x + z\dot{\theta}'_y - y\dot{\theta}'_z, \\ \dot{\gamma}_{xy} &= -\dot{\theta}_z + \dot{u}'_y - z\dot{\theta}'_x, \\ \dot{\gamma}_{xz} &= \dot{\theta}_y + \dot{u}'_z + y\dot{\theta}'_x, \\ \dot{\varepsilon}_y &= \dot{\varepsilon}_z = \dot{\gamma}_{yz} = 0. \end{aligned} \tag{6.5}$$

Following (6.4), the derivatives of the interpolated variables should be constants. Therefore, (6.5) is expressed as

$$\begin{aligned} \dot{\varepsilon}_x &= (\dot{u}_{x2} + z\dot{\theta}_{y2} - y\dot{\theta}_{z2} - \dot{u}_{x1} - z\dot{\theta}_{y1} + y\dot{\theta}_{z1})/L, \\ \dot{\gamma}_{xy} &= -\dot{\theta}_{z1}(1-\xi)/2 - \dot{\theta}_{z2}(1+\xi)/2 + (\dot{u}_{y2} - z\dot{\theta}_{x2} - \dot{u}_{y1} + z\dot{\theta}_{x1})/L, \\ \dot{\gamma}_{xz} &= \dot{\theta}_{y1}(1-\xi)/2 + \dot{\theta}_{y2}(1+\xi)/2 + (\dot{u}_{z2} + y\dot{\theta}_{x2} - \dot{u}_{z1} - y\dot{\theta}_{x1})/L. \end{aligned} \tag{6.6}$$

Recall the calculation for nodal force and moment, which is for the shell element based on R-M theory, we have, similarly

$$\begin{aligned} F_{i\alpha}\delta\dot{u}_{i\alpha} + M_{i\alpha}\delta\dot{\theta}_{i\alpha} &= -\int \sigma_{ij}\delta\dot{\varepsilon}_{ij}\mathrm{d}\Omega \\ &= -\int (\sigma_x\delta\dot{\varepsilon}_x + \kappa(\tau_{xy}\delta\dot{\gamma}_{xy} + \tau_{xz}\delta\dot{\gamma}_{xz}))\mathrm{d}\Omega, \end{aligned} \tag{6.7}$$

where κ is the shear correction factor.

It is observed that the shear locking is also a problem for Timoshenko beam element. For example, consider the case of elastic pure bending deformation, with $-\theta_{y1} = \theta_{y2} = \theta$ and all the other displacement components vanishing. From (6.6), this displacement field results in the following strain components:

$$\begin{aligned} \varepsilon_x &= z(\theta_{y2} - \theta_{y1})/L = 2z\theta/L, \\ \gamma_{xy} &= 0, \\ \gamma_{xz} &= \theta_{y1}(1-\xi)/2 + \theta_{y2}(1+\xi)/2 = \xi\theta. \end{aligned}$$

It has the same pattern as what is observed in the R-M plate-bending situation. The shear strain energy will dominate when the thickness in z-direction becomes small. Like the R-M plate element, shear-locking issue of Timoshenko beam element is also of research interest. Theoretical assessment can be found, for example, in Chen et al. (1997) .

To control the shear locking, we adopt the concept of reduced integration (at center point along the element axial direction, $\xi = 0$) for the shear strain components. As a matter of fact, all other terms can be explicitly integrated. For elastic beam, we can directly derive the nodal forces and moments from (6.6) to (6.7):

$$\begin{aligned}
-F_{x2} &= F_{x1} = EAu'_x, \\
-F_{y2} &= F_{y1} = \kappa GA\left(u'_y - \bar{\theta}_z\right), \\
-F_{z2} &= F_{z1} = \kappa GA\left(u'_z + \bar{\theta}_y\right), \\
-M_{x2} &= M_{x1} = G\int\left(y^2 + z^2\right)\mathrm{d}A\theta'_x = GJ\theta'_x, \\
-M_{y2} &= M_{y1} = \int\left(Ez^2\theta'_y - \kappa GL\left(u'_z + \bar{\theta}_y\right)\right)\mathrm{d}A = EI_y\theta'_y - F_{z1}L, \\
-M_{z2} &= M_{z1} = \int\left(Ey^2\theta'_z + \kappa GL\left(u'_y - \bar{\theta}_z\right)\right)\mathrm{d}A = EI_z\theta'_z + F_{y1}L.
\end{aligned} \tag{6.8}$$

Here, we denote $\bar{\theta}_y$ and $\bar{\theta}_z$ for the middle point values or the average nodal values of the rotation. As usual, we define the moment of inertia of the cross section by $I_y = \int z^2\mathrm{d}A$, $I_z = \int y^2\mathrm{d}A$, and $J = I_y + I_z = \int(y^2 + z^2)\mathrm{d}A$.

For nonlinear materials, integration in the cross section should be performed like the integration through thickness for shell elements.

In practical application of large deformation analysis, usually the connection modeled by beam element is quite strong and only small deformation is expected. In many cases, the beam element is a simplified modeling rather than a real beam type of structure. Under such assumption, the linear elasticity model is appropriate and the stress resultants can be explicitly expressed. When large distortion of the cross section along with nonlinear material behavior happens, the determination of deformation and stress becomes a much more complex subject, which is beyond our scope. In fact, the usual beam theory does not include the arbitrary large deformation of the cross section. In such situation, reliability becomes questionable when utilizing the nonlinear beam element. Other type of elements is therefore to be recommended.

6.3 SPRING ELEMENT

6.3.1 One Degree of Freedom Spring Element

The usage of a two-node spring element in large deformation analysis is most likely for a simplified modeling of the connection between two structural parts. It is used to transmit the loads, determined by the change of distance between the connected parts and the given spring stiffness. The material property is defined in the application of truss and beam elements. The response force is calculated by element stiffness method based on the concept of continuum mechanics. The spring element is a type of discrete element and the computation is straightforward. The usage, however, varies from software to software with different specifications. Here, we describe some basic applications; see Hallquist (2006) and ALTAIR (2009) for more details.

For a linear spring with stiffness K, the force to be transmitted is $F = K \times (d - d_0)$ with current distance d and the neutral distance d_0 between the two nodes. Nonlinear

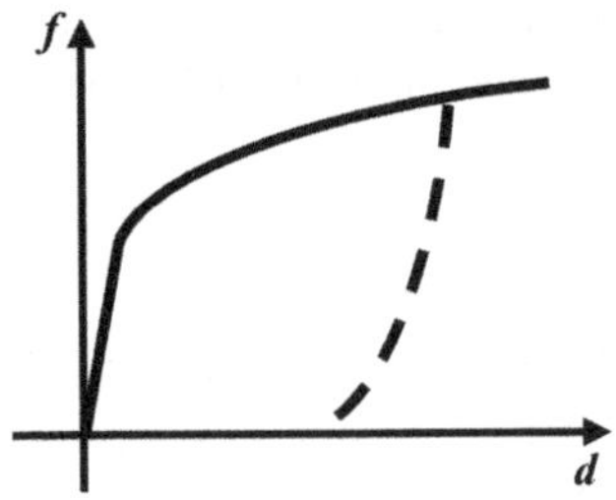

FIGURE 6.3 Spring element with different loading and unloading behavior.

spring is also useful, for which a force–deflection function F_k is prescribed. Then $F = F_k(d - d_0)$. The damping effect can be included to make it as a spring-damper. The damping force can also be linear with a given viscosity or nonlinear with a viscosity function. In general, the spring force is expressed as

$$F = F_k(d - d_0) + F_c(\dot{d}). \tag{6.9}$$

In nonlinear applications, the spring element is sometimes designed to represent a structural component's large deformation behavior. Therefore, certain property such as plastic unloading, buckling, and postbuckling can be modeled by implementing an unloading path which is different from the loading path, shown in Figure 6.3. This type of application is of course using the existing knowledge of the components for simplified modeling. It is also useful for sensitivity study to investigate the effect of certain components to overall response in the other cases.

6.3.2 Six Degrees of Freedom Spring Element

The usage of spring element can be extended to more general application with six degrees of freedom. A possible approach is to simply use the difference in the nodal positions, that is, the difference in the nodal displacement components and the nodal rotations, to determine the components of forces and moments, for example,

$$\begin{aligned} F_j &= F_{kj}\left(u_j^2 - u_j^1\right), \\ M_j &= M_{kj}\left(\theta_j^2 - \theta_j^1\right), \end{aligned} \qquad j = 1, 2, 3. \tag{6.10}$$

Note that with this algorithm, the displacement would happen accordingly when rotation occurs as shown in Figure 6.4. The force would then be generated even with a rigid body type of rotation. This is unexpected and will possibly mislead people. Therefore, its usage is limited to certain controllable situations.

Another choice is to let the spring perform as a beam but with the beam property directly provided in the form of a force–deflection function. For instance, we can still use the concept of Timoshenko beam and use (6.5) to calculate the increment of

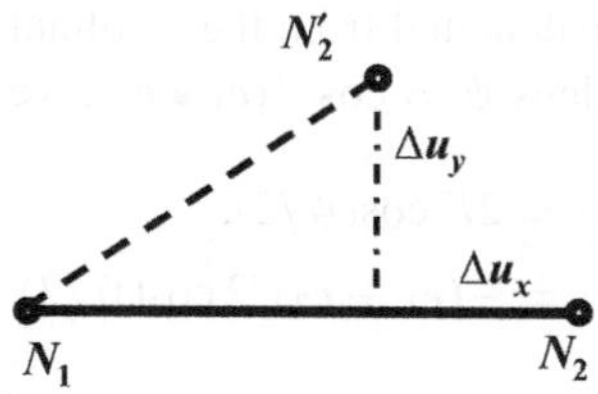

FIGURE 6.4 A 6-dof spring element experiencing a motion with rotation.

deformation. Then get the nodal forces and moments by using predefined functions, instead of the constitutive equations and material properties.

$$\begin{aligned} F_j &= F_j(u_j), \\ M_j &= M_j(\theta_j), \end{aligned} \quad j = 1, 2, 3. \tag{6.11}$$

In this way, the rigid body rotation of the element will not produce force or moment.

6.3.3 Three-node Spring Element

As another extension, we can use a three-node spring element to model a pulley or a similar mechanism. The concept is illustrated in Figure 6.5. The third node N_3 acts like a pulley. The spring element acts like the rope around the pulley. As a compression is not meaningful to the rope, exam is needed before using the three-node spring. The positions of the nodes determine the lengths of the two legs, $L_1 = |x^1 - x^3|$ and $L_2 = |x^2 - x^3|$. The deformation (elongation) of the spring is the change of the total length and the force is determined by a given function, just like the situation discussed above:

$$\begin{aligned} d &= L_1 + L_2 - L_0, \\ F &= F_k(d). \end{aligned} \tag{6.12}$$

The damping force can be added into the formula if needed. Node N_1 and N_2 will have force $F_1 = F_2 = F$ along the direction of the legs $\boldsymbol{e}_\alpha = (\boldsymbol{x}^3 - \boldsymbol{x}^\alpha)/L_\alpha$. The

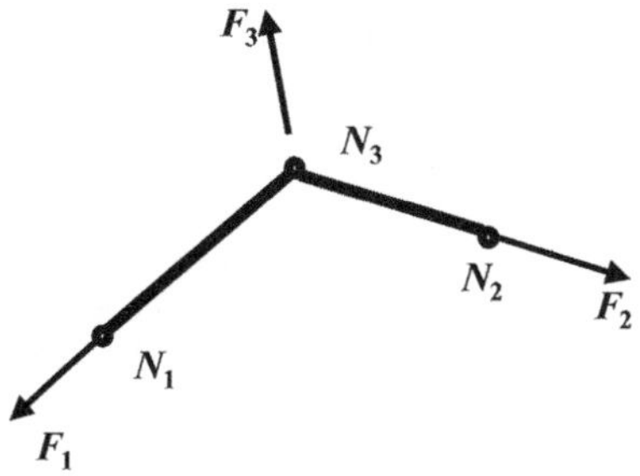

FIGURE 6.5 A three-node spring element simulating a pulley.

force applied on node N_3 is calculated from the resultant $\boldsymbol{F}_3 = -(\boldsymbol{F}_1 + \boldsymbol{F}_2)$. Using the angle formed by the two legs $\psi = \cos^{-1}(\boldsymbol{e}_1 \bullet \boldsymbol{e}_2)$, we obtain

$$\begin{aligned} F_3 &= 2F\cos(\psi/2), \\ \boldsymbol{e}_3 &= -(\boldsymbol{e}_1 + \boldsymbol{e}_2)/2\cos(\psi/2). \end{aligned} \tag{6.13}$$

6.4 SPOT WELD ELEMENT

Spot weld plays an important role. The spot weld holds two or multiple pieces of sheet metals together, as depicted in Figure 6.6a. The weld is formed by the compression of electrodes with certain level of current and holding time. The sheet metals are melting together under high current and then cool down to form a circular shaped nugget embedded in the sheets, shown in Figure 6.6b. Surrounding the nugget perimeter, there is a thin layer of material called heat affected zone. Generally speaking, the spot welds are designed to strongly hold a structure composed of the sheet metals without breaking off. Under certain static or dynamic loading conditions, however, the spot weld connection has the possibility to fail. Therefore great efforts are devoted to investigate the mechanism of failure, the conditions regarding the possibility of welds separation, and the failure criterion for analysis. Meanwhile, a kind of simplified method to model the spot weld was attempted.

6.4.1 Description of Spot Weld Separation

It is well known that the spot welding needs to follow certain specification or weld schedule in order to produce good quality welds. The weld quality will deteriorate if the electric current is not appropriately set or the electrodes do not hold with proper time duration. Examining the load capacity of spot weld has been an active research topic. Kaga et al. (1976) designed cross-tension test (also known as normal tension test) and tensile-shear test (also known as lap shear test). The cross-tension test used

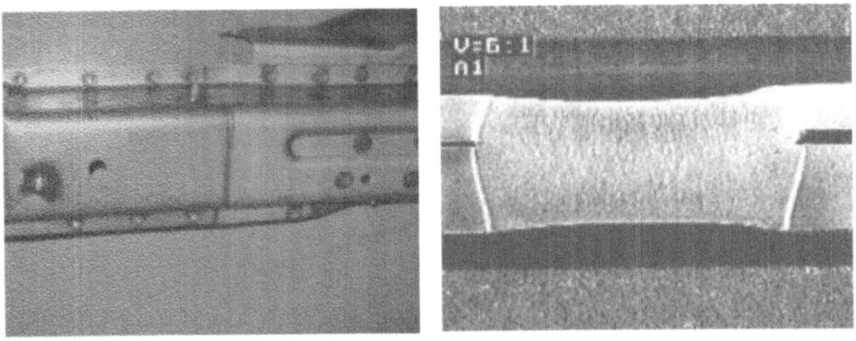

(a) (b)

FIGURE 6.6 Configuration of spot weld: (a) a structural component composed by several pieces welded together; (b) the cross section of a spot weld with two steel sheets of different thicknesses.

two strips welded together lying in a cross position. The load was applied on one strip, with the other one fixed at both ends, to pull apart the weld. In the tensile-shear test, two strips were welded together in a head on tail position. The load was applied in the longitudinal direction shearing the weld. Their objective was to find the yield point P_y and the ultimate load P_u for Aluminum with 1.6 mm thickness and mild steel with 1 mm thickness. The test apparatus was designed to pull the weld specimen in arbitrary direction. The main findings from tests were: (1) the ultimate load P_u was not greatly affected by loading direction due to the local large deformation of the aluminum sheets occurred before fracture; and (2) the yield point P_y was clearly affected by the loading direction. The yield point P_y decreased when the load angle increased from 0° (corresponding to lap shear) to 90° (corresponding to normal tension).

Ewing et al. (1982) added a coach peel test, with two L-shaped strips welded together in a symmetric position, involving bending effect. Specimens of low-carbon steel SAE-1006 and high-strength steel SAE-960X, with bare and galvanized surface treatment, were tested. Specimens were made with thickness ranging from 1.2 to 1.5 mm. Dynamic test was also performed using drop-weight tower, with impact speed up to 24 km/h. Along with material, the weld nugget size, and the parameters set by welding schedule were investigated for their effects to the load capacity. It was observed that the pull out failure under normal tension test occurred through the thickness of the base material by shearing at the nugget perimeter. It was found that in general the failure load increased with the loading speed.

6.4.2 Failure Criterion

Based on the tests similar to Ewing et al. (1982), Riesner et al. (2000) reported an empirical formula postulated for the failure strength of normal tension:

$$F_N = C_\sigma \sigma_{\text{haz}} \pi D \zeta, \tag{6.14}$$

where D is the weld diameter; ζ is the thickness; σ_{haz} is the yield stress of the heat affected zone; and C_σ is a coefficient, 0.577 for Mises yield condition and 0.5 for Tresca yield condition, respectively (see discussion in Chapter 7). The material property in the narrow heat affected zone is not a constant though because it links the weld nugget and the base material. Figure 6.7 presents a sample test result of hardness distribution in the cross section of a weld of the high strength steel. The hardness is usually tested as reference of the yield stress of material. It is higher in the weld nugget area than that of the base material. The hardness changes from the nugget through the heat affected zone to the base material.

For lap shear, the failure strength was proposed in Riesner et al. (2000) :

$$F_S = 0.5\sigma_{\text{haz}} \pi D \zeta + 2\frac{0.577}{\cos 35.3^{\circ}} \sigma_u \zeta W_a, \tag{6.15}$$

where σ_u is the maximum tensile stress of the base material, and W_a is the active flange width, that is, the distance from weld nugget to the edge of the specimen.

The heat-affected zone has variable strength and relatively lower ductility than what the base material has. If we assume a perfectly plastic model for the heat

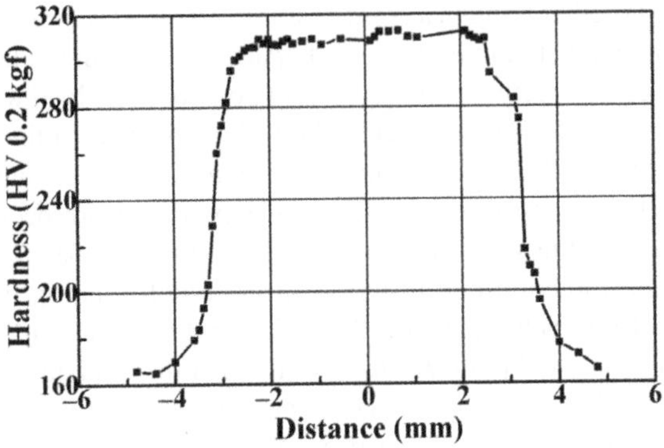

FIGURE 6.7 Hardness test in the spot weld.

affected zone, (6.14) represents the maximum load. After the nugget starts to move out, the load begins to drop until complete separation. This rupture process seems to be a kind of unstable structural softening, which happens very fast in the test.

Several types of simple failure criterion have been implemented in commercial software. For instance, a formula with combination of responses in six degrees of freedom (like a beam) is proposed in LS-DYNA; see Hallquist (2006):

$$\left(\frac{N_{rr}}{N_{rrF}}\right)^2 + \left(\frac{N_{rs}}{N_{rsF}}\right)^2 + \left(\frac{N_{rt}}{N_{rtF}}\right)^2 + \left(\frac{M_{ss}}{M_{ssF}}\right)^2 + \left(\frac{M_{tt}}{N_{ttF}}\right)^2 + \left(\frac{M_{rr}}{M_{rrF}}\right)^2 \leq 1. \tag{6.16}$$

Here, $\boldsymbol{N}$ and $\boldsymbol{M}$ are the calculated resultant force and moment transmitted by the spot weld. $\boldsymbol{N}_F$ and $\boldsymbol{M}_F$ are the failure loads obtained from tests in a single direction. When a large number of different combinations of sheet metals to be welded in the product, it becomes a burden to get the information and prepare the input data.

Lin et al. (2001, 2002a, 2002b) extended these studies and investigated static and dynamic tests under loading with an orientation angle from 0° for normal tension to near 90° (lap shear). An improved specimen was designed for the tests to obtain better consistency. The failure load was found dependent of the thickness and yield stress of the base material, the loading direction, and the loading speed. The failure load was formulated as

$$\begin{aligned} p_0 &= \pi D \zeta \tau_0, \\ P_1 &= C_v p_0, \\ \bar{N} &= N/p_1, \\ \bar{S} &= S/p_1, \\ \bar{N}^2 &+ \Lambda \bar{S}^2 \leq 1. \end{aligned} \tag{6.17}$$

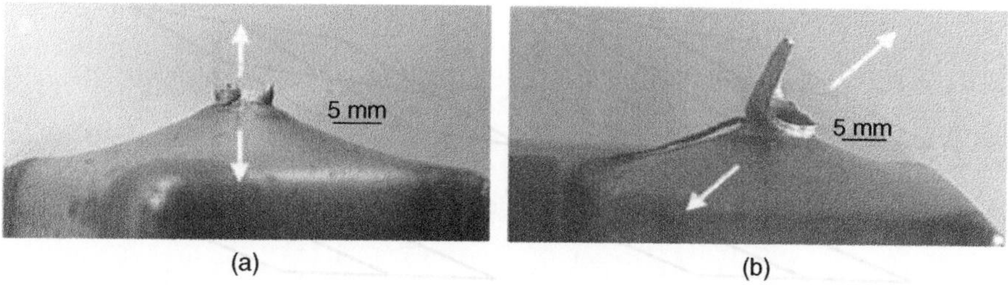

FIGURE 6.8 Rupture of spot weld: (a) with a normal tension load; (b) with a load in a given direction. (Lin S.H., Pan J., Wu S.R., Tyan T., and Wung P., Failure loads of spot welds under combined opening and shear static loading conditions. Elsevier 2001.)

Here, p_0 represents a shear force in the peripheral of the nugget with the diameter D of the nugget, the thickness ζ, and the shear yield stress τ_0 of the base material. p_0 serves as a basic estimate of separation load in static normal pull test, an extension of Riesner et al. (2000) . C_v is a function of loading rate. The scalar Λ is a function related to the nugget geometry parameters. C_v and Λ are determined from simple specimen tests. N and S are the normal and tangential components of the force transmitted by the spot weld. Ideally, in this way the engineers need only to define the diameter of the welds and the connected nodes or elements as input. The software will be able to find the material property and establish the rupture strength of the welds.

It was observed, for example, by Lin et al. (2002a) that under normal tension, the rupture started from the heat affected zone and was quite uniform around the nugget, as shown in Figure 6.8a. When loading in the direction with an angle from the normal, the rupture started from the heat affected zone and extended into the base material, as shown in Figure 6.8b. This distinguishes from the rupture mode of normal tension, shown in Figure 6.8a. Having observed these, we understand that the failure of spot weld is a very complex process. It does not seem to be practical to formulate the failure criterion purely based on mechanics rule. The experimental results are essentially leading to an applicable solution.

6.4.3 Finite Element Representation of Spot Weld

Many tests on spot weld are in fact designed to find the failure criterion for finite element applications. The finite element model that represents spot weld as connecting two metal sheets can have quite many choices. Xiang et al. (2006) presented a collection of 16 approaches of spot weld definition. These approaches were evaluated by correlating a component impact test. Several of them worked well.

The failure criteria discussed in Section 6.4.2 are basically suitable for a discrete type of element. The two-node spring element has been used to model spot weld for many applications. The two sheets to be welded together can lie on each other without gap. Shell elements are used to model the thin sheets, whose position is defined at the mid-surfaces. Hence, the mid-surfaces have a gap of the average thickness. The spot weld element connecting the two sheets looks like a beam element (or solid element in view of the physical presentation).

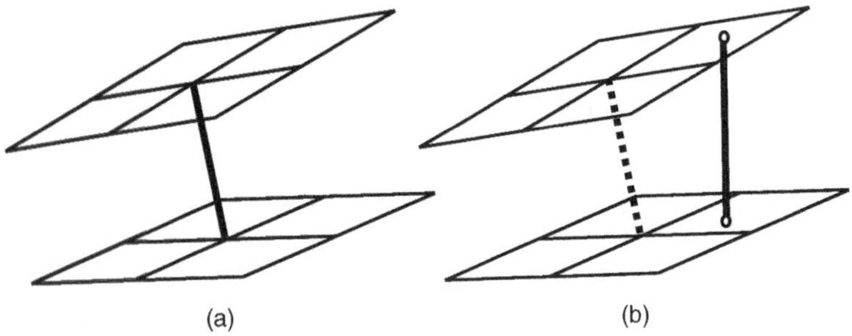

FIGURE 6.9 Representation of spot weld by a spring element: (a) connected at the grid points in the mesh; (b) connected at the designed location with constraint conditions.

However, only a small number of the welds (as designed) are located within the small neighborhood of the nodes in the mesh when generating the mesh. Then, a pair of nodes nearby instead is found for defining the spot weld element. The two nodes are very possibly at locations away from the designed locations. The line direction of the spot weld thus defined may not be normal to the surfaces to be welded, as depicted in Figure 6.9a. Improving from this method, Xiang et al. (2006) proposed a multinodes connection method where four or more pairs of nodes were to be found from the two surfaces. Its shortcomings are: (1) more time consuming; (2) the welds are still not guaranteed to be normal to the surfaces; and (3) the rupture of one connection does not mean failure and the applicable failure criterion becomes more complicated. Using solid element is another approach because the weld nugget looks like a solid structure. The model with a solid element or several solid elements by refinement, however, not only needs smaller time steps but also is still hard to accurately represent the weld. It will be more difficult to accurately simulate the rupture process if we consider the fact that the failure mainly occurs in the heat affected zone but not the nugget self.

A new connection method has been available in the commercial software in recent years. To have a good connection at the designed location and in the normal direction, an artificial beam or six degrees of freedom spring element is created, which can be done automatically by the preprocessor software, for example, HYPERMESH. The two end points of the beam or spring element are tied to the corresponding shell elements on both sides by the constraint conditions, shown in Figure 6.9b.

All these connection methods depend on the shell mesh. When mesh refinement is considered, the spot weld becomes a mesh dependent type of element. For single point connection, rigid or deformable, the spot weld has nothing to refine, but approaches a singularity in the mesh. For multipoint connection, the spot weld is redefined but not necessarily refined. We are facing another unexplored question about the convergence by mesh refinement with spot weld. On the other hand, the real weld nugget has a diameter about 5–6 mm or so. When element size of the mesh is near or even smaller than this scale, we may have to consider the effect of the geometric representation of the spot weld. Overall, the accurate representation of spot welds remains to be a challenging task.

PART III

MATERIAL MODELS

CHAPTER 7

MATERIAL MODEL OF PLASTICITY

Plasticity is an important behavior of metal materials. After 100 years and more of development, plasticity has evolved into an active research field of solid mechanics. Interesting topics include plastic analysis of structures, plastic wave propagation, elastic–plastic buckling, elastic–plastic damage and fracture, metal forming, transient dynamic response, impact, crashworthiness, etc. Valuable theories have been developed for large plastic deformation, for example, yield criterion, hardening rules, flow rules, constitutive relations, slip line methods, limit analysis, min–max principles, as well as various numerical methods. Successful stories have been reported with many practical applications. A great number of publications recorded the historical development of plasticity theory and its applications. For influential texts, we refer to some examples in chronological order, Nádai (1931), Mikhlin (1934), Sokolovskii (1945), Il'yushin (1948), Hill (1950), Prager and Hodge (1951), Prager (1955a, 1955b), Kachanov (1956), Hodge (1959), Koiter (1960), Perzyna (1966), Martin (1975), Lemaitre and Chaboche (1990), Simo and Hughes (1997), Han and Reddy (1999), and Lubarda (2002). A detailed survey is beyond the scope of this book, we refer to Jones (2009) for historical commentary.

Large deformation of metal materials always accompanies plasticity. For a durable and stable structure, plastic deformation and buckling should be avoided, whereas controllable plasticity for optimal usage of materials is allowed in some cases. On the other hand, in metal forming analysis, the plastic deformation is the desired deformation and forms the core process. In impact engineering, the energy involved in the large plastic deformation is critically important.

Introduction to the Explicit Finite Element Method for Nonlinear Transient Dynamics, First Edition.
Shen R. Wu and Lei Gu.

In the precomputer era, simplified mathematical models such as elastic–perfect plasticity and rigid–perfect plasticity were developed to derive analytical solutions. These models and solution methods are still incredibly valuable today even with the aid of modern computers and sophisticated nonlinear programming. Of course, many applications have brought many more challenges, which are solvable only by numerical approximations such as finite element methods. Our objective is to summarize the well-established basics of plasticity theory and the algorithms implemented in the explicit finite element software. Examples of plastic deformation are provided to evaluate the performance of the shell elements discussed in Chapters 3 and 4.

7.1 FUNDAMENTALS OF PLASTICITY

7.1.1 Tensile Test

The plastic behavior can be observed from the tensile test of a standard material coupon. A stress–strain curve is presented in Figure 7.1. Here, the stress $\sigma = f/A$, with the load f and the current cross-section area A. The strain $\varepsilon = \ln(L/L_0)$, with L_0 and L being the original and current gauge length, respectively. During the loading process, the first piece of stress–strain curve is nearly linear up to point A with a slope E, which is called Young's modulus. Point A is called the proportionality limit. Next is the elastic limit, point B, beyond which deformation is no longer completely recoverable after unloading. If the load is removed at point C somewhere after B with strain ε, the unloading path then follows a nearly straight line with slope E. A portion of the strain is recovered, denoted by ε_e, and the rest part ε_p represents the permanent deformation. When the coupon is reloaded, the curve reverses the unloading path and goes up with nearly the same slope as E, until reaching point C. Thus, ε_e represents the elasticity behavior and ε_p represents the plasticity behavior. The total strain is decomposed:

$$\varepsilon = \varepsilon_e + \varepsilon_p. \tag{7.1}$$

If it keeps on loading after point C, the curve becomes nonlinear again and looks like a natural continuation before point C, therefore, looks just like the unloading had

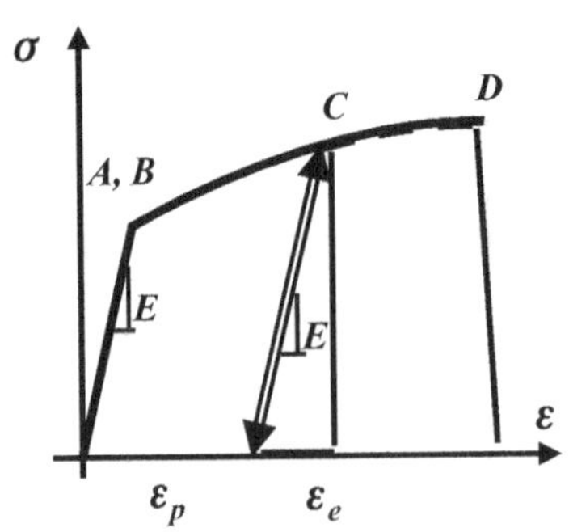

FIGURE 7.1 Tensile test.

not happened. The curve runs up to a point D where the coupon breaks with a strain ε_u and a stress σ_u. The magnitude of ε_u varies from material to material and depends on manufacturing conditions, for example, about 15–35% elongation for mild steel, 15–25% for aluminum alloys AA5456, and about 8% for magnesium AM60.

When depicting in real scale, the linear part of stress–strain curve in Figure 7.1 looks almost vertical. Points A and B are close to each other and assume no difference hereafter in our interest of nonlinear applications. Some textbooks provide more descriptions of details such as upper and lower yield points, which are not to be discussed here as well. Traditionally, a strain of 0.2% is adopted as the elasticity limit for steel. The corresponding stress σ_y^0 is called the initial yield stress. The stress at point C has the similar meaning for reloading, is therefore called a subsequent yield stress, which is denoted by $\sigma_y = \sigma_y^c$. The curve is now described by

$$\sigma_y = f_1(\varepsilon). \tag{7.2}$$

For metal materials, the curve after the initial yield is generally nondecreasing, $\sigma_y \geq \sigma_y^0$. Along the unloading and reloading paths, only the elastic part of strain changes

$$\varepsilon_e = \sigma/E, \quad \sigma \leq \sigma_y. \tag{7.3}$$

The plastic part ε_p does not change in this process. Hence, the yield stress can also be expressed as a function of plastic strain

$$\sigma_y = f_2(\varepsilon_p), \quad \varepsilon_p = \varepsilon - \sigma_y/E. \tag{7.4}$$

In Figure 7.1, the area below the graph (ε, σ_y) represents the elastic behavior, unloading or reloading (including the initial loading) up to the yield stress. The region above the graph is not physically admissible. The stress–strain curve serves as a boundary between the admissible and nonadmissible zones. The curve itself is admissible. Only when ε_p increases, can the response path keep advancing along the curve with increasing σ_y. Note that $\sigma(\varepsilon)$ is not a single valued state due to the unloading and reloading process, but $\sigma_y(\varepsilon_p)$ is.

For a pair of (ε, σ) in the admissible region (including the stress–strain curve), assume that a small increment of strain $\Delta\varepsilon$ happens during the process. If $\sigma + E\Delta\varepsilon \leq f_1(\varepsilon + \Delta\varepsilon)$, the process is elastic and includes the unloading from the curve $\sigma = f_1(\varepsilon)$ with $\Delta\varepsilon < 0$, as shown in Figure 7.2a. For $\sigma = f_1(\varepsilon)$ and $\Delta\varepsilon > 0$, we have $\sigma + E\Delta\varepsilon > f_1(\varepsilon + \Delta\varepsilon)$, which is not admissible. The material must have higher composition of plasticity with $\Delta\varepsilon_p > 0$ and $\sigma + \Delta\sigma = f_1(\varepsilon + \Delta\varepsilon)$ or $\sigma + \Delta\sigma = f_2(\varepsilon_p + \Delta\varepsilon_p)$, as shown in Figures 7.2b and 7.2c. Using the tangent of the yield stress, $E_t = f_1'(\varepsilon)$ or $E_p = f_2'(\varepsilon_p)$, we obtain

$$\Delta\sigma = E_t\Delta\varepsilon = E\Delta\varepsilon_e = E_p\Delta\varepsilon_p, \tag{7.5}$$

$$\frac{1}{E_t} = \frac{1}{E} + \frac{1}{E_p}, \quad E_t = \frac{EE_p}{E + E_p}, \quad E_p = \frac{EE_t}{E - E_t}. \tag{7.6}$$

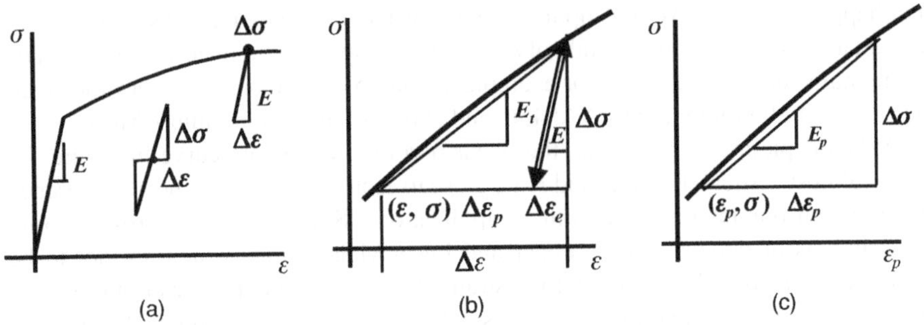

FIGURE 7.2 Increment of strain: (a) elastic loading and unloading; (b) expressed as $\sigma_y = f_1(\varepsilon)$; (c) expressed as $\sigma_y = f_2(\varepsilon^p)$.

We call E_t the tangent modulus and E_p the plastic modulus. Thus the relation between strain and stress is established in the incremental form.

For this type of materials, the above discussion can be summarized as: for any admissible $\tau = \sigma + \Delta\sigma$,

$$(\tau - \sigma)\Delta\varepsilon_p \geq 0. \tag{7.7}$$

The meaning of inequality (7.7) has two folds. For elastic unloading or reloading, $\Delta\varepsilon_p = 0$. On the other hand, for continuous plastic loading from a yield point, $\Delta\varepsilon_p > 0$, we have an increased stress, so $\tau - \sigma > 0$. Note that, due to possibility of unloading and reloading, the admissible stress (or strain) cannot be determined by the strain (or stress respectively) alone without loading history.

7.1.2 Hardening

For a lot of metal materials, continued loading after the initial yield point results in increasing of stress, as shown in Figure 7.1 ($\sigma^C \geq \sigma^A = \sigma_y^0$). This is called hardening, or softening otherwise. In case if the hardening effect is not significant, then the stress is approximately constant after the initial yield. This is called perfect plasticity.

On the other hand, in compression test, same amount of initial yield stress as that in tension test, is observed (in absolute value) for a lot of metal materials. Even after unloading from tension (or compression respectively) at a stress state and loading in the opposite direction for compression (or tension respectively), materials may exhibit nearly the same subsequent yield stress, as shown in Figure 7.3. This model is called isotropic hardening; see Hill (1950).

Some of the other material types, however, exhibit smaller reversal yield stress after unloading and loading into the opposite direction. This is known as Bauschinger effect, which was first reported by Bauschinger (1886). Also refer to early study by Bader (1927). One idealization assumes that after unloading from any point on the

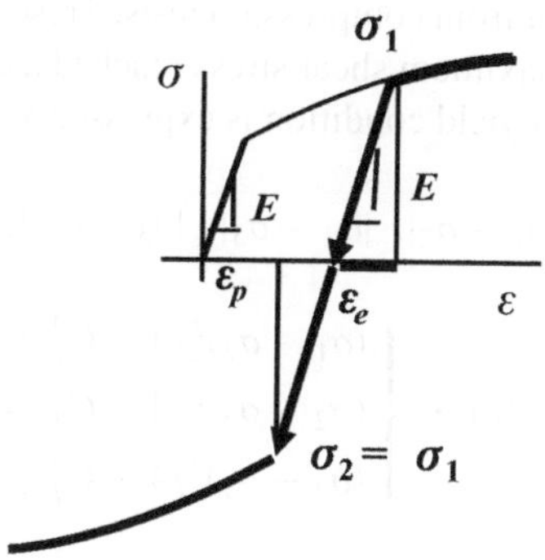

FIGURE 7.3 Isotropic hardening.

stress–strain curve, as shown in Figure 7.4, the reference state has moved from zero to a back-stress σ_b:

$$\sigma_b = \sigma_y - \sigma_y^0 = f_2(\varepsilon_p) - \sigma_y^0 \tag{7.8}$$

Reversal yield will happen when $|\sigma - \sigma_b| = \sigma_y^0$. This model was introduced by Melan (1938a) and Prager (1955a, 1956), called kinematic hardening. The decomposition of strain is still valid.

A better model may assume that the recovered stress is not a constant σ_y^0, but varies as a function of ε_p. A simplification may use the combination of these two hardening models. It is available in most of the commercial software now.

7.1.3 Yield Surface

The first step to extend the previous discussion for the uniaxial loading case to the general three-dimensional (3D) loading case is to define the yield criterion for the complex loading conditions.

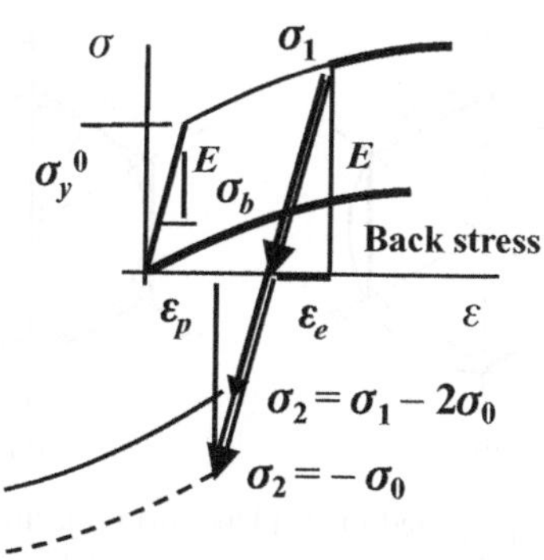

FIGURE 7.4 Kinematic hardening.

Based on the observation from compression tests, Tresca (1864) assumed that the yield happened when the maximum shear stress reached a critical value C_T. In terms of principal stresses, Tresca yield condition is expressed as

$$F(\boldsymbol{\sigma}, C_T) = \max\{|\sigma_1 - \sigma_2|,\ |\sigma_2 - \sigma_3|,\ |\sigma_3 - \sigma_1|\}/2 - C_T \le 0, \tag{7.9}$$

or

$$F(\boldsymbol{\sigma}, C_T) = \begin{cases} (\sigma_1 - \sigma_2)^2/4 - C_T^2 \le 0 \\ (\sigma_2 - \sigma_3)^2/4 - C_T^2 \le 0 \\ (\sigma_3 - \sigma_1)^2/4 - C_T^2 \le 0 \end{cases}. \tag{7.9a}$$

Here, the function F is called yield function. $F = 0$ represents a surface in the principal stress space, which is called the yield surface. $F < 0$ means an elastic state. The projection of the yield surface in the π-plane (passing the origin with normal direction (1, 1, 1)) of the principal stress space is a regular hexagon. In fact, it is easy to see that the projection in O-$\sigma_1\sigma_2$ plane is a hexagon. By symmetry, the yield surface should be a hexagonal prism normal to the π-plane. Consider a unit cubic in the principal stress space, depicted in Figure 7.5a. From the cross-section containing

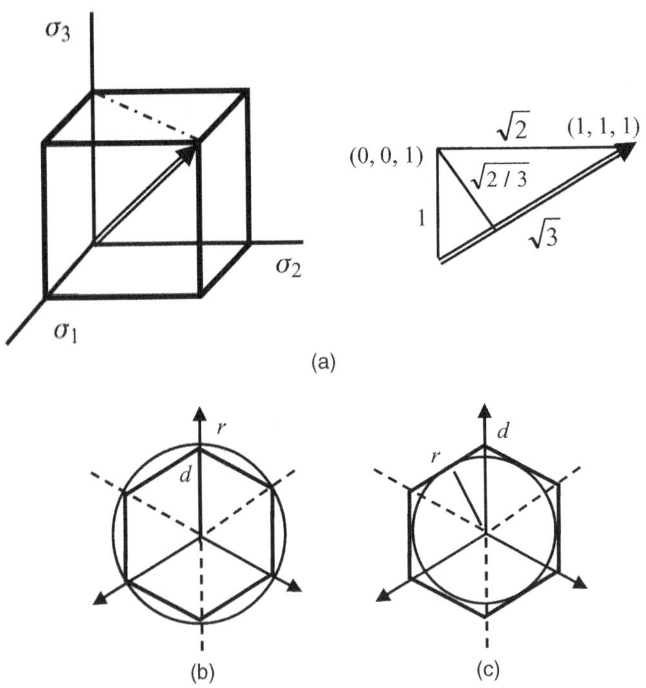

FIGURE 7.5 Yield surface projected on π-plane: (a) projection on the π-plane; (b) yield condition determined by uniaxial tension test: $r = d$; (c) yield condition determined by pure shear test.

the σ_3-axis and the major diagonal, we discover that the projection of (0, 0, 1) in the π-plane has a distance $\sqrt{2/3}$ to the origin. Stress point ($\sigma_1 = 0, \sigma_2 = 0, \sigma_3 = 2C_T$) is on the yield surface, whose projection in the π-plane is a corner of the hexagon. The distance to the origin is $d = 2\sqrt{2/3}C_T$, as shown in Figures 7.5b and 7.5c.

It is observed from the multiaxial tests of metal materials, that the yield and plastic behavior are independent of the hydrostatic pressure and determined by the distortional energy. von Mises (1913) assumed that the yield happened when J_2, the second invariant of the deviatoric stress tensor, reached the critical value $C = (C_M)^2$, where

$$J_2 = S_{ij}S_{ij}/2,\ S_{ij} = \sigma_{ij} - \bar{\sigma}\delta_{ij},\ \bar{\sigma} = -p = \sigma_{kk}/3. \tag{7.10}$$

Here, the summation convention on the repeated indices is applied. It is expressed alternatively in terms of the stress components or the principal stresses:

$$\begin{aligned} J_2 &= \left(\sigma_{11}^2 + \sigma_{22}^2 + \sigma_{33}^2 - (\sigma_{11}\sigma_{22} + \sigma_{22}\sigma_{33} + \sigma_{33}\sigma_{11})\right)/3 \\ &\quad + \sigma_{12}^2 + \sigma_{23}^2 + \sigma_{31}^2 \\ &= ((\sigma_1 - \sigma_2)^2 + (\sigma_2 - \sigma_3)^2 + (\sigma_3 - \sigma_1)^2)/6 \\ &= \left(\sigma_1^2 + \sigma_2^2 + \sigma_3^2 - (\sigma_1\sigma_2 + \sigma_2\sigma_3 + \sigma_3\sigma_1)\right)/3. \end{aligned} \tag{7.11}$$

The von Mises yield condition is defined as

$$F(\boldsymbol{\sigma}, C_M) = \sqrt{J_2} - C_M = \varphi(\boldsymbol{\sigma}) - C_M = 0. \tag{7.12}$$

Its projection on O-$\sigma_1\sigma_2$ plane is an ellipse. Therefore, this yield surface in the principal stress space should be a cylindrical surface. The projection of the yield surface $F = 0$ in the π-plane is a circle by symmetry. If stress point ($\sigma_1 = 0, \sigma_2 = 0, \sigma_3 = \sigma$) is on the yield surface, $\sigma = \sqrt{3}C_M$. Its projection in the π-plane has a distance $r = \sqrt{2/3}\sqrt{3}C_M = \sqrt{2}C_M$ to the origin.

Hencky (1924) considered J_2 to be related to the distortional energy. In fact, using distortional energy for yield condition was published earlier by Huber (1904). The concept might trace back to James Maxwell's informal letters to William Thomson (later Lord Kelvin) in 1850s; see Larmor (1937).

When using tensile test to determine the critical values for the yield stress σ_y, we have $\sigma_1 = \sigma_y, \sigma_2 = \sigma_3 = 0$, and obtain

$$\begin{aligned} C_T &= \sigma_y/2, \quad & d &= \sqrt{2/3}\sigma_y, \\ C_M &= \sigma_y/\sqrt{3}, \quad & r &= \sqrt{2/3}\sigma_y. \end{aligned} \tag{7.13}$$

In the π-plane, depicted in Figure 7.5b, the Tresca hexagon inscribes the Mises circle with $d = r$. When using pure shear test to determine the critical values for the yield stress τ_y, we have $\sigma_{12} = \tau_y$ and other stress components $\sigma_{ij} = 0$; or $\sigma_1 = -\sigma_2 = \tau_y$ and $\sigma_3 = 0$, which are expressed in terms of principal stresses. Then

$$\begin{aligned} C_T &= \tau_y, \quad d = 2\sqrt{2/3}\tau_y, \\ C_M &= \tau_y, \quad r = \sqrt{2}\tau_y. \end{aligned} \tag{7.14}$$

In the π-plane, depicted in Figure 7.5c, the Tresca hexagon circumscribes the Mises circle in this case. Multiaxial tests for metal materials, such as a thin-walled tube subjected to tension with torsion and tension with internal pressure, have demonstrated that Mises criterion fits better than Tresca criterion. We can see that the difference of calculation between Mises circle and Tresca hexagon is within 15.5%. For isotropic hardening plasticity, due to the independence of the coordinate system and the hydrostatic pressure, the projection of the yield surface in π-plane has 12 symmetry axes.

From (7.3), we have $\sigma_y = \sqrt{3}C_M = \sqrt{3J_2}$ for the yield under tension. A measure for stress is defined as the equivalent stress, also named Mises stress

$$\sigma_M = \sqrt{3J_2} = \sqrt{\frac{3}{2}S_{ij}S_{ij}}. \tag{7.15}$$

It indicates that, by Mises criterion, the material yields when σ_M reaches the tensile yield stress σ_y. The yield condition (7.12) can be rewritten as

$$\begin{aligned} F(\boldsymbol{\sigma}, C_M) &= \varphi(\boldsymbol{\sigma}) - \sigma_y = 0, \\ \varphi(\boldsymbol{\sigma}) &= \sigma_M. \end{aligned} \tag{7.12a}$$

This form can also be applied to Tresca criterion, except that the function $\varphi(\sigma)$ needs another definition.

The yield criteria (7.9) and (7.12) can represent the isotropic hardening if C_T or C_M grows with the loading. On the other hand, they can be modified to represent the kinematic hardening. For example, Melan (1938a) and Prager (1955a, 1956) assumed a translation of the yield surface with the same shape. For 3D case with the back-stress σ_{ij}^b, the yield functions are defined below for Tresca and Mises criterion, respectively,

$$\begin{aligned} \hat{F}_T(\boldsymbol{\sigma}, \boldsymbol{\sigma}^b, C_T) &= F_T(\boldsymbol{\sigma} - \boldsymbol{\sigma}^b, C_T) \\ &= \max\left\{ \left|\sigma_1 - \sigma_2 - \left(\sigma_1^b - \sigma_2^b\right)\right|, \left|\sigma_2 - \sigma_3 - \left(\sigma_2^b - \sigma_3^b\right)\right| \right. \\ &\qquad \left. \left|\sigma_3 - \sigma_1 - \left(\sigma_3^b - \sigma_1^b\right)\right| \right\}/2 - C_T, \end{aligned} \tag{7.16a}$$

$$\hat{F}_M(\boldsymbol{\sigma}, \boldsymbol{\sigma}^b, C_M) = \varphi(\boldsymbol{\sigma} - \boldsymbol{\sigma}^b) - C_M = \sqrt{\left(s_{ij} - s_{ij}^b\right)\left(s_{ij} - s_{ij}^b\right)/2} - C_M. \tag{7.16b}$$

Pure kinematic hardening model uses constant C_T or C_M corresponding to the initial yield. Variable C_T or C_M can represent the mixed hardening model.

The diagram in the π-plane is a good representation for yield condition, which is independent of the hydrostatic pressure. Looking at the π-plane, the area inside the yield surface, represented by Tresca hexagon or Mises circle, describes the elastic behavior including unloading from the yield surface. For the hardening material, subsequent loading after the initial (or current) yield results in increasing yield stress as demonstrated in the uniaxial case. The isotropic hardening is characterized by an expanding hexagon or circle with increasing parameter C_T or C_M. The kinematic hardening is represented by the moving hexagon or circle with their center representing the back-stress σ_b, depicted in Figure 7.6a. Viewing at the higher space of stresses with hardening parameters, hardening is a move along a surface. The cross section with the hardening parameter is moving apart from the origin and the projection in the π-plane is expanding and/or moving, as conceptually illustrated in Figure 7.6b. The yield surface serves as the boundary of the elastic zone and itself is the plastic

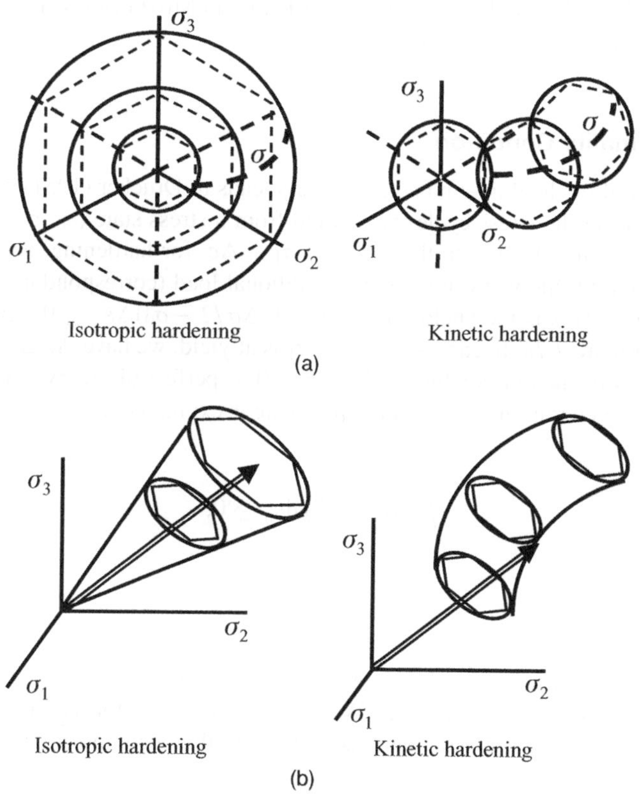

FIGURE 7.6 Hardening representation: (a) hardening projected in π-plane; (b) hardening presented in principal stress space.

zone. The exterior of the surface is still not admissible. We summarize below as the yield criterion (e.g., Mises) in general:

$$
\begin{array}{lll}
F(\boldsymbol{\sigma}, \boldsymbol{\sigma}^b, C_M) = \varphi(\boldsymbol{\sigma}, \boldsymbol{\sigma}^b) - C_M & & \\
F < 0 \Leftrightarrow \varphi < C_M & \text{elastic} & \\
F = 0 \Leftrightarrow \varphi(\boldsymbol{\sigma}) = C_M & \text{plastic yield} & \\
F = 0, \left\{ \begin{array}{l} \mathrm{d}F < 0 \Leftrightarrow \mathrm{d}\varphi(\boldsymbol{\sigma}) < 0 \\ \mathrm{d}F = 0, \mathrm{d}\varphi(\boldsymbol{\sigma}) = 0 \\ \mathrm{d}F = 0, \mathrm{d}\varphi(\boldsymbol{\sigma}) > 0 \end{array} \right. & \begin{array}{l} \text{elastic unloading} \\ \text{plastic neutral loading} \\ \text{plastic hardening.} \end{array} &
\end{array}
\tag{7.17}
$$

Tresca criterion is a somewhat convenient method for obtaining analytical solutions by applying straight lines of the yield conditions, but its corners bring in difficulty in mathematical assessment. Koiter (1953), Prager (1953), and Sanders (1954) presented the treatment with singular yield surface. On the other hand, Mises criterion provides a convenient mathematical platform. There also are other yield criteria being investigated to model complex material behaviors, for example, the anisotropic yield criteria in Hill (1950), multisurface yield/hardening in Mroz (1967), and nonconvex yield surface in Kim and Oden (1984).

7.1.4 Normality Condition

Investigation of the loading–unloading process leads to Drucker (1951) postulate. It states that during the stress cycle that started from a stress state (ε_0, σ_0), loaded to yield point σ_1, then loaded further to $\sigma_2 = \sigma_1 + \Delta\sigma$ for hardening, and unloaded back to σ_0, the plastic work done by the additional load (corresponding to $\sigma - \sigma_0$) is nonnegative. This is the enclosed area $(\sigma_1 + \Delta\sigma/2 - \sigma_0)\Delta\varepsilon^p \geq 0$ schematically depicted in Figure 7.7a. In case when $\sigma_0 = \sigma_1$ is at yield, we have $\Delta\sigma\,\Delta\varepsilon^p \geq 0$. The equality holds for neutral loading with $\Delta\varepsilon^p = 0$ or perfect plasticity with $\Delta\sigma = 0$. For 3D loading conditions, the Drucker postulate is expressed as

$$\oint_{\sigma_0} \left(\sigma_{ij} - \sigma_{ij}^0\right) \mathrm{d}\varepsilon_{ij}^p \geq 0, \tag{7.18}$$

$$\mathrm{d}\sigma_{ij}\mathrm{d}\varepsilon_{ij}^p \geq 0. \tag{7.18a}$$

During the stress cycle, the elastic part of deformation is recovered and it obeys the generalized Hooke's law of elasticity. We consider that the deformation is composed of two parts: elastic and plastic, and also introduce the decomposition of the strain increment:

$$\mathrm{d}\varepsilon_{ij} = \mathrm{d}\varepsilon_{ij}^e + \mathrm{d}\varepsilon_{ij}^p. \tag{7.19}$$

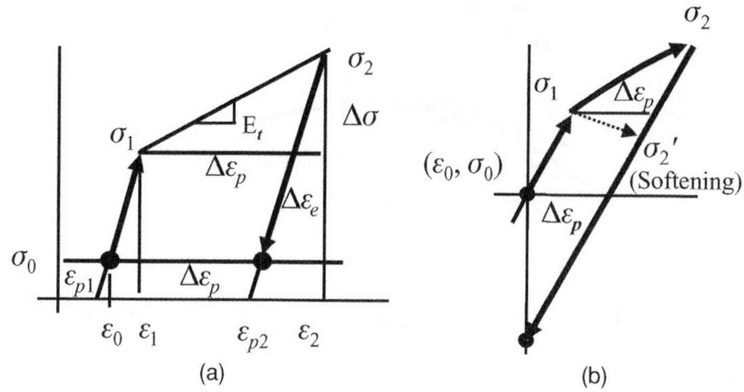

FIGURE 7.7 Postulates of plastic work: (a) Drucker postulate; (b) I'Lyusin postulate.

The elastic part can be expressed with the elasticity tensor and compliance tensor:

$$
\begin{aligned}
E_{ijkl} &= 2\mu\delta_{ik}\delta_{jl} + 2\mu\nu/(1-2\nu)\delta_{ij}\delta_{kl}, \\
C_{ijkl} &= \delta_{ik}\delta_{jl}/2\mu - \delta_{ij}\delta_{kl}\nu/E, \\
\mathrm{d}\varepsilon_{ij}^{e} &= C_{ijkl}\mathrm{d}\sigma_{kl} = \mathrm{d}\sigma_{ij}/2\mu - \delta_{ij}\mathrm{d}\sigma_{kk}\nu/E, \\
\mathrm{d}\sigma_{ij} &= E_{ijkl}\mathrm{d}\varepsilon_{kl}^{e} = 2\mu\mathrm{d}\varepsilon_{ij}^{e} + \delta_{ij}\mathrm{d}\varepsilon_{kk}^{e}2\mu\nu/(1-2\nu).
\end{aligned}
\tag{7.20}
$$

The inequality (7.18) is valid for hardening material, which is also called a stable material. For softening material, moving along the stress–strain curve could indicate a decreasing of load. Unloading back to σ_0 may lose meaning in such situation. The concept was extended by Il'yushin (1961) postulate, which described a strain cycle with unloading to strain ε_0 instead of σ_0, as shown in Figure 7.7b. It applies to both types of materials.

For an infinitesimal stress cycle (small $\Delta\sigma_{ij}$), (7.18) reduces to a differential form

$$\left(\sigma_{ij} - \sigma_{ij}^{0}\right)\mathrm{d}\varepsilon_{ij}^{p} \geq 0. \tag{7.21}$$

As stress state σ_{ij} is considered, (7.21) indicates that $\mathrm{d}\varepsilon_{ij}^{p}$ is a property of σ_{ij} and is independent of the choice of σ_{ij}^{0}. It has the same meaning of (7.7). The geometrical representation can be viewed in the stress space shown in Figure 7.8, with the plastic strain space-aligned. The curve is the projection of the yield surface with the given hardening parameter. Here σ_{ij}, σ_{ij}^{0}, and $\mathrm{d}\varepsilon_{ij}^{p}$ are all represented by points as vectors in the stress space. If the yield surface is differentiable at σ_{ij}, (7.21) leads to a conclusion that $\mathrm{d}\varepsilon_{ij}^{p}$ is parallel to the normal of the yield surface at σ_{ij}. This fact is expressed in the rate form or in the incremental form with the gradient of the yield surface:

$$\mathrm{d}\varepsilon_{ij}^{p} = \mathrm{d}\lambda\partial F/\partial\sigma_{ij}. \tag{7.22}$$

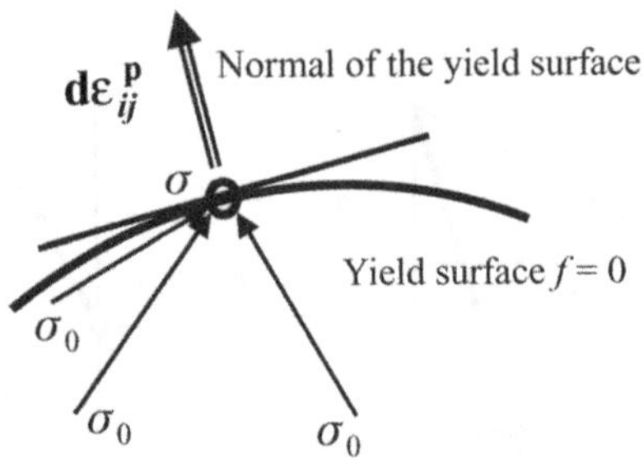

FIGURE 7.8 Normality of plasticity expressed in stress space.

This property is known as the normality. Furthermore, all the admissible stresses σ_{ij}^{0} (in the elastic zone or on the yield surface) have to lie on one side of the tangent superplane and the yield surface is convex, which explains the meaning of convexity. In fact, the assumption of normality or convexity can lead to Drucker postulate. They are equivalent hypotheses. The uniqueness of solution to the elastic–plastic problem can be proved based on this type of postulates. In fact, Melan (1938a, 1938b) at the first time established the uniqueness for elastic–plastic problem with incremental form. More discussions about the uniqueness and stability of plasticity can be found, for example, in Drucker (1956) and Hill (1958).

For material with piecewise smooth yield surface, such as Tresca yield surface with corners, the convexity still holds, but the normality involves a set of subdifferentials instead of the single regular normal. The theory of convex analysis has therefore found its importance in applications for plasticity; see Han and Reddy (1999).

7.1.5 Strain Rate Effect/Viscoplasticity

Plasticity is often accompanied with viscosity effect. The dynamic tensile test shows that the strength is intensified, and significantly intensified in some cases when the tensile strain rate is increased, as illustrated in Figure 7.9. In practical applications,

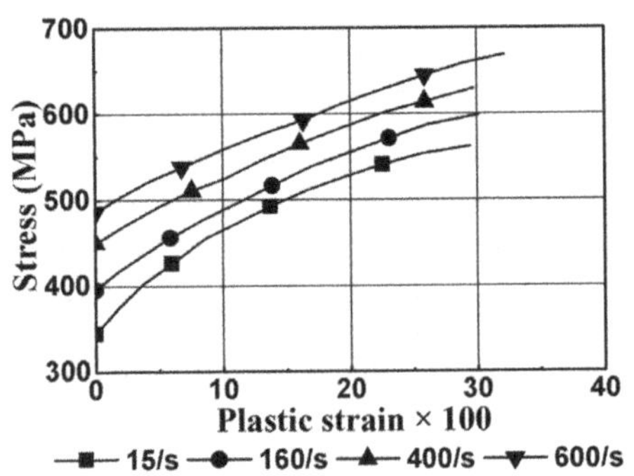

FIGURE 7.9 Tensile test of a steel material under various strain rates.

this phenomenon is called strain rate effect, which is important to many dynamic- and impact-related problems. Early studies on strain rate effect can be found in Manjoine (1944), and Clark and Duwez (1950). Also refer to Perzyna (1963) and Rice (1970) for more development.

It could be costly to formulate a general functional form for describing such material behavior and implementing in the finite element software. We refer to Perzyna (1966, 1971) for the fundamental theories of viscoplasticity, including viscosity in the constitutive equation. Empirical formulas based on test data have been implemented in software for applications. The basic approach is to modify the stress–strain curve(s) by including the strain rate as a variable. It recovers the static behavior when the rate is low. The following are a few examples that are available in the commercial explicit software:

Cowper and Symonds (1957)

$$\sigma_y^d = \sigma_y^s(1 + (\dot{\varepsilon}/\dot{\varepsilon}_0)^{1/\gamma}). \tag{7.23}$$

Johnson and Cook (1983)

$$\sigma_y^d = \sigma_y^s(1 + c\ln(\dot{\varepsilon}/\dot{\varepsilon}_0))\,(1 - (T - 298/T_{\text{melt}} - 298)^m). \tag{7.24}$$

Zerilli and Armstrong (1987)

$$\sigma_y^d = \sigma_y^s + c_1 e^{(-c_3 T + c_4 T\ln(\dot{\varepsilon}/\dot{\varepsilon}_0))}. \tag{7.25}$$

Zhao (1997)

$$\sigma_y^d = \sigma_y^s + \left(c_1 - c_2\varepsilon_p^m\right)\ln(\dot{\varepsilon}/\dot{\varepsilon}_0) + c_3\dot{\varepsilon}^k. \tag{7.26}$$

Here, $\dot{\varepsilon}$ is the current strain rate. $\dot{\varepsilon}_0$ is a reference strain rate, whose value is determined by the material property as well as the characteristics of the model. Each of these formulas contains several parameters to be determined from the laboratory tests with various constant strain rates. Some formulas include temperature effect. The commonly used test equipment is the Hopkinson's (1905, 1914) splitting compression bar, which records the stress–strain response of the specimen with near constant strain rate. The technique of high-speed tensile tests also belongs to an active research field that includes the design of test apparatus, test specimens, and the test methods, as well as the post process of the test data.

7.2 CONSTITUTIVE EQUATIONS

7.2.1 Relations between Stress Increments and Strain Increments

As discussed in Section 7.1.1, the stress–strain relation in plastic deformation is not uniquely determined, but depends on the loading history. It is convenient to use the rate form or the incremental form. With decomposition of the strain (7.19), we use the elastic relation (7.20) with the normality condition (7.22) and (7.17),

$$\begin{aligned} d\sigma_{ij} &= E_{ijkl} d\varepsilon^e_{kl} = E_{ijkl}\left(d\varepsilon_{kl} - \gamma d\varepsilon^p_{kl}\right) = E_{ijkl}(d\varepsilon_{kl} - \gamma d\lambda \partial\varphi/\partial\sigma_{kl}), \\ d\varepsilon_{ij} &= d\varepsilon^e_{kl} + d\varepsilon^p_{kl} = C_{ijkl}\sigma_{kl} + \gamma d\lambda \partial\varphi/\partial\sigma_{ij}, \end{aligned} \tag{7.27}$$

where $\gamma = 0$ is for elastic behavior and $\gamma = 1$ is for plastic behavior. The key task now is to determine $d\lambda$.

We proceed with the concept of a plastic work rate. For this purpose, we introduce a measure for the equivalent plastic strain rate (or increment), which corresponds to the equivalent stress (Mises stress) defined in (7.15),

$$d\varepsilon^p = \alpha\sqrt{d\varepsilon^p_{ij} d\varepsilon^p_{ij}}. \tag{7.28}$$

In the sense of plastic power, we have

$$dW^p = \sigma_{ij} d\varepsilon^p_{ij} = \sigma_M d\varepsilon^p. \tag{7.29}$$

For plastic loading in the uniaxial test, we have $\sigma_{11} = \sigma_y$, and other stress components $\sigma_{ij} = 0$. Hence, $\sigma_M = \sigma_y$. Meanwhile, $d\varepsilon^p_{22} = d\varepsilon^p_{33} = -0.5 d\varepsilon^p_{11}$ by plastic incompressibility, and the shear components are zeroes. In this case, we have $d\varepsilon^p = \alpha\sqrt{1.5} d\varepsilon^p_{11}$. From (7.29), $dW^p = \sigma_{11} d\varepsilon^p_{11} = \sigma_M d\varepsilon^p$. It follows that $\alpha = \sqrt{2/3}$ and

$$d\varepsilon^p = \sqrt{\frac{2}{3} d\varepsilon^p_{ij} d\varepsilon^p_{ij}}. \tag{7.28a}$$

Note that the total amount of plastic strain is expressed by $\varepsilon^p = \int d\varepsilon^p$, but not by $\varepsilon^p = \sqrt{\frac{2}{3}\varepsilon^p_{ij}\varepsilon^p_{ij}}$, which has no physical meaning except for special situation.

Then, the normality condition leads to

$$d\varepsilon^p = d\lambda\sqrt{\frac{2}{3}\frac{\partial\varphi}{\partial\sigma_{ij}}\frac{\partial\varphi}{\partial\sigma_{ij}}}, \quad d\lambda = d\varepsilon^p \Bigg/ \sqrt{\frac{2}{3}\frac{\partial\varphi}{\partial\sigma_{ij}}\frac{\partial\varphi}{\partial\sigma_{ij}}}. \tag{7.30}$$

We further assume that at yield, the relation between the equivalent stress and the equivalent plastic strain represents material characterization. Therefore, the yield function is uniquely determined by the simple tensile test, $\sigma_M = \sigma_y = f_2(\varepsilon^p)$. Now we are ready to derive the constitutive relation for plasticity, and to express it by a general form of a smooth yield function (7.12a).

For plastic loading, the stress remains on the yield surface. Hence, the consistency condition $\mathrm{d}F = 0$ applies:

$$\begin{aligned} \mathrm{d}F &= \mathrm{d}\varphi - \mathrm{d}\sigma_y = 0, \\ (\partial\varphi/\partial\sigma_{ij})\mathrm{d}\sigma_{ij} &- E_p \mathrm{d}\varepsilon^p = 0. \end{aligned} \tag{7.31}$$

This gives

$$\mathrm{d}\varepsilon^p = \frac{\partial\varphi}{\partial\sigma_{ij}}\frac{\mathrm{d}\sigma_{ij}}{E_p}. \tag{7.32}$$

Thus, (7.30) and (7.32) lead to

$$\mathrm{d}\lambda = \frac{\dfrac{\partial\varphi}{\partial\sigma_{ij}}\mathrm{d}\sigma_{ij}}{E_p\sqrt{\dfrac{2}{3}\dfrac{\partial\varphi}{\partial\sigma_{ij}}\dfrac{\partial\varphi}{\partial\sigma_{ij}}}}. \tag{7.33}$$

We then obtain the constitutive relation in rate form from (7.27) and (7.33)

$$\mathrm{d}\varepsilon_{ij} = \mathrm{d}\varepsilon_{ij}^e + \mathrm{d}\varepsilon_{ij}^p = \left(C_{ijkl} + \gamma \frac{\dfrac{\partial\varphi}{\partial\sigma_{ij}}\dfrac{\partial\varphi}{\partial\sigma_{kl}}}{E_p\sqrt{\dfrac{2}{3}\dfrac{\partial\varphi}{\partial\sigma_{mn}}\dfrac{\partial\varphi}{\partial\sigma_{mn}}}} \right)\mathrm{d}\sigma_{kl}, \tag{7.34}$$

where $\gamma = 0$ indicates an elastic state and $\gamma = 1$ indicates a plastic state.

The equation for stress increment expressed in terms of strain increment can be obtained by using the elasticity (7.27) and normality (7.33) again

$$\mathrm{d}\sigma_{ij} = E_{ijkl}\mathrm{d}\varepsilon_{kl}^e = E_{ijkl}\left(\mathrm{d}\varepsilon_{kl} - \mathrm{d}\varepsilon_{kl}^p\right) = E_{ijkl}\left(\mathrm{d}\varepsilon_{kl} - \gamma \frac{\dfrac{\partial\varphi}{\partial\sigma_{kl}}\dfrac{\partial\varphi}{\partial\sigma_{mn}}\mathrm{d}\sigma_{mn}}{E_p\sqrt{\dfrac{2}{3}\dfrac{\partial\varphi}{\partial\sigma_{mn}}\dfrac{\partial\varphi}{\partial\sigma_{mn}}}} \right).$$

This is equivalent to applying a multiplication with the elasticity tensor to (7.34). A tensor contraction operation with the above equation leads to

$$\frac{\partial\varphi}{\partial\sigma_{mn}}\mathrm{d}\sigma_{mn} = \frac{E_{ijkl}\dfrac{\partial\varphi}{\partial\sigma_{ij}}\mathrm{d}\varepsilon_{kl}}{1 + \gamma\dfrac{E_{ijkl}\dfrac{\partial\varphi}{\partial\sigma_{ij}}\dfrac{\partial\varphi}{\partial\sigma_{kl}}}{E_p\sqrt{\dfrac{2}{3}\dfrac{\partial\varphi}{\partial\sigma_{mn}}\dfrac{\partial\varphi}{\partial\sigma_{mn}}}}}. \tag{7.35}$$

From this, it is followed by

$$d\sigma_{ij} = \left(E_{ijkl} - \gamma \frac{E_{ijmn}\dfrac{\partial\varphi}{\partial\sigma_{mn}} E_{stkl}\dfrac{\partial\varphi}{\partial\sigma_{st}}}{E_p\sqrt{\dfrac{2}{3}\dfrac{\partial\varphi}{\partial\sigma_{mn}}\dfrac{\partial\varphi}{\partial\sigma_{mn}}} + E_{mnst}\dfrac{\partial\varphi}{\partial\sigma_{mn}}\dfrac{\partial\varphi}{\partial\sigma_{st}}} \right) d\varepsilon_{kl}. \tag{7.36}$$

From (7.32), (7.33), and (7.35), we obtain $d\varepsilon^p$ and $d\lambda$, expressed in terms of $d\varepsilon_{ij}$,

$$d\varepsilon^p = \frac{E_{ijkl}\dfrac{\partial\varphi}{\partial\sigma_{ij}} d\varepsilon_{kl}}{E_p + \dfrac{E_{ijkl}\dfrac{\partial\varphi}{\partial\sigma_{ij}}\dfrac{\partial\varphi}{\partial\sigma_{kl}}}{\sqrt{\dfrac{2}{3}\dfrac{\partial\varphi}{\partial\sigma_{mn}}\dfrac{\partial\varphi}{\partial\sigma_{mn}}}}}, \tag{7.37}$$

$$d\lambda = \frac{E_{ijkl}\dfrac{\partial\varphi}{\partial\sigma_{ij}} d\varepsilon_{kl}}{E_p\sqrt{\dfrac{2}{3}\dfrac{\partial\varphi}{\partial\sigma_{mn}}\dfrac{\partial\varphi}{\partial\sigma_{mn}}} + E_{mnst}\dfrac{\partial\varphi}{\partial\sigma_{mn}}\dfrac{\partial\varphi}{\partial\sigma_{st}}}. \tag{7.37a}$$

With plastic incompressibility and its independence of hydrostatic pressure, we have $d\varepsilon^p_{kk} = d\lambda\partial\varphi/\partial\sigma_{kk} = 0$. Therefore, with the symmetry of elasticity tensors, we obtain

$$E_{ijkl} d\varepsilon^p_{kl} = \lambda d\varepsilon^p_{kk}\delta_{ij} + 2\mu d\varepsilon^p_{ij} = 2\mu d\varepsilon^p_{ij},$$

$$E_{ijkl}\frac{\partial\varphi}{\partial\sigma_{ij}} = \lambda\frac{\partial\varphi}{\partial\sigma_{jj}}\delta_{kl} + 2\mu\frac{\partial\varphi}{\partial\sigma_{kl}} = 2\mu\frac{\partial\varphi}{\partial\sigma_{kl}}.$$

The constitutive equations (7.36) and (7.37) are reduced to

$$d\varepsilon^p = \frac{2\mu\dfrac{\partial\varphi}{\partial\sigma_{kl}} d\varepsilon_{kl}}{E_p + \sqrt{6}\mu\sqrt{\dfrac{\partial\varphi}{\partial\sigma_{ij}}\dfrac{\partial\varphi}{\partial\sigma_{ij}}}}, \tag{7.38a}$$

$$d\lambda = \frac{2\mu\dfrac{\partial\varphi}{\partial\sigma_{kl}} d\varepsilon_{kl}}{E_p\sqrt{\dfrac{2}{3}\dfrac{\partial\varphi}{\partial\sigma_{ij}}\dfrac{\partial\varphi}{\partial\sigma_{ij}}} + 2\mu\dfrac{\partial\varphi}{\partial\sigma_{ij}}\dfrac{\partial\varphi}{\partial\sigma_{ij}}}, \tag{7.38b}$$

$$d\sigma_{ij} = \left(E_{ijkl} - \gamma \frac{2\mu \dfrac{\partial\varphi}{\partial\sigma_{ij}} \dfrac{\partial\varphi}{\partial\sigma_{kl}}}{\dfrac{E_p}{2\mu}\sqrt{\dfrac{2}{3}\dfrac{\partial\varphi}{\partial\sigma_{mn}}\dfrac{\partial\varphi}{\partial\sigma_{mn}}} + \dfrac{\partial\varphi}{\partial\sigma_{pq}}\dfrac{\partial\varphi}{\partial\sigma_{pq}}} \right) d\varepsilon_{kl}. \tag{7.39}$$

Note that there is nothing to be simplified for (7.32)–(7.34) expressed in terms of $d\sigma_{ij}$.

7.2.2 Constitutive Equations for Mises Criterion

Using (7.15) and (7.12a) for Mises criterion, we now have

$$\begin{aligned} \frac{\partial\varphi}{\partial\sigma_{ij}} &= \frac{\partial\varphi}{\partial S_{ij}} = \frac{3S_{ij}}{2\varphi}, \\ \frac{\partial\varphi}{\partial\sigma_{ij}}\frac{\partial\varphi}{\partial\sigma_{ij}} &= \frac{3}{2}. \end{aligned} \tag{7.40}$$

Then obtain $d\lambda = d\varepsilon^p$ from (7.30). We also have

$$\begin{aligned} E_{ijkl}\frac{\partial\varphi}{\partial\sigma_{kl}} &= E_{ijkl}\frac{3S_{kl}}{2\varphi} = \frac{3\mu S_{ij}}{\varphi}, \\ E_{ijkl}\frac{\partial\varphi}{\partial\sigma_{ij}}\frac{\partial\varphi}{\partial\sigma_{kl}} &= \frac{9\mu S_{ij}S_{ij}}{2\varphi^2} = 3\mu. \end{aligned} \tag{7.41}$$

Equation (7.35) is reduced to

$$\frac{\partial\varphi}{\partial\sigma_{ij}} d\sigma_{ij} = \frac{3\mu S_{ij} d\varepsilon_{ij}/\varphi}{1 + \gamma 3\mu/E_p}. \tag{7.42}$$

By using (7.40)–(7.42), the relations (7.32) and (7.33) are reduced to

$$d\lambda = d\varepsilon^p = \frac{3S_{ij} d\sigma_{ij}}{2E_p\sigma_y} = \frac{S_{ij} d\varepsilon_{ij}}{(1 + E_p/3\mu)\sigma_y}. \tag{7.43}$$

The constitutive equations (7.34) and (7.39) are reduced to

$$\begin{aligned} d\varepsilon_{ij} &= \left(C_{ijkl} + \gamma \frac{9S_{ij}S_{kl}}{4E_p\sigma_y^2} \right) d\sigma_{kl}, \\ d\sigma_{ij} &= \left(E_{ijkl} - \gamma \frac{3\mu S_{ij}S_{kl}}{(1 + E_p/3\mu)\sigma_y^2} \right) d\varepsilon_{kl}. \end{aligned} \tag{7.44}$$

For isotropic hardening material obeying von Mises criterion with yield function defined in (7.12), the normality condition (7.22) leads to

$$d\varepsilon_{ij}^{p} = d\lambda \frac{\partial \varphi}{\partial \sigma_{ij}} = d\lambda \frac{3S_{ij}}{2\varphi}. \tag{7.45}$$

It states that the plastic strain rate tensor is parallel to the deviatoric stress. This is named flow rule. In fact, Lévy (1870) and von Mises (1913) independently proposed the relation $d\varepsilon_{ij} = d\lambda S_{ij}$, named as Lévy–Mises equation. Considering rigid-plastic material, ignoring the elastic part, this is the same as (7.22). This is an extension from St. Vénant's (1870) model for rigid-perfectly plastic material in plane stress, where the axes of strain increment were supposed to coincide with the axes of principle stress. The idea was extended to elastoplasticity of plane problem in Prandtl (1924) and general 3D problem in Reuss (1930). The system combining (7.20) and (7.22) is named as Prandtl–Reuss equation:

$$d\varepsilon_{ij} = \frac{1}{2\mu} d\sigma_{ij} - \delta_{ij} \frac{\nu}{E} d\sigma_{kk} + d\lambda S_{ij}. \tag{7.46}$$

It is worth noting that the process to derive incremental constitutive equations discussed so far can be applied to the case including additional variables, such as internal variable and damage variable in general.

7.2.3 Application to Kinematic Hardening

For kinematic hardening, we consider a modification of the yield function F with the same function φ and constant σ_y as in (7.12a), based on the idea of Prager (1956):

$$F(\boldsymbol{\sigma}, \boldsymbol{\sigma}^b, C_M) = \varphi(\boldsymbol{\sigma} - \boldsymbol{\sigma}^b) - \sigma_y. \tag{7.47}$$

This is a rigid translation of the yield surface. An incremental model of motion in the direction of plastic strain increment is suggested:

$$d\sigma_{ij}^{b} = \alpha d\varepsilon_{ij}^{p}. \tag{7.48}$$

See Shield and Ziegler (1958), Xyaн (1958), and Ziegler (1959) for more discussions published in that time period. Considering the stress cycle from the yield surface and an increment of plastic deformation, as discussed in Section 7.1.4, we make further assumption that the equivalence of $d\sigma_{ij}^{b} d\varepsilon_{ij}^{p} = d\sigma^{b} d\varepsilon^{p} = E_p (d\varepsilon^{p})^2 = 2E_p d\varepsilon_{ij}^{p} d\varepsilon_{ij}^{p}/3$. Then

$$\alpha = 2E_p/3. \tag{7.49}$$

The consistency of plastic loading gives $dF = 0$. The normality condition leads to $(\partial\varphi/\partial\sigma_{ij})d\sigma_{ij} = (\partial\varphi/\partial\sigma_{ij})d\sigma^b_{ij} = \alpha d\lambda(\partial\varphi/\partial\sigma_{ij})(\partial\varphi/\partial\sigma_{ij})$. Then

$$d\lambda = \frac{\dfrac{\partial\varphi}{\partial\sigma_{ij}}d\sigma_{ij}}{\alpha\dfrac{\partial\varphi}{\partial\sigma_{ij}}\dfrac{\partial\varphi}{\partial\sigma_{ij}}}. \tag{7.50}$$

Similar to the previous discussion, we obtain the constitutive relations

$$d\varepsilon_{ij} = d\varepsilon^e_{ij} + d\varepsilon^p_{ij} = \left(C_{ijkl} + \gamma\frac{\dfrac{\partial\varphi}{\partial\sigma_{ij}}\dfrac{\partial\varphi}{\partial\sigma_{kl}}}{\alpha\dfrac{\partial\varphi}{\partial\sigma_{mn}}\dfrac{\partial\varphi}{\partial\sigma_{mn}}}\right)d\sigma_{kl}, \tag{7.51}$$

$$d\sigma_{ij} = \left(E_{ijkl} - \gamma\frac{E_{ijmn}\dfrac{\partial\varphi}{\partial\sigma_{mn}}E_{stkl}\dfrac{\partial\varphi}{\partial\sigma_{st}}}{\alpha\dfrac{\partial\varphi}{\partial\sigma_{mn}}\dfrac{\partial\varphi}{\partial\sigma_{mn}} + E_{ijkl}\dfrac{\partial\varphi}{\partial\sigma_{ij}}\dfrac{\partial\varphi}{\partial\sigma_{kl}}}\right)d\varepsilon_{kl}. \tag{7.52}$$

For Mises criterion, with (7.40) and (7.49), the constitutive relations (7.51) and (7.52) have the same appearance as (7.44), with s_{ij} replaced by $s_{ij} - s^b_{ij}$ only.

Ways to define back-stress and shape of the subsequent yield surface lead to establishment of various theories. Detailed discussion is however beyond the scope of this book.

The above discussion is for the incremental theory of plasticity. There is also a theory regarding total strain proposed by Hencky (1924). It is named deformation theory; see Nádai (1931), Sokolovskii (1945), Il'yushin (1948), and Budiansky (1959) for continuing development in this subject.

7.3 SOFTWARE IMPLEMENTATION

Plasticity usually occurs with large deformation. The nonlinearity creates substantial difficulties in seeking analytical solutions of plastic deformations. Some problems with simple geometry and loading conditions have been solved analytically, such as perfectly plastic thin-walled cylindrical shell subjected to uniform tension and torsion; rigid–perfectly-plastic beam under uniform bending; axisymmetric deformation of elastic–perfectly-plastic thick cylinder under uniform pressure; axisymmetric deformation of elastic–plastic (with hardening) thick cylinder. In addition to these successes, slip line method has been developed based on the characteristic method of partial differential equations, cf. Hencky (1924) for the fundamental theory. The slip line method has been applied to many rigid–perfectly-plastic plane problems, which have more general geometry than the problems mentioned above have. These contributions are considered great achievements of the precomputer era. Many of the plasticity theories have been developed along with the validation of these solutions. Their limitation to simple geometry and simple loading is however obvious. Some of those solutions can now serve as benchmarks for evaluating the software implementation of plasticity theory.

7.3.1 Explicit Finite Element Procedure with Plasticity

As discussed in Section 2.3.4, a major step of computation in the explicit finite element procedure is to calculate the internal nodal forces, illustrated in Figure 2.8. The computation is accomplished by integration of all the elements, described in (2.19):

$$\left(F_i^M\right)_{int} = -\int_{\Omega} (\sigma_{ij}\Phi_{M,j})\, \mathrm{d}\Omega = -\sum_e \int_{\Omega_e} (\sigma_{ij(e)}\varphi_{M(e),j})\, \mathrm{d}\Omega_e.$$

The element procedure is discussed in Part 2 of this book, with an assumption that the stress can be obtained from the material procedure. Here, the discussion extends to the material model of plasticity. The main procedure is illustrated in Figure 7.10, as the extension of Figure 2.8 representing elasticity only.

The calculation is for all the elements at each integration point.

7.3.2 Normal (Radial) Return Scheme

To implement the constitutive equations discussed in Section 7.2, we need to replace the rate form with difference or increment form. At time t_n, consider a stress state

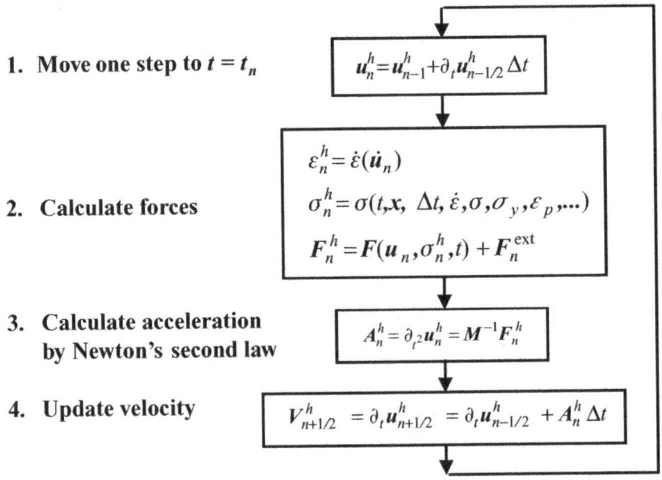

FIGURE 7.10 Explicit finite element procedure—with plasticity (extended from Figure 2.8).

$(\sigma_{ij})_n$. The strain increment $\Delta\varepsilon_{ij} = \dot{\varepsilon}_{ij}\Delta t = (\dot{u}_{i,j} + \dot{u}_{j,i})\Delta t/2$ is given by marching one time step. The values of the functions and variables are calculated at time t_n. First, we assume that this action results in an elastic state. We then proceed with a trial stress, which is assigned to be a pure elastic response:

$$\begin{aligned} \Delta\sigma_{ij}^t &= E_{ijkl}\Delta\varepsilon_{kl}, \\ \sigma_{ij}^t &= (\sigma_{ij})_n + \Delta\sigma_{ij}^t. \end{aligned} \tag{7.53}$$

Then, we test the yield condition with

$$F(\boldsymbol{\sigma}^t, \sigma_y) = \varphi(\boldsymbol{\sigma}^t) - \sigma_y. \tag{7.54}$$

If $F < 0$, then the trial state is elastic and the trial is correct. If $F \geq 0$, then the new state must be plastic. First, consider the current state as a yield point with $\varphi(\boldsymbol{\sigma}^t) - \sigma_y((\varepsilon_p)_n) \geq 0$. Using (7.19), (7.20), (7.22), and (7.30), we obtain

$$\begin{aligned} &\Delta\sigma_{ij} = E_{ijkl}\Delta\varepsilon_{kl}^e = E_{ijkl}(\Delta\varepsilon_{kl} - \Delta\lambda\partial\varphi/\partial\sigma_{kl}) = \Delta\sigma_{ij}^t - \theta(N_{ij})_n, \\ &(\sigma_{ij})_{n+1} = \sigma_{ij}^t - \theta(N_{ij})_n, \end{aligned} \tag{7.55}$$

where we denote by $\boldsymbol{N}_n$ for the unit vector normal to the yield surface at t_n, and θ for a scalar:

$$\begin{aligned} (N_{ij})_n &= \frac{\partial\varphi/\partial\sigma_{ij}}{\sqrt{(\partial\varphi/\partial\sigma_{ij})(\partial\varphi/\partial\sigma_{ij})}}, \\ \theta &= \sqrt{6}\mu\Delta\varepsilon^p. \end{aligned} \tag{7.56}$$

$\Delta\varepsilon^p$ can be calculated from (7.38). Equation (7.55) means a correction in the normal direction.

We continue the discussion for Mises criterion with (7.43) from Section 7.2.2. Replacing the rate form by the incremental form, we can directly use (7.44) to obtain the results:

$$\Delta\lambda = \Delta\varepsilon^p = \frac{3S_{ij}\Delta\sigma_{ij}}{2E_p\sigma_y} = \frac{S_{ij}\Delta\varepsilon_{ij}}{(1 + E_p/3\mu)\sigma_y}, \tag{7.57}$$

$$\begin{aligned} \Delta\varepsilon_{ij} &= \left(C_{ijkl} + \gamma\frac{9S_{ij}S_{kl}}{4E_p\sigma_y^2}\right)\Delta\sigma_{kl}, \\ \Delta\sigma_{ij} &= \left(E_{ijkl} - \gamma\frac{3\mu S_{ij}S_{kl}}{(1 + E_p/3\mu)\sigma_y^2}\right)\Delta\varepsilon_{kl}. \end{aligned} \tag{7.58}$$

Then, we obtain the stress at the new time point t_{n+1} after one time step:

$$(\sigma_{ij})_{n+1} = (\sigma_{ij})_n + \Delta\sigma_{ij}. \tag{7.59}$$

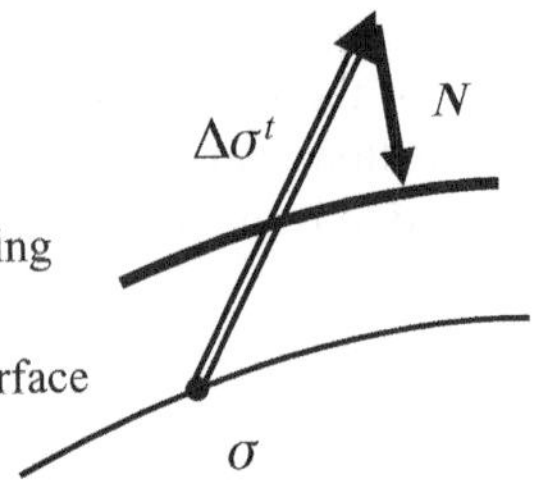

FIGURE 7.11 Normal return.

Also, we have the updated equivalent plastic strain and yield stress:

$$
\begin{aligned}
(\varepsilon^p)_{n+1} &= (\varepsilon^p)_n + \Delta\varepsilon^p, \\
(\sigma_y)_{n+1} &= \sigma_y((\varepsilon^p)_{n+1}).
\end{aligned}
\tag{7.60}
$$

Note that we have the normal direction:

$$
N_{ij} = \frac{S_{ij}}{\sqrt{S_{mn}S_{mn}}} = \sqrt{\frac{3}{2}}\frac{S_{ij}}{\sigma_M}. \tag{7.61}
$$

The correction now is in the direction of ***S*** // ***N***, and hence bares the name of radial return, illustrated in Figure 7.11. The method was first studied by Wilkins (1964). The coincidence of radial and normal direction is only for Mises criterion. In general, it is a return scheme along the normal direction as in (7.55), but not along the radial direction. It also applies to other smooth yield surfaces such as the individual piece of Tresca surface.

The procedure with Mises criterion is summarized below:

1. Assign a trial stress as an elastic prediction by (7.53).
2. Calculate the deviatoric stress of the trial stress.
3. Check the yield condition (7.54).
4. If it is elastic, the computation is done; if it is plastic, make the correction, that is, normal return, using (7.58) and (7.59).
5. Calculate $\Delta\varepsilon_p$ from (7.57) and update the equivalent plastic strain and yield stress with (7.60).

It seems straightforward. But it does not guarantee that σ_{ij}^{n+1} can be just on the yield surface with the updated plastic strain due to the incremental procedure. The balance $\sigma_M^{n+1} = \varphi(\sigma^{n+1}) = \sigma_y((\varepsilon^p)^{n+1})$ should be checked at least every few steps of computation. Appropriate adjustment may be needed if the equation is deviated.

Now let us consider the case starting with the elastic stress state and with $F = \varphi(\sigma^t) - \sigma_y > 0$ ($F \leq 0$ is trivial). We can divide the increment into two subprocesses. The first one advances to the yield surface elastically and the second one is the case discussed earlier. The task of the first process is to find a $\xi \in (0, 1)$ such that $\varphi(\boldsymbol{\sigma} + \xi \Delta\boldsymbol{\sigma}^t) - \sigma_y(\varepsilon_p) = 0$. This is a simple algebra equation for simple form of yield function. For instance, with Mises criterion and isotropic hardening, we have $\sqrt{3(S_{ij} + \xi \Delta S_{ij}^t)(S_{ij} + \xi \Delta S_{ij}^t)/2} - \sigma_y = 0$ and

$$\xi = \left(-S_{ij}\Delta S_{ij}^t + \sqrt{\left(S_{ij}\Delta S_{ij}^t\right)^2 + \left(2\sigma_y^2/3 - S_{ij}S_{ij}\right)\left(\Delta S_{mn}^t \Delta S_{mn}^t\right)}\right) \Big/ \Delta S_{st}^t \Delta S_{st}^t .$$

7.3.3 A Generalized Plane Stress Model

As discussed in Chapters 3 and 4, the plate and shell adopt the generalized plane stress concept with $\sigma_{33} = 0$. The calculation described in Section 7.2.2, however, does not promise satisfaction of this condition. Even it is satisfied at the last step, it might be violated due to the possibility of $\Delta\sigma_{33} \neq 0$ when using 3D computation. In this situation, $\sigma_{33} = 0$ can be treated as a constraint condition. An iterative procedure is implemented in some software. An alternative approach is using the plane stress model to solve $\Delta\sigma_{ij}$ directly without iteration. The method was first discussed in Wilkins (1964); see Krieg (1977) and Schreyer et al. (1979) for more discussion. In fact, the procedure to determine $\Delta\lambda$ and θ (of the return scheme) still holds. As a special case of the general 3D problems, we use the generalized plane stress model for elasticity to obtain:

$$\begin{bmatrix} \Delta\sigma_{11}^t \\ \Delta\sigma_{22}^t \\ \Delta\sigma_{ij}^t \end{bmatrix} = E_1 \begin{bmatrix} 1 & \nu & 0 \\ \nu & 1 & 0 \\ 0 & 0 & 1-\nu/2 \end{bmatrix} \begin{bmatrix} \Delta\varepsilon_{11} \\ \Delta\varepsilon_{22} \\ 2\Delta\varepsilon_{ij} \end{bmatrix} \quad i,j = 1,2,3, \quad i \neq j, \tag{7.62}$$

where $E_1 = E/(1-\nu^2)$. Similar to the procedure of the general 3D case, we have

$$\begin{aligned} (\sigma_{11})_{n+1} &= \sigma_{11}^t - E_1\left(\Delta\varepsilon_{11}^p + \nu\Delta\varepsilon_{22}^p\right) = \sigma_{11}^t - E_1\Delta\lambda(\partial\varphi/\partial\sigma_{11} + \nu\partial\varphi/\partial\sigma_{22}), \\ (\sigma_{22})_{n+1} &= \sigma_{22}^t - E_1\left(\Delta\varepsilon_{22}^p + \nu\Delta\varepsilon_{11}^p\right) = \sigma_{22}^t - E_1\Delta\lambda(\partial\varphi/\partial\sigma_{22} + \nu\partial\varphi/\partial\sigma_{11}). \\ (\sigma_{ij})_{n+1} &= \sigma_{ij}^t - 2\mu\Delta\lambda\partial\varphi/\partial\sigma_{ij}, \quad i,j = 1,2,3, \quad i \neq j. \end{aligned} \tag{7.63}$$

For the example of Mises criterion with isotropic hardening, we use a reduced form of the Mises stress $\varphi = \sqrt{\sigma_{11}^2 + \sigma_{22}^2 - \sigma_{11}\sigma_{22} + 3(\sigma_{12}^2 + \sigma_{23}^2 + \sigma_{31}^2)}$ with five components. From (7.63), we have

$$\begin{aligned} (\sigma_{11})_{n+1} &= \sigma_{11}^t - \theta\left(S_{11}^t + \nu S_{22}^t\right)/(1-\nu), \\ (\sigma_{22})_{n+1} &= \sigma_{22}^t - \theta\left(S_{22}^t + \nu S_{11}^t\right)/(1-\nu), \\ (\sigma_{ij})_{n+1} &= \sigma_{ij}^t - \theta S_{ij}^t, \quad i,j = 1,2,3, \quad i \neq j, \\ \theta &= 3\mu\Delta\varepsilon_p/\sigma_M^t. \end{aligned} \tag{7.64}$$

This is also a return scheme. Note that $\Delta\varepsilon_{33}^{p}$ does not participate in these computations. It is recovered by the plastic incompressibility $\Delta\varepsilon_{33}^{p} = -\Delta\varepsilon_{11}^{p} - \Delta\varepsilon_{22}^{p}$.

7.3.4 Stress Resultant Approach

For plate and shell, the membrane forces, the transverse shear forces, and the bending moments are of particular interest. As discussed in Chapter 3, they are defined below as the resultant of stress components integrated through the thickness ζ:

$$
\begin{aligned}
N_{\alpha\beta} &= \int_{-\zeta/2}^{\zeta/2} \sigma_{\alpha\beta}\mathrm{d}z, \\
M_{\alpha\beta} &= \int_{-\zeta/2}^{\zeta/2} z\sigma_{\alpha\beta}\mathrm{d}z, \\
Q_{\alpha} &= \int_{-\zeta/2}^{\zeta/2} \sigma_{\alpha 3}\mathrm{d}z.
\end{aligned}
\tag{7.65}
$$

According to the hypothesis that the normal keeps straight in Reissner–Mindlin plate theory, the deformation consists of contributions from membrane stretch and rotation of the normal:

$$
\varepsilon_{\alpha\beta} = e_{\alpha\beta} + z\kappa_{\alpha\beta}. \tag{7.66}
$$

For linear elasticity, the stresses are also linear in the thickness. Then (7.65) can be integrated explicitly, and the stress can be viewed as the composition of the contributions from membrane actions and bending/twisting actions. In fact, using the constitutive relation of (7.62), we have

$$
\begin{aligned}
\sigma_{\alpha\beta} &= E_{\alpha\beta\gamma\delta}^{1}(e_{\gamma\delta} + z\kappa_{\gamma\delta}), \\
N_{\alpha\beta} &= E_{\alpha\beta\gamma\delta}^{1}e_{\gamma\delta}\zeta, \quad M_{\alpha\beta} = E_{\alpha\beta\gamma\delta}^{1}\kappa_{\gamma\delta}\zeta^{3}/12, \\
\sigma_{\alpha\beta} &= \frac{1}{\zeta}N_{\alpha\beta} + \frac{12z}{\zeta^{3}}M_{\alpha\beta}.
\end{aligned}
\tag{7.67}
$$

Here, $\boldsymbol{E}^{1}$ represents the elasticity tensor in (7.62).

Now Mises stress is expressed in terms of the stress resultants, neglecting the transverse shear terms:

$$
\sigma_{M} = \sqrt{S_{ij}S_{ij}3/2} = \frac{1}{\zeta}\sqrt{I_{N} + (12z/\zeta^{2})^{2}I_{M} + (24z/\zeta)I_{MN}}, \tag{7.68}
$$

where

$$
\begin{aligned}
I_{N} &= N_{11}^{2} + N_{22}^{2} - N_{11}N_{22} + 3N_{12}^{2}, \\
I_{M} &= M_{11}^{2} + M_{22}^{2} - M_{11}M_{22} + 3M_{12}^{2}, \\
I_{MN} &= N_{11}M_{11} + N_{22}M_{22} - 0.5(N_{11}M_{22} + N_{22}M_{11}) + 3N_{12}M_{12}.
\end{aligned}
\tag{7.69}
$$

Il'yushin (1948) applied these items to define a yield function for perfect plasticity. The concept has been extended for hardening plasticity and dynamic applications, cf. Chou et al. (1993, 1994, 1996), where a yield function was proposed:

$$F(\boldsymbol{N},\boldsymbol{M},\sigma_y) = \varphi(\boldsymbol{N},\boldsymbol{M}) - \Sigma_y,$$
$$\varphi(\boldsymbol{N},\boldsymbol{M}) = \sqrt{I_N + \alpha I_M + \beta I_{MN}}, \quad \Sigma_y = \zeta \eta \sigma_y(\varepsilon^p). \tag{7.70}$$

Here, parameters α, β, and η were introduced, defined as

$$\alpha = \alpha_1 \left(\frac{4}{\zeta}\right)^2, \quad \beta = \beta_1 \frac{8}{\zeta}. \tag{7.71}$$

When $\alpha_1 = (3z/\zeta)^2$ and $\beta_1 = z/\zeta$, (7.68) was recovered. η was chosen to be 1. These parameters were used to approximate for the whole range of z in the thickness.

The stress resultant approach seems to be an appealing intuition. For linear elasticity, it is straightforward to integrate the stress through the thickness to obtain the membrane forces and bending moments. It becomes a complex situation however, when plasticity is involved. We give a brief description here, basically following Chou et al. (1993, 1994, and 1996). Note that the quoted references used dimensionless parameters. The following assumptions are introduced then:

(A1) Proportional straining: the proportions between the plastic strain components stay constant during the loading process

(A2) Deep plasticity: the material deforms with very large plasticity so that the elastic strain is negligible and the plastic strain is considered to be the total strain.

(A3) The power law of plasticity

$$\sigma = A\varepsilon^{1/n}. \tag{7.72}$$

The following is sometimes used for power law instead of (7.72):

$$\sigma = \begin{cases} E\varepsilon & \text{if } \varepsilon \le \sigma_y/E, \\ \sigma_y(E\varepsilon/\sigma_y)^{1/n}, & \quad \varepsilon > \sigma_y/E. \end{cases} \tag{7.72a}$$

The difference of these two expressions is mainly in the elastic zone and small plastic zone. For simplicity, we adopt (7.72) for the illustration.

A combined loading of N_{11} and M_{11} associated with plastic strain $\varepsilon_{11} = e_p + z\kappa_p$ leads to a yield condition in the form of $N_{11}^2 + \alpha_1(\frac{4}{\zeta})^2 M_{11}^2 + 2\beta_1 \frac{4}{\zeta} N_{11} M_{11} = (\zeta\sigma_y)^2$. By using the complementary potential with Assumption (A1) for the proportional straining and Assumption (A3) for power law of plasticity, an approximation was used in Chou et al. (1993):

$$\alpha_1 = \frac{1}{4\left(\dfrac{n}{2n+1}\right)^{2n/(n+1)}}. \tag{7.73}$$

Another parameter β_1 was approximated for each of the power index n, ranging from 0 for $n = 1$ to 0.6064 for $n = \infty$.

To simplify the notations, we define the generalized stress

$$\boldsymbol{S} = \{N_{11}, N_{22}, N_{12}, M_{11}, M_{22}, M_{12}\}. \tag{7.74}$$

The yield function, as a generalized Mises criterion, is expressed by

$$\varphi(\boldsymbol{S}) = \sqrt{\mu_{ij} S_i S_j / 2}\,(\text{summation on } i \text{ and } j = 1 \text{ to } 6). \tag{7.75}$$

From (7.70) and (7.71), we have (μ_{ij}) and its inverse directly:

$$\boldsymbol{\mu} = \begin{bmatrix} 2\boldsymbol{A}_3 & \beta_1 \boldsymbol{A}_3 \\ \beta_1 \boldsymbol{A}_3 & 2\alpha_1 \boldsymbol{A}_3 \end{bmatrix}_{6\text{x}6}, \quad \boldsymbol{\mu}^{-1} = \frac{4}{3\left(4\alpha_1 - \beta_1^2\right)} \begin{bmatrix} 2\alpha_1 \boldsymbol{B}_3 & -\beta_1 \boldsymbol{B}_3 \\ -\beta_1 \boldsymbol{B}_3 & 2\boldsymbol{B}_3 \end{bmatrix}_{6\text{x}6},$$

$$\boldsymbol{A}_3 = \begin{bmatrix} 1 & -1/2 & 0 \\ -1/2 & 1 & 0 \\ 0 & 0 & 3 \end{bmatrix}, \quad \boldsymbol{B}_3 = \begin{bmatrix} 1 & 1/2 & 0 \\ 1/2 & 1 & 0 \\ 0 & 0 & 1/4 \end{bmatrix}. \tag{7.76}$$

On the other hand, we can use the membrane strain and curvature for the associated generalized strain, corresponding to the generalized stress:

$$\boldsymbol{E} = \{e_{11}, e_{22}, 2e_{12}, \kappa_{11}, \kappa_{22}, 2\kappa_{12}\}. \tag{7.77}$$

The plastic flow rule gives

$$\mathrm{d}E_i^p = \mathrm{d}\Lambda \frac{\partial F}{\partial S_i} = \mathrm{d}\Lambda \frac{\mu_{ij} S_j}{2\varphi}. \tag{7.78}$$

Like the process discussed in Section 7.2, we introduce the generalized equivalent plastic strain rate and the plastic work rate:

$$\mathrm{d}W_p = S_i \mathrm{d}E_i^p = \Sigma_y \mathrm{d}E^p. \tag{7.79}$$

From (7.78), $\mathrm{d}\Lambda \mu_{ij} S_j S_i / 2\varphi = \varphi \mathrm{d}E^p$, we obtain

$$\mathrm{d}\Lambda = \mathrm{d}E^p. \tag{7.80}$$

On the other hand, from the inverse of (7.78), $S_i = 2\varphi(\boldsymbol{\mu}^{-1})_{ij}\mathrm{d}E_j^p/\mathrm{d}\Lambda$, we obtain, $S_i\mathrm{d}E_i^p = 2\varphi(\boldsymbol{\mu}^{-1})_{ij}\mathrm{d}E_i^p\mathrm{d}E_j^p/\mathrm{d}\Lambda$. This suggests defining the generalized equivalent plastic strain rate as

$$\mathrm{d}E^p = \sqrt{2(\boldsymbol{\mu}^{-1})_{ij}\mathrm{d}E_i^p\mathrm{d}E_j^p}. \tag{7.81}$$

Direct calculation gives

$$\mathrm{d}E^p = \zeta\sqrt{\frac{4\alpha_1}{3\left(\alpha_1-\beta_1^2/4\right)}}\sqrt{I_{\dot{e}} + \frac{1}{\alpha_1}I_{\dot{\kappa}} - \frac{\beta_1}{2\alpha_1}I_{\dot{e}\dot{\kappa}}}. \tag{7.82}$$

Here, we denote

$$\begin{aligned}
I_{\dot{e}} &= \left(\mathrm{d}e_{11}^p\right)^2 + \left(\mathrm{d}e_{22}^p\right)^2 + \mathrm{d}e_{11}^p\mathrm{d}e_{22}^p + \left(\mathrm{d}e_{12}^p\right)^2, \\
I_{\dot{\kappa}} &= \left(\mathrm{d}\kappa_{11}^p\right)^2 + \left(\mathrm{d}\kappa_{22}^p\right)^2 + \mathrm{d}\kappa_{11}^p\mathrm{d}\kappa_{22}^p + \left(\mathrm{d}\kappa_{12}^p\right)^2, \\
I_{\dot{e}\dot{\kappa}} &= 2\mathrm{d}e_{11}^p\mathrm{d}\kappa_{11}^p + 2\mathrm{d}e_{22}^p\mathrm{d}\kappa_{22}^p + \mathrm{d}e_{11}^p\mathrm{d}\kappa_{22}^p + \mathrm{d}e_{22}^p\mathrm{d}\kappa_{11}^p + 2\mathrm{d}e_{12}^p\mathrm{d}\kappa_{12}^p.
\end{aligned} \tag{7.83}$$

We can verify that $\mathrm{d}\varepsilon^p = \sqrt{4/3}\sqrt{I_{\dot{e}} + z^2 I_{\dot{\kappa}} + 2zI_{\dot{e}\dot{\kappa}}}$ by the standard definition. $\alpha_1/(\alpha_1-\beta_1^2/4)$ was illustrated to be near 1 in Chou et al. (1993). Hence, consider (7.82) as a representation for the entire thickness.

To derive the constitutive relation in terms of the generalized items, we start with the elastic plane stress model for plate and shell (7.67). Expressed in matrix notation

$$\begin{aligned}
\boldsymbol{N} &= \zeta\boldsymbol{E}_1\boldsymbol{e}^e, \\
\boldsymbol{M} &= \zeta^3\boldsymbol{E}_1\boldsymbol{\kappa}^e/12,
\end{aligned} \tag{7.84}$$

where the elasticity tensor $\boldsymbol{E}_1$ for plane stress is shown in (7.62). Hence, we can use the following relations for the generalized stress and the generalized strain:

$$\mathrm{d}\boldsymbol{S} = \boldsymbol{D}\mathrm{d}\boldsymbol{E}^e, \quad \boldsymbol{D}_{6x6} = \begin{bmatrix} \zeta\boldsymbol{E}_1 & \boldsymbol{O}_{3x3} \\ \boldsymbol{O}_{3x3} & \dfrac{\zeta^3}{12}\boldsymbol{E}_1 \end{bmatrix}, \quad \boldsymbol{C}_{6x6} = \boldsymbol{D}^{-1} = \begin{bmatrix} \dfrac{1}{\zeta}\boldsymbol{C}_1 & \boldsymbol{O}_{3x3} \\ \boldsymbol{O}_{3x3} & \dfrac{12}{\zeta^3}\boldsymbol{C}_1 \end{bmatrix}. \tag{7.85}$$

The yield stress is expressed as $\bar{\Sigma}_y(E^p)$. By using the plasticity consistency, we have $\partial\varphi/\partial S_i\mathrm{d}S_i = (\bar{\Sigma}_y)'\mathrm{d}E^p$. Note that $(\bar{\Sigma}_y(E^p))'\mathrm{d}E^p = \zeta\eta(\Sigma_y)'\mathrm{d}\varepsilon^p$. With power law, $\Sigma_y(\varepsilon^p) = \zeta a(\varepsilon^p)^{1/n}$ and $\bar{\Sigma}_y(E^p) = a_1(E^p)^{1/n}$, we consider $(\bar{\Sigma}_y)' = \eta_1(\Sigma_y)'$. It was illustrated in Chou et al. (1993) that the coefficient $\eta_1 = 1$ could be chosen by approximation. Thus,

$$\mathrm{d}\Lambda = \mathrm{d}E^p = \frac{\partial\varphi}{(\bar{\Sigma}_y)'\partial S_i}\mathrm{d}S_i. \tag{7.86}$$

Following the steps developed for the rate form of constitutive equations in Section 7.2, we have

$$\mathrm{d}E_i = \mathrm{d}E_i^e + \mathrm{d}E_i^p = C_{ij}\mathrm{d}S_j + \gamma\mathrm{d}\Lambda\frac{\partial\varphi}{\partial S_i} = \left(C_{ij} + \gamma\frac{\mu_{im}\mu_{jn}S_m S_n}{4\Sigma_y^2(\Sigma_y)'}\right)\mathrm{d}S_j. \quad (7.87)$$

The reverse side can be derived

$$\mathrm{d}S_i = D_{ij}\mathrm{d}E_j^e = D_{ij}\left(\mathrm{d}E_j - \mathrm{d}E_j^p\right) = D_{ij}\left(\mathrm{d}E_j - \gamma\frac{\partial\varphi}{\partial S_k}\frac{\mathrm{d}S_k}{(\Sigma_y)'}\frac{\partial\varphi}{\partial S_j}\right),$$
$$\frac{\partial\varphi}{\partial S_i}\mathrm{d}S_i = \frac{\dfrac{\partial\varphi}{\partial S_i}D_{ij}\mathrm{d}E_j}{1 + \gamma\dfrac{1}{(\Sigma_y)'}\dfrac{\partial\varphi}{\partial S_i}D_{ij}\dfrac{\partial\varphi}{\partial S_j}}. \quad (7.88)$$

Then

$$\mathrm{d}\Lambda = \mathrm{d}E^p = \frac{\dfrac{\partial\varphi}{\partial S_i}D_{ij}\mathrm{d}E_j}{(\Sigma_y)' + \dfrac{\partial\varphi}{\partial S_m}D_{mn}\dfrac{\partial\varphi}{\partial S_n}} = \frac{2\varphi D_{ij}\mu_{ik}S_k}{4\varphi^2(\Sigma_y)' + D_{mn}\mu_{ms}\mu_{nt}S_s S_t}\mathrm{d}E_j, \quad (7.89)$$

$$\mathrm{d}S_i = D_{ij}\mathrm{d}E_j - \gamma\frac{D_{im}\dfrac{\partial\varphi}{\partial S_n}D_{nj}\dfrac{\partial\varphi}{\partial S_m}}{D_p + \dfrac{\partial\varphi}{\partial S_m}D_{mn}\dfrac{\partial\varphi}{\partial S_n}}\mathrm{d}E_j$$
$$= \left(D_{ij} - \gamma\frac{D_{im}D_{jn}\mu_{ms}\mu_{nt}S_s S_t}{4\varphi^2(\Sigma_y)' + D_{mn}\mu_{ms}\mu_{nt}S_s S_t}\right)\mathrm{d}E_j. \quad (7.90)$$

This also suggests a return scheme:

$$\mathrm{d}S_i = \mathrm{d}S_i^t - \gamma\mathrm{d}\Lambda D_{ij}\partial\varphi/\partial S_j,$$
$$\mathrm{d}S_i^t = D_{ij}\mathrm{d}E_j. \quad (7.91)$$

Numerical experiments show that the computing speed of the stress resultant approach is close to that of the 3D plasticity model. Some difference in results is expected for the situation under general complex loading conditions.

7.4 EVALUATION OF SHELL ELEMENTS WITH PLASTIC DEFORMATION

The examples discussed in Sections 3.8.1 and 4.4.1 for shell elements are about small elastic deformation. Here, we continue the study for shell elements with large plastic deformation. The computation is performed using commercial software LS-DYNA V971 and the plasticity material model.

Example 7.1 Simulation of tensile test Strain rate dependence is an important property for many materials. Several dynamic test procedures have been designed to obtain the rate dependent stress–strain relation. One type of high-speed tensile test uses a small specimen for better resolution. The specimen has an effective gauge length of 8 mm, with cross section of 4 × 2 mm. Both ends of the specimen are clipped by test machines. A mesh mainly composed of quadrilateral elements for the specimen is presented in Figure 7.12. The right end is fixed. Several rows of elements with high density are added to the left end, modeling the fixture with an attached weight of 10 kg. The initial velocity of 8 mm/ms is assigned to the left end. This is to simulate the test condition. The stress–strain relation from test data is used as the material property.

The plasticity material model uses stress–strain relation from static test and Cowper–Symonds model of strain rate effect, with parameters $\dot{\varepsilon}_0 = 40$ and $\gamma = 5$ in (7.23). In this case, we expect to see strain rate to be around 1,000/s. To obtain more information about the strain rate effect, we also use lower loading speed $v = 0.8$ and 0.08 mm/ms, respectively. Meanwhile, we redefine mass density for the attached weight by 10^2 and 10^4 times larger, respectively, so that the tests can have the same initial energy to load the specimen.

The calculated stress–plastic strain relation of an element at the center of the specimen is depicted in Figure 7.13a. It is supposed to simulate material behavior under test conditions. The quasi-static test data is also included for comparison. It is observed that the calculated element stress is higher than that from the static test data due to the strain rate effect. With higher loading speed the response is stronger. The resultant force integrated from a cross section, depicted in Figure 7.13b, represents the load. The element stress and the resultant force have oscillatory nature, particularly in the short time interval at beginning. Higher loading speed results in stronger

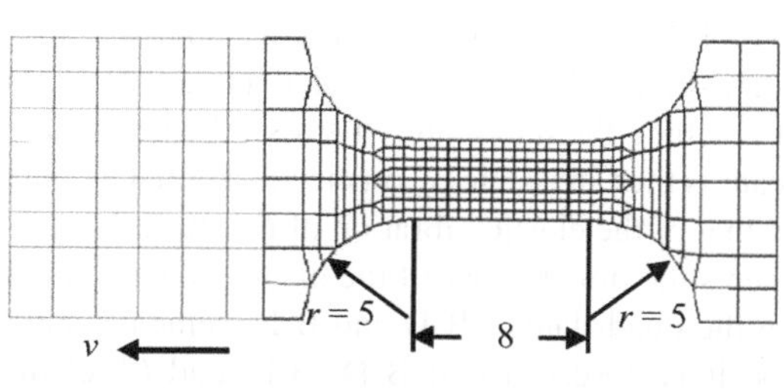

FIGURE 7.12 Model of tensile test specimen.

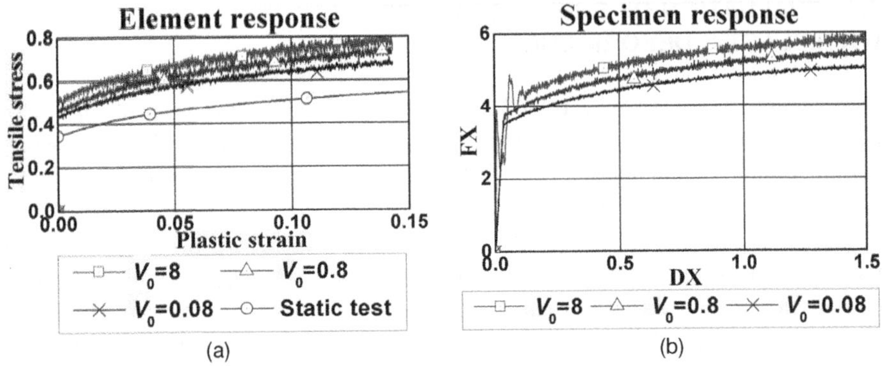

FIGURE 7.13 Results of the simulation of tensile test: (a) stress–strain of the element at the center area; (b) force–deformation of the tensile test.

oscillation. Such oscillation is usually observed in the material tests. We consider it dynamic response of the specimen but not material property. The element uses the smooth stress–strain curve from static test and strain rate effect without oscillation. The numerical results present oscillation.

Example 7.2 Twisted beam with plastic deformation When plasticity is considered, heavier load will cause the twisted beam discussed in Section 3.8 to have plastic deformation. Assume that the yield stress equals 0.1% of the Young's modulus E. We apply a load 60 times of that in Examples 3.1 and 4.1 so that the stress of elements subjected to large deformation will surpass the yield stress. With significant warping of the twisted beam, B-T element does not provide good solution for the elastic case. For the plastic case, B-T element does not provide reasonable solution either.

We use a set of four quadrilateral meshes with 6×1, 12×2, 24×4, and 48×8 elements respectively for B-D and B-L elements. The corresponding triangular meshes are generated by splitting the quadrilateral elements in the same way as in Example 4.1. The evaluation includes C^0 and DKT triangular elements as discussed in Example 4.1. As an illustration, the deformation at $t = 0.01$ calculated by using B-D element with the 12×2-elements mesh is presented in Figure 7.14a. The time history of Mises stress of the element at the second row from the clamped end is presented in Figure 7.14b. It is the largest stress calculated at the three integration points of the Gauss quadrature. The beam is loaded into a state of plasticity with large deformation. After reaching the maximum of stress, the beam starts to bounce back and unload. The beam is then in a state of elastic vibration. Of course, the magnitude is much larger than that of the elastic vibration of Example 3.1.

Figure 7.15 presents the time history of the displacement at the loaded end point, calculated by using the quadrilateral B-D and B-L elements; and the triangular C^0 and DKT elements. It is observed that B-D, B-L, and C^0 elements do not have clear tendency to converge yet. The response of DKT element is acceptable. For fine

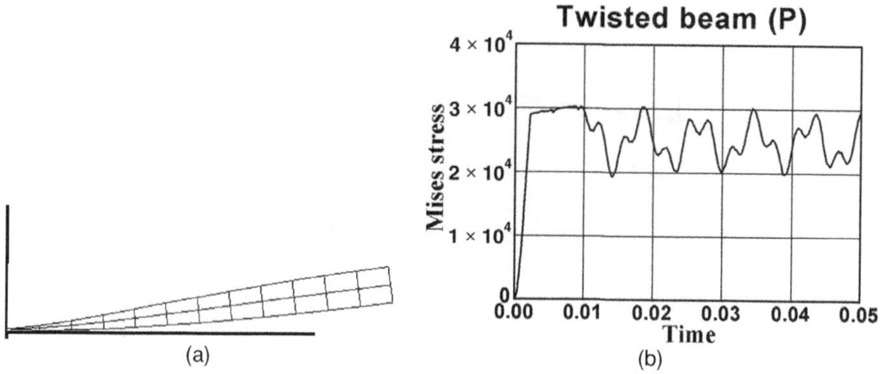

FIGURE 7.14 Result of twisted plastic beam by B-D element, with 12 × 2 mesh: (a) deformation at $t = 0.01$; (b) Mises stress of an element near the clamped end.

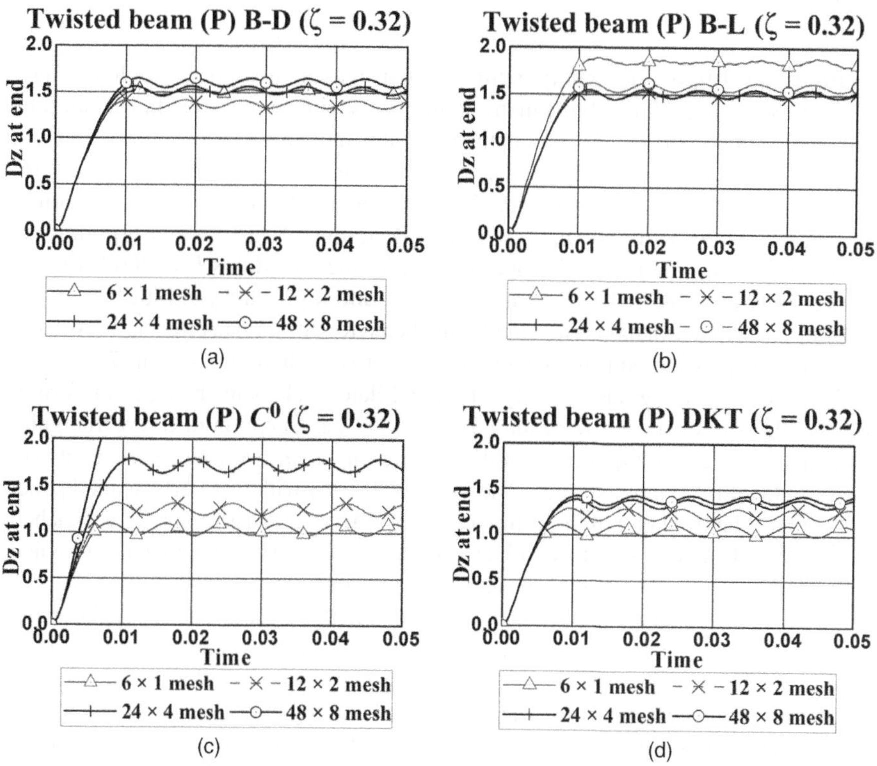

FIGURE 7.15 Displacement at the loaded end at $t = 0.01$: (a) results of B-D element; (b) results of B-L element; (c) results of C^0 element; (d) results of DKT element.

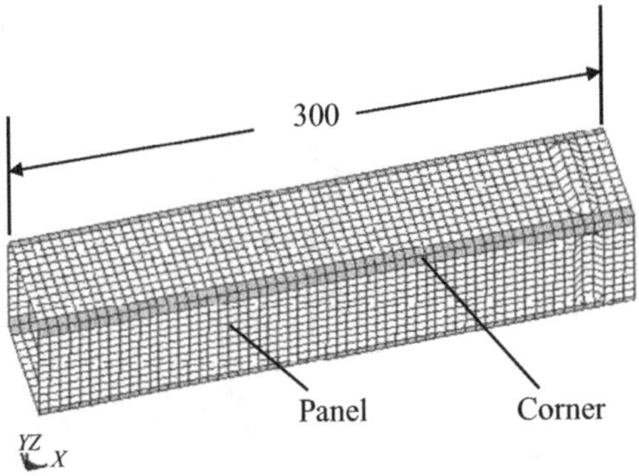

FIGURE 7.16 Mesh of a steel component.

meshes, the results of B-D and B-L elements are close. The result of C^0 element is questionable.

Example 7.3 Impact of a component In many cases, the impact problem is analyzed at a system level, involving many components with complex loading conditions and constraint conditions. As a fundamental study, the plastic large deformation of a simple component is investigated using the shell elements. As depicted in Figure 7.16, the box-beam shaped steel component has a length = 300 mm and a cross section = 60 mm × 60 mm, with thickness = 1 mm and radius of curvature = 5 mm at the folding line area. The material properties are: Young's modulus $E = 200$ kN/mm^2, mass density $\rho = 7.85 \times 10^{-6}$ kg/mm^3, Poisson ratio = 0.3, and yield stress $\sigma_y = 0.2$ kN/mm^2. The material has certain hardening behavior. Strain rate effect is considered by using Cowper-Symonds model with parameter $\dot{\varepsilon}_0 = 40$ and $\gamma = 5$ in (7.23).

The component is graded mainly with quadrilateral elements being as uniformly as possible with element size around 5 mm, with the exception of a few triangular elements in the convolution area. In the folding line area, two rows of elements are used to model the curved surface. The mesh is defined with two parts, named panel and corner. The front end of the component is fixed. A big mass of 500 kg is attached at the end of the component to simulate a heavy structure. The component is assigned with an initial velocity of 15 mm/ms and moves ahead.

Large deformation at the front end happens immediately and triggers the buckling in the convolution area, which is designed to induce the dynamic buckling in the desired mode. The deformation at 0.6 ms calculated by using B-T element is presented in Figure 7.17a, showing severe warping in the buckling area. As B-T element does not provide good solution to the twisted beam test due to warping, we need further investigation. The mesh is refined with element size around 2.5 mm and further refined to have size of 1.25 mm. The deformation is depicted in Figures 7.17b and 7.17c, respectively. The warpage in the buckling area is mainly located in two rows

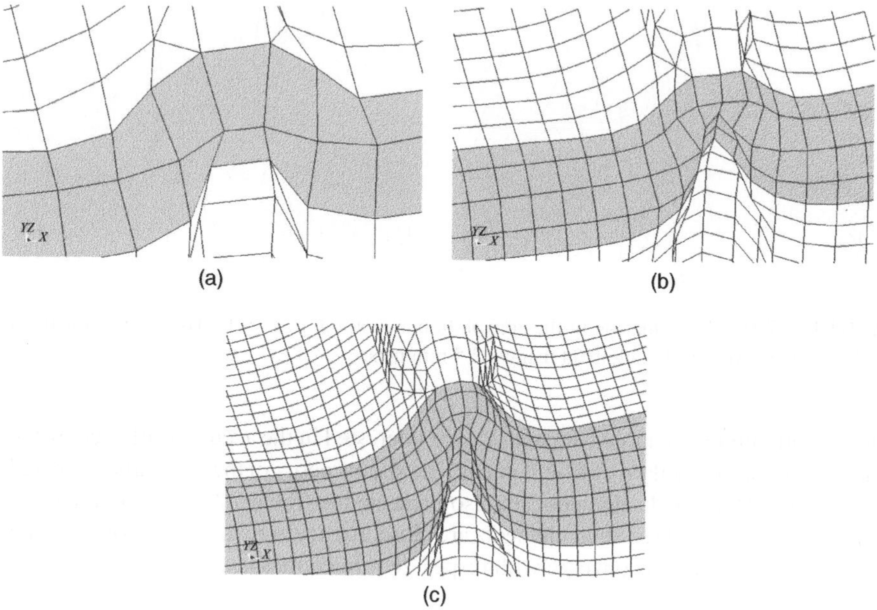

FIGURE 7.17 Deformation in the convolution area at 0.6 ms, by B-T element: (a) using a 5 mm mesh; (b) using a 2.5 mm mesh; (c) using a 1.25 mm mesh.

of elements around the corners. The internal energy induced by hourglass control method is presented in Table 7.1. When using fine mesh, the warpage affected area shrinks. The internal energy due to hourglass control method is mainly contributed from the corner area. Furthermore, the deformation calculated by using B-D and B-L elements with the 2.5 mm mesh is presented in Figure 7.18. The main features of the deformation are similar to those of B-T element. In this case, warping occurs due to large deformation under heavy load, and happens to all three types of elements.

TABLE 7.1 Strain Energy of the Component at 0.6 ms

Mesh	BT-hg	BT-IE	BD-IE	BL-IE	C^0-IE	DKT-IE
Part 1 (Panel)						
5	0.52464	83.559	89.168	88.872	108.38	90.580
2.5	0.13239	73.577	76.047	74.537	81.160	79.568
1.25	0.06962	70.212	71.735	70.861	71.922	72.305
Part 2 (Corner)						
5	0.78751	127.14	141.21	132.31	138.95	114.20
2.5	0.43334	102.45	109.04	106.02	108.39	103.05
1.25	0.21901	95.816	99.590	97.782	102.66	103.12

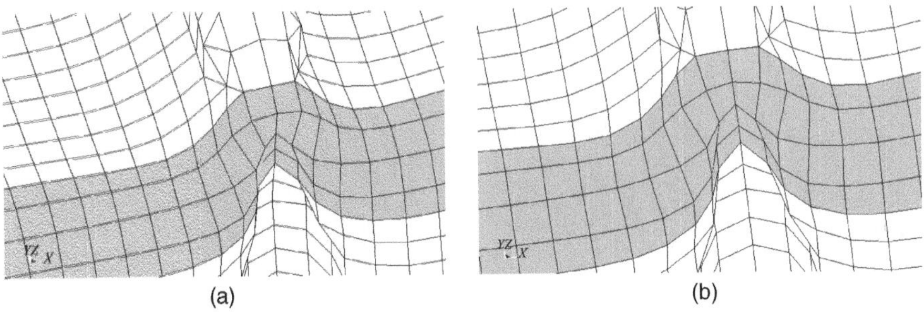

FIGURE 7.18 Deformation in the convolution area at 0.6 ms: (a) by B-D element using a 2.5 mm mesh; (b) by B-L element using a 2.5 mm mesh.

For deformation with large warpage, we are interested in the accuracy affected by the hourglass control method of B-T element. The strain energy of the two parts computed by using B-D and B-L elements are also presented in Table 7.1. It is observed that the results of these three shell elements are close, particularly in the case of 1.25 mm mesh. They all tend to converge with respect to refined meshes.

CHAPTER 8

CONTINUUM MECHANICS MODEL OF DUCTILE DAMAGE

Plasticity implies that the material is ductile. The strain that the specimen finally breaks at under tensile test is a measure of the ductility. Materials like cast aluminum and magnesium are usually considered brittle due to their lower ductility. Mild steel and many high-strength steels are considered ductile due to their higher ductility. Generally, for example in metal forming and in certain areas of impact problem, rupture is considered a failure and is prevented. Some material rupture may eventually happen during severe impact loading. Unlike structural design that uses a safety factor in its design process to prevent material failure, here it is necessary to know precisely whether the rupture will happen under certain loading condition. Furthermore, knowing where and when the rupture is going to happen is needed. In some situations, material rupture at certain location is inevitable. In some other cases, rupture may not happen by a simple load, but by accumulation of material's internal changes during the complex repeated loading process. It is a challenging task to predict the possibility of material ruptures under complex loading condition with acceptable accuracy.

8.1 CONCEPT OF DAMAGE MECHANICS

Several simple rupture criteria are available in the software. For example, as discussed in Wu et al. (2001), with the simple tensile test data, we can use the critical value of von Mises stress, the equivalent plastic strain, or the total strain energy. These

Introduction to the Explicit Finite Element Method for Nonlinear Transient Dynamics, First Edition.
Shen R. Wu and Lei Gu.

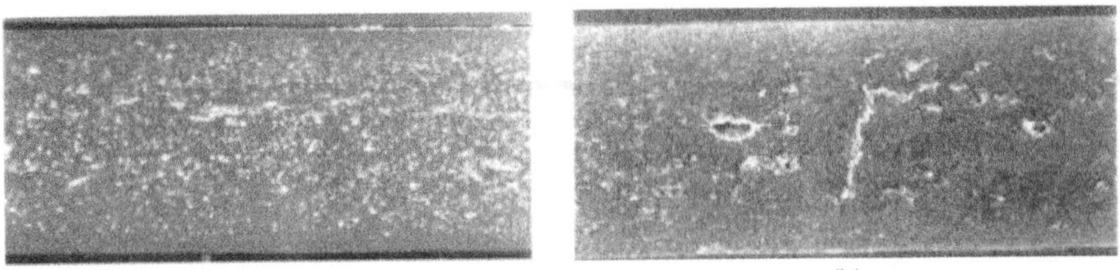

(a) (b)

FIGURE 8.1 Porosity found in the specimen: (a) before test; (b) after test. (Reprinted from Wu et al. (2001). Copyright (2001), with permission from ASME.)

simple criteria do not distinguish the difference between rupture caused by tension or by compression. It may be improved by using the principal tensile (or compressive) stress or strain. Whereas it is applicable to cases with loading dominated in one dimension, its deficiency in multiaxial complex loading is evident. In general, results of biaxial tests by these criteria do not match the pattern shown in the failure limit diagram. In addition, these approaches do not reflect the strain rate effect on ductility or the rupture. The material strength test shows that higher strain rate does not only increase the yield stress but also reduce the ductility.

On the other hand, the theory of damage mechanics may provide better approaches to describe and analyze the mechanism of material rupture. Since Kachanov (1958) introduced the concept of damage mechanics, research field has been very active with rich development of theories and applications.

It is observed from tensile test that when unloading after yielding the Young's modulus may decrease from that of the original material. Loading further into plasticity may cause further reduction of the Young's modulus. Metalloscopy analysis has found imperfections in the form of microvoids, inclusions, or cracks in the material, as illustrated in Figure 8.1. Generally, the materials used to produce the components and structures are not perfect. It is therefore assumed that imperfection either exists in the material or initiates when loading develops. When the material keeps on being loaded, the imperfection, which is considered as damage grows gradually. Material rupture will happen when damage reaches the critical condition.

Assume that the voids are distributed uniformly and the volume fraction of voids is ω. If the tensile specimen is subjected to a load F, then the measured Cauchy stress is the nominal stress $\sigma = F/A$, as if the material were perfect. The effective material has a cross-sectional area $\tilde{A} = A(1-\omega)$ and the effective stress or the true stress measured from the effective material is

$$\tilde{\sigma} = \frac{F}{(1-\omega)A} = \frac{\sigma}{1-\omega}. \tag{8.1}$$

Due to the reduced material section area, the effective stress is higher than the nominal stress.

Conceptually, this also applies to other forms of imperfections. Kachanov (1958) introduced the concept of continuity defined by $\psi = 1 - \omega$ for the effective material.

Along with the increasing of load, microdefect grows and effective material shrinks. Robotnov (1963, 1969) then introduced the damage parameter using ω. It has a reverse meaning of ψ. ω is zero for perfect material and grows with increasing load for imperfect material. The parameter ω represents the severity of damage. Janson and Hult (1977) was the first to use the terminology of damage mechanics.

Note that the damage mechanics approach deals with the situation where the microdefects may exist in the material but not visible in the macroscale. On the other hand, the traditional fracture mechanics approach deals with the situation where the cracks are identified and visible under the macroscale. The fracture mechanics, focusing on the crack, investigates the stress distribution due to the existence of a crack and the conditions of crack growth. In many applications, crack is considered a defect and is not supposed to exist in the product, but initiates under certain loading conditions. When using numerical method to solve these problems, the fracture mechanics model usually needs to mesh the details of the existing cracks; but the damage mechanics model does not have crack information in the mesh. Instead, it uses some internal variables, which are added to the material models to describe the process. The theory of damage mechanics draws a line apart from the methods of fracture mechanics.

Microcracks usually grow fast, which is similar to macrocrack investigated in fracture mechanics. Material rupture may happen with smaller ductility and therefore a research field focus on brittle damage is formed. On the other hand, microvoids go through the process of nucleation, growth, and coalescence. This may take a little bit longer time and result in slightly larger ductility at rupture. Based on this, a research field focus on ductile damage mechanics is formed.

Analysis of the classical damage elements, including the growth, nucleation, and coalescence of the voids, is the foundation of micromechanics-based damage mechanics. As an example, the Gurson's model is described in Section 8.2. Another approach uses the continuum mechanics-based theory to assume that the voids are uniformly distributed in the materials and to only concern about the macroprocess without microdetails. The later approach is also named phenomenological model of damage mechanics. Chow's isotropic damage model and anisotropic model are discussed in Sections 8.3 and 8.4, respectively. There are also approaches based on the combination of these two methods. Other methods use stochastic approaches for the distribution of voids, which are of random nature. For more discussion, we refer to Lemaitre and Chaboche (1990), Lemaitre (2001), and the references within. To find a generally acceptable and applicable approach, however, remains to be a challenging task.

8.2 GURSON'S MODEL

Gurson (1975, 1977) studied the plastic deformation of the cylindrical void contained in the finite cylinder and the plastic deformation of the spherical void contained in the finite sphere. Earlier analytical solutions of these types of problems for rigid-plastic material can be found, for example, in McClintock (1968) for the cylindrical void, and in Rice and Tracey (1969) for the spherical void. Based on these studies, a yield function with the void volume fraction as the internal variable was proposed and the evolution rule of the void volume fraction was established. Gurson's model has been applied to many cases. Several researchers have also contributed to the improvement of the model; for examples, Chu and Needleman (1980), Tvergaard (1990), and Needleman et al. (1992). Here, we describe the improved model, which was summarized in Tvergaard and Needleman (2001).

8.2.1 Damage Variables and Yield Function

As an extension from the classical isotropic hardening, the yield function is defined below:

$$F(\boldsymbol{\sigma}, \tilde{\sigma}_y, f) = \varphi(\boldsymbol{\sigma}) - \tilde{\sigma}_y(1 - 2q_1 f^* \cosh(q_2\sigma_{kk}/2\tilde{\sigma}_y) + q_3 f^{*2}), \tag{8.2}$$

where $\boldsymbol{\sigma}$ is the nominal Cauchy stress tensor and $\varphi(\boldsymbol{\sigma})$ is the yield function defined as in Chapter 7. $\tilde{\sigma}_y$ is the current yield stress of the effective material. Now, $\sigma_{kk} = 0$ is not necessary. q_i are constants. f^* is defined below as a modification of the void volume fraction f, representing the effect of void coalescence to rupture:

$$f^* = \begin{cases} f & \text{if } f < f_c \\ f_c + \dfrac{f_u - f_c}{f_F - f_c}(f - f_c) & \text{if } f \geq f_c. \end{cases} \tag{8.3}$$

When f reaches the critical value f_c, coalescence of voids starts. When $f \geq f_c$, we have $\mathrm{d}f^* \geq \mathrm{d}f$. The evolution of f^* speeds up. When f reaches the rupture value f_F, $f^* = f_u$, the material rupture starts. In summary

$$\mathrm{d}f^* = k_f \mathrm{d}f \geq \mathrm{d}f; \quad k_f = \begin{cases} 1 & \text{if } f < f_c \\ (f_u - f_c)/(f_F - f_c) & \text{if } f \geq f_c. \end{cases} \tag{8.4}$$

The evolution of f consists of two parts: the growth of the volume fraction of the existing voids and the nucleation of new voids:

$$\begin{aligned} \mathrm{d}f &= \mathrm{d}f_g + \mathrm{d}f_n, \\ \mathrm{d}f_g &= (1 - f)\mathrm{d}\varepsilon^p_{kk}, \\ \mathrm{d}f_n &= A\mathrm{d}\tilde{\sigma}_y + B\mathrm{d}\sigma_{kk}/3, \\ A &= \frac{f_N}{E_p S_N \sqrt{2\pi}} \exp\left(-\frac{1}{2}\left(\frac{\tilde{\varepsilon}_p - \varepsilon_N}{S_N}\right)^2\right), \\ B &= \begin{cases} A, & \text{if nucleation is stress controlled} \\ 0, & \text{if nucleation is plastic strain controlled.} \end{cases} \end{aligned} \tag{8.5}$$

The growth of void volume fraction depends on the plastic strain. Because of the plastic incompressibility of the effective material, the nominal volume change is due to the void growth. From $V = V_{\text{eff. mat.}} + V_{\text{voids}}$ and $V_{\text{voids}} = fV$, we obtain $\mathrm{d}V = \mathrm{d}V_{\text{void}} = V\mathrm{d}f_g + f\mathrm{d}V$. The nominal volume change is mainly due to the nominal plastic strain $\mathrm{d}\varepsilon_{kk}^p = \mathrm{d}V/V$. Thus, $\mathrm{d}f_g = (1-f)\mathrm{d}\varepsilon_{kk}^p$ and we derived the second equation of (8.5). The form of A accounts for the stochastic nature of void nucleation. ε_N is the mean strain of nucleation and S_N is the standard deviation. f_N is the volume fraction of the material by void nucleation.

8.2.2 Constitutive Equation and Damage Growth

It is assumed that the global behavior of the imperfect material obeys normality condition and that the consistency of plastic loading is valid:

$$\begin{aligned} \mathrm{d}\varepsilon_{ij}^p &= \mathrm{d}\lambda \partial F/\partial\sigma_{ij}, \\ \mathrm{d}F &= \partial F/\partial\sigma_{ij}\mathrm{d}\sigma_{ij} + \partial F/\partial\tilde{\sigma}_y \mathrm{d}\tilde{\sigma}_y + \partial F/\partial f^* \mathrm{d}f^* = 0. \end{aligned} \tag{8.6}$$

The derivatives involved are:

$$\begin{aligned} \partial F/\partial\sigma_{ij} &= \partial\varphi/\partial\sigma_{ij} + q_1 q_2 f^* \sinh(q_2\sigma_{kk}/2\tilde{\sigma}_y)\delta_{ij}, \\ \partial F/\partial\tilde{\sigma}_y &= -q_1 q_2 f^* \sigma_{kk} \sinh(q_2\sigma_{kk}/2\tilde{\sigma}_y)/\tilde{\sigma}_y \\ &\quad -(1 - 2q_1 f^* \cosh(q_2\sigma_{kk}/2\tilde{\sigma}_y) + q_3 f^{*2}), \\ \partial F/\partial f^* &= \tilde{\sigma}_y(2q_1 \cosh(q_2\sigma_{kk}/2\tilde{\sigma}_y) - 2q_3 f^*), \end{aligned} \tag{8.7}$$

$$\mathrm{d}f^* = k_f \mathrm{d}f = k_f((1-f)\mathrm{d}\varepsilon_{kk}^p + A\mathrm{d}\tilde{\sigma}_y + B\mathrm{d}\sigma_{kk}/3). \tag{8.8}$$

For simplicity, we denote

$$\begin{aligned} \alpha &= q_1 q_2 f^* \sinh(q_2\sigma_{kk}/\tilde{\sigma}_y), \\ p_{ij} &= \partial F/\partial\sigma_{ij} = \partial\varphi/\partial\sigma_{ij} + \alpha\delta_{ij}. \end{aligned} \tag{8.9}$$

From (8.6) and (8.8), we obtain

$$\mathrm{d}\tilde{\sigma}_y = -\frac{p_{ij}\mathrm{d}\sigma_{ij} + k_f \partial F/\partial f^*((1-f)\mathrm{d}\varepsilon_{kk}^p + B\mathrm{d}\sigma_{kk}/3)}{\partial F/\partial\tilde{\sigma}_y + k_f A \partial F/\partial f^*}. \tag{8.10}$$

We have the equivalence of plastic work measured from the nominal observation and the effective material:

$$\begin{aligned} \mathrm{d}W^p &= \sigma_{ij}\mathrm{d}\varepsilon_{ij}^p = \sigma_{ij}\mathrm{d}\lambda\partial F/\partial\sigma_{ij} = \mathrm{d}\lambda p_{ij}\sigma_{ij} \\ &= (1-f)\tilde{\sigma}_y \mathrm{d}\tilde{\varepsilon}^p = (1-f)\tilde{\sigma}_y \mathrm{d}\tilde{\sigma}_y/E_p. \end{aligned}$$

Using this equation to eliminate $\mathrm{d}\tilde{\sigma}_y$ from (8.10); and using the normality condition $\mathrm{d}\varepsilon_{kk}^p = \mathrm{d}\lambda \partial F/\partial \sigma_{kk}$, we then have

$$\mathrm{d}\lambda = q_{ij}\mathrm{d}\sigma_{ij}/H, \tag{8.11}$$

where we define:

$$\begin{aligned} q_{ij} &= p_{ij} + \delta_{ij} k_f \frac{B}{3}\frac{\partial F}{\partial f^*}, \\ H &= -\left(k_f(1-f)\frac{\partial F}{\partial f^*}\frac{\partial F}{\partial \sigma_{kk}} + \frac{E_p}{(1-f)\tilde{\sigma}_y} p_{ij}\sigma_{ij}\left(\frac{\partial F}{\partial \tilde{\sigma}_y} + Ak_f\frac{\partial F}{\partial f^*}\right)\right). \end{aligned} \tag{8.12}$$

This leads to the constitutive relation:

$$\mathrm{d}\varepsilon_{ij} = \mathrm{d}\varepsilon_{ij}^e + \mathrm{d}\varepsilon_{ij}^p = (C_{ijkl} + p_{ij}q_{kl}/H)\mathrm{d}\sigma_{kl}. \tag{8.13}$$

The inverse can be obtained in a similar way as described in Section 7.2:

$$\begin{aligned} \mathrm{d}\sigma_{ij} &= E_{ijkl}\left(\mathrm{d}\varepsilon_{kl} - \mathrm{d}\varepsilon_{kl}^p\right) = E_{ijkl}\mathrm{d}\varepsilon_{kl} - E_{ijkl}p_{kl}q_{mn}\mathrm{d}\sigma_{mn}/H, \\ q_{ij}\mathrm{d}\sigma_{ij} &= E_{ijkl}q_{ij}\mathrm{d}\varepsilon_{kl}/(1 + E_{ijkl}q_{ij}p_{kl}/H), \\ \mathrm{d}\sigma_{ij} &= \left(E_{ijkl} - \frac{E_{ijst}p_{st}E_{mnkl}q_{mn}}{H + E_{stmn}q_{st}p_{mn}}\right)\mathrm{d}\varepsilon_{kl}, \\ \mathrm{d}\lambda &= \frac{q_{ij}\mathrm{d}\sigma_{ij}}{H} = \frac{E_{ijkl}q_{ij}\mathrm{d}\varepsilon_{kl}}{H + E_{ijkl}q_{ij}p_{kl}}. \end{aligned} \tag{8.14}$$

The last equation is from (8.11).

About the parameters, Tvergaard and Needleman (2001) recommended using $1.25 < q_1 < 2, 0.9 < q_2 < 1, q_3 = q_1^2$, and $0.03 < f_c < 0.15$.

8.3 CHOW'S ISOTROPIC MODEL OF CONTINUUM DAMAGE MECHANICS

The fundamental concept of Kachanov's damage mechanics model is extended to the three-dimensional (3D) loading conditions. Various types of damage models under the framework of continuum mechanics have been developed. For example, Lemaitre and Chaboche (1990) used a single parameter D for the damage effect to the three principal stresses individually. Based on the experimental observation of varying Poisson's ratio along with the damage growth, Chow and Wei (1999), and Chow et al. (2000) developed an isotropic model with an additional parameter μ and several anisotropic models. The later will be discussed in the next section. The additional parameter μ was also prompted by the elasticity theory that the isotropic elastic material has two independent material parameters: Young's modulus E and

Poisson's ratio ν. We briefly discuss the concept and methods illustrated in Chow and Wei (1999), and Chow et al. (2000) in the following sections.

8.3.1 Damage Effect Tensor

The conceptual relation (8.1) between the effective stress $\tilde{\sigma}$ (the real stress of the effective material) and the nominal Cauchy stress σ (measured in the usual way) is extended to the 3D stress states:

$$\tilde{\boldsymbol{\sigma}} = \boldsymbol{M}\boldsymbol{\sigma}, \quad \tilde{\sigma}_{ij} = M_{ijkl}\sigma_{kl},$$
$$M_{ijkl} = \frac{1}{1-\hat{d}}((1-\hat{\mu})\delta_{ik}\delta_{jl} + \hat{\mu}\delta_{ij}\delta_{kl}). \tag{8.15}$$

Here, a symmetric damage effect tensor $\boldsymbol{M}$ of fourth order is introduced. Taking the symmetry of stress tensor into account, (8.15) can be expressed in a matrix form with the six-component representation of stress:

$$\boldsymbol{\sigma}^T = [\sigma_{11}, \sigma_{22}, \sigma_{33}, \sigma_{12}, \sigma_{23}, \sigma_{31}], \tilde{\boldsymbol{\sigma}}^T = [\tilde{\sigma}_{11}, \tilde{\sigma}_{22}, \tilde{\sigma}_{33}, \tilde{\sigma}_{12}, \tilde{\sigma}_{23}, \tilde{\sigma}_{31}], \tag{8.16}$$

$$\boldsymbol{M} = \frac{1}{1-\hat{d}}\begin{bmatrix} 1 & \hat{\mu} & \hat{\mu} & & & \\ \hat{\mu} & 1 & \hat{\mu} & & & \\ \hat{\mu} & \hat{\mu} & 1 & & & \\ & & & 1-\hat{\mu} & & \\ & & & & 1-\hat{\mu} & \\ & & & & & 1-\hat{\mu} \end{bmatrix}. \tag{8.17}$$

Here $\hat{d}$ and $\hat{\mu}$ are two damage parameters, denoted by D_1 and D_2, respectively. When $\hat{\mu} = 0$, (8.15) reduces to a simple isotropic model discussed in Lemaitre and Chaboche (1990). The parameter $\hat{\mu}$ represents the damage effect on the lateral and shear deformation related to Poisson's ratio.

The development of the constitutive relation is casted in the framework of irreversible thermodynamics. The Helmholtz free energy is postulated by Lemaitre and Chaboche (1990):

$$\rho\Psi = W_E + \rho\Psi_p(q), \tag{8.18}$$

where ρ is the density. W_E is the elastic energy. Ψ_p is the plastic part of the free energy due to strain hardening with q representing the parameter(s) to describe plastic strain hardening. Lemaitre (1971) proposed a hypothesis of strain equivalence, also known as the principle of equivalent elastic energy. The principle assumed that the elastic energy of the damaged material be written in the same form as that of an undamaged material except that the stress was replaced by the effective stress. Based on this concept, we have

$$W_E = C^0_{ijkl}\tilde{\sigma}_{ij}\tilde{\sigma}_{kl}/2. \tag{8.19}$$

Here, C^0 is the elastic compliance tensor for the isotropic undamaged material. From (8.15), we can interpret (8.19) in terms of nominal stress and the nominal elastic compliance:

$$W_E = C_{ijkl}\sigma_{ij}\sigma_{kl}/2, \tag{8.20}$$

$$C_{ijkl} = M_{ijmn}M_{klpq}C^0_{mnpq}. \tag{8.21}$$

With the definition of M in (8.17), direct calculation gives

$$C_{ijkl} = \frac{1}{E}((1+\nu)\delta_{ik}\delta_{jl} - \nu\delta_{ij}\delta_{kl}), \tag{8.22}$$

$$\begin{aligned} E &= \frac{E_0(1-\hat{d})^2}{1-4\nu_0\hat{\mu}+2(1-\nu_0)\hat{\mu}^2}, \\ \nu &= \frac{\nu_0 - 2(1-\nu_0)\hat{\mu} - (1-3\nu_0)\hat{\mu}^2}{1-4\nu_0\hat{\mu}+2(1-\nu_0)\hat{\mu}^2}, \end{aligned} \tag{8.23}$$

where E_0 and ν_0 are the Young's modulus and Poisson's ratio of the undamaged material, respectively. E and ν are for the nominal ones. The inverse $C^{-1} = E$ has the same form as the regular elasticity tensor. E and ν serve as the elasticity constants in the generalized Hook's law for the nominal measurement. Note that E and ν are variables, which are dependent of the damage parameters. Now, we can express (8.18) as

$$\rho\Psi = C_{ijkl}\sigma_{ij}\sigma_{kl}/2 + \rho\Psi_p(q). \tag{8.24}$$

The nominal elastic strain is expressed based on the free energy (8.24) and (8.20):

$$\begin{aligned} \varepsilon^e_{ij} &= \rho\partial\Psi/\partial\sigma_{ij} = C_{ijkl}\sigma_{kl}, \\ W_E &= \sigma_{ij}\varepsilon^e_{ij}/2. \end{aligned} \tag{8.25}$$

Comparing to the elastic energy expressed in terms of the effective stress and effective strain $W_E = \tilde{\sigma}_{ij}\tilde{\varepsilon}^e_{ij}/2 = M_{ijkl}\sigma_{kl}\tilde{\varepsilon}^e_{ij}/2$, we obtain

$$\varepsilon^e_{ij} = M_{klij}\tilde{\varepsilon}^e_{kl}, \quad \tilde{\varepsilon}^e_{ij} = (M^{-1})_{klij}\varepsilon^e_{kl}. \tag{8.26}$$

Due to the symmetry of M, we will not distinguish M and M^T in the later discussion.

From (8.18) and (8.24), we consider that the Helmholtz free energy is expressed in a form that separates the damage parameters and the plasticity parameters. Some damage mechanics models use the hypothesis of strain equivalence or stress equivalence. In fact, the damage growth involves both stress and strain distributions, for example, about the crack tip growth. Here, the hypothesis of energy equivalence is adopted. When heat effect is negligible, postulate (8.18) means that there are two major parts of energy dissipation: plastic flow and microcavitations of microcracking. The former associates with distortion and the later associates with dilatation.

We can write the energy equation in the rate or the incremental form:

$$\begin{aligned} \mathrm{d}w &= \mathrm{d}\tilde{w}, \\ \mathrm{d}w &= \mathrm{d}w^e + \mathrm{d}w^p + \mathrm{d}w^d, \\ \mathrm{d}\tilde{w} &= \mathrm{d}\tilde{w}^e + \mathrm{d}\tilde{w}^p, \\ \mathrm{d}w^p &= \mathrm{d}\tilde{w}^p. \end{aligned} \tag{8.27}$$

From $\mathrm{d}w^p = \sigma_{ij}\mathrm{d}\varepsilon_{ij}^p$ and $\mathrm{d}\tilde{w}^p = \tilde{\sigma}_{ij}\mathrm{d}\tilde{\varepsilon}_{ij}^p$, we obtain

$$\mathrm{d}\varepsilon_{ij}^p = M_{ijkl}\mathrm{d}\tilde{\varepsilon}_{kl}^p, \quad \mathrm{d}\tilde{\varepsilon}_{ij}^p = (\boldsymbol{M}^{-1})_{ijkl}\mathrm{d}\varepsilon_{kl}^p. \tag{8.28}$$

8.3.2 Yield Function and Constitutive Equation

We need to derive the constitutive equations for the nominal stress and strain. The rate of nominal strain will be available in the finite element procedure at each time step. The nominal stress will be used for computing nodal force and moment. Note that with the phenomenological approach, only the quantities measured in a macrosense are accessible and interested to us. The yield function of a general form (7.12a) can still be used, with the notion that the Cauchy stress and the yield stress are the stresses of the effective material. We then have the yield function, the normality condition, and the equivalent plastic strain rate as described in Section 7.1. We will use Mises criterion here. Following the steps described in Section 7.2, also using both expressions by the effective stress and nominal stress with damage variables, we can write

$$\begin{aligned} F(\tilde{\boldsymbol{\sigma}}, \tilde{\sigma}_y) &= \varphi(\tilde{\boldsymbol{\sigma}}) - \tilde{\boldsymbol{\sigma}}_y = \hat{\varphi}(\boldsymbol{\sigma}, \boldsymbol{D}) - \tilde{\sigma}_y, \\ \varphi(\tilde{\boldsymbol{\sigma}}) &= \sigma_M(\tilde{\boldsymbol{\sigma}}) = \sqrt{3\tilde{S}_{ij}\tilde{S}_{ij}/2}, \end{aligned} \tag{8.29}$$

$$\begin{aligned} \mathrm{d}\tilde{\varepsilon}_{ij}^p &= \mathrm{d}\tilde{\lambda}\frac{\partial F}{\partial \tilde{\sigma}_{ij}} = \mathrm{d}\tilde{\lambda}\frac{3\tilde{S}_{ij}}{2\varphi(\tilde{\boldsymbol{\sigma}})}, \\ \mathrm{d}\tilde{\varepsilon}^p &= \mathrm{d}\tilde{\lambda}. \end{aligned} \tag{8.30}$$

From (8.15), we verify $\tilde{\sigma}_{ij} = ((1-\hat{\mu})\sigma_{ij} + \hat{\mu}\sigma_{kk}\delta_{ij})/(1-\hat{d})$, $\tilde{\sigma}_{kk} = (1+2\hat{\mu})\sigma_{kk}/(1-\hat{d})$, and then

$$\tilde{S}_{ij} = \tilde{\sigma}_{ij} - \tilde{\sigma}_{kk}/3 = \frac{1-\hat{\mu}}{1-\hat{d}}S_{ij}. \tag{8.31}$$

Hence, we obtain

$$\begin{aligned}
\varphi(\tilde{\boldsymbol{\sigma}}) &= \frac{1-\hat{\mu}}{1-\hat{d}}\sqrt{\frac{3}{2}S_{ij}S_{ij}} = \frac{1-\hat{\mu}}{1-\hat{d}}\varphi(\boldsymbol{\sigma}) = \hat{\varphi}(\boldsymbol{\sigma}, \boldsymbol{D}), \\
\frac{\partial\hat{\varphi}}{\partial\sigma_{ij}} &= \frac{1-\hat{\mu}}{1-\hat{d}}\frac{3S_{ij}}{2\varphi(\boldsymbol{\sigma})} = \frac{3\tilde{S}_{ij}}{2\varphi(\boldsymbol{\sigma})}.
\end{aligned} \tag{8.32}$$

The consistency condition of plastic hardening yields

$$\begin{aligned}
\mathrm{d}F &= \frac{\partial\hat{\varphi}}{\partial\sigma_{ij}}\mathrm{d}\sigma_{ij} + \frac{\partial\hat{\varphi}}{\partial D_\alpha}\mathrm{d}D_\alpha - \tilde{E}_p\mathrm{d}\tilde{\varepsilon}^p = 0, \\
\mathrm{d}\tilde{\varepsilon}^p &= \frac{1}{\tilde{E}_p}\left(\frac{\partial\hat{\varphi}}{\partial\sigma_{ij}}\mathrm{d}\sigma_{ij} + \frac{\partial\hat{\varphi}}{\partial D_\alpha}\mathrm{d}D_\alpha\right).
\end{aligned} \tag{8.33}$$

From (8.15), (8.25), and (8.28)–(8.33), for the nominal variables we have

$$\begin{aligned}
\mathrm{d}\varepsilon_{ij} &= \mathrm{d}\varepsilon^e_{ij} + \mathrm{d}\varepsilon^p_{ij} = C_{ijkl}\mathrm{d}\sigma_{kl} + \mathrm{d}C_{ijkl}\sigma_{kl} + M_{ijkl}\mathrm{d}\tilde{\varepsilon}^p_{kl} \\
&= \left(C_{ijkl} + \frac{9\tilde{S}_{ij}\tilde{S}_{kl}}{4\tilde{E}_p\varphi^2(\boldsymbol{\sigma})}\right)\mathrm{d}\sigma_{kl} + \frac{\partial C_{ijkl}}{\partial D_\alpha}\mathrm{d}D_\alpha\sigma_{kl} + \frac{3\tilde{S}_{ij}}{2\tilde{E}_p\varphi(\boldsymbol{\sigma})}\frac{\partial\hat{\varphi}}{\partial D_\alpha}\mathrm{d}D_\alpha.
\end{aligned} \tag{8.34}$$

On the other hand, we have

$$\begin{aligned}
\mathrm{d}\sigma_{ij} &= \mathrm{d}E_{ijkl}\varepsilon^e_{kl} + E_{ijkl}\left(\mathrm{d}\varepsilon_{kl} - \mathrm{d}\varepsilon^p_{kl}\right) \\
&= \frac{\partial E_{ijkl}}{\partial D_\alpha}\mathrm{d}D_\alpha\varepsilon^e_{kl} + E_{ijkl}\left(\mathrm{d}\varepsilon_{kl} - M_{klpq}\mathrm{d}\tilde{\varepsilon}^p_{pq}\right) \\
&= \frac{\partial E_{ijkl}}{\partial D_\alpha}\mathrm{d}D_\alpha C_{klpq}\sigma_{pq} + E_{ijkl}\mathrm{d}\varepsilon_{kl} \\
&\quad - \frac{3\mu\tilde{S}_{ij}}{\varphi(\boldsymbol{\sigma})}\frac{\dfrac{\partial\hat{\varphi}}{\partial\sigma_{mn}}\mathrm{d}\sigma_{mn} + \dfrac{\partial\hat{\varphi}}{\partial D_\alpha}\mathrm{d}D_\alpha}{\tilde{E}_p}.
\end{aligned} \tag{8.35}$$

From (8.35), using (8.15 and 8.30) and the nominal shear modulus μ, we find:

$$\frac{\partial\hat{\varphi}}{\partial\sigma_{mn}}\mathrm{d}\sigma_{mn} = \frac{\dfrac{\partial\hat{\varphi}}{\partial\sigma_{ij}}\left(\dfrac{\partial E_{ijkl}}{\partial D_\alpha}\mathrm{d}D_\alpha C_{klpq}\sigma_{pq} + E_{ijkl}\mathrm{d}\varepsilon_{kl} - \dfrac{3\mu}{\tilde{E}_p\varphi(\boldsymbol{\sigma})}\tilde{S}_{ij}\dfrac{\partial\hat{\varphi}}{\partial D_\alpha}\mathrm{d}D_\alpha\right)}{1 + \dfrac{3\mu}{\tilde{E}_p\varphi(\boldsymbol{\sigma})}\dfrac{\partial\hat{\varphi}}{\partial S_{ij}}\tilde{S}_{ij}}. \tag{8.36}$$

Plugging (8.36) back into (8.35) leads to the rate form

$$
\mathrm{d}\sigma_{ij} = E_{ijkl}\mathrm{d}\varepsilon_{kl} + \frac{\partial E_{ijkl}}{\partial D_\alpha}\mathrm{d}D_\alpha C_{klpq}\sigma_{pq}
- \frac{\dfrac{3\mu\tilde{S}_{ij}\tilde{S}_{kl}\mathrm{d}\varepsilon_{kl}}{\varphi^2(\boldsymbol{\sigma})} + \dfrac{\tilde{S}_{ij}}{\varphi(\boldsymbol{\sigma})}\dfrac{\partial\hat{\varphi}}{\partial D_\alpha}\mathrm{d}D_\alpha + \dfrac{3\tilde{S}_{ij}\tilde{S}_{st}}{2\varphi^2(\boldsymbol{\sigma})}\dfrac{\partial E_{stkl}}{\partial D_\alpha}\mathrm{d}D_\alpha C_{klpq}\sigma_{pq}}{\dfrac{\tilde{E}_p}{3\mu} + \dfrac{\varphi^2(\tilde{\boldsymbol{\sigma}})}{\varphi^2(\boldsymbol{\sigma})}}. \quad (8.37)
$$

The yield stress needs an update after the calculation of plasticity, involving the update of the equivalent plastic strain of the effective material. From (8.15), (8.28), and the plastic incompressibility, we obtain

$$
\mathrm{d}\varepsilon_{ij}^p = \left((1-\hat{\mu})d\tilde{\varepsilon}_{ij}^p + \hat{\mu}\mathrm{d}\tilde{\varepsilon}_{kk}^p\delta_{ij}\right)/(1-\hat{d}) = (1-\hat{\mu})\mathrm{d}\tilde{\varepsilon}_{ij}^p/(1-\hat{d}),
$$
$$
\mathrm{d}\tilde{\varepsilon}_{ij}^p = \frac{1-\hat{d}}{1-\hat{\mu}}\mathrm{d}\varepsilon_{ij}^p, \quad \mathrm{d}\tilde{\varepsilon}^p = \frac{1-\hat{d}}{1-\hat{\mu}}\mathrm{d}\varepsilon^p. \quad (8.38)
$$

Now the elasticity modulus tensor is not constant, but dependent of the damage parameters. From (8.22) and (8.23), we can verify

$$
\begin{aligned}
\partial C_{ijkl}/\partial\hat{d} &= 2C_{ijkl}/(1-\hat{d}),\\
\partial C_{ijkl}/\partial\hat{\mu} &= 2A_{ijkl}/(1-\hat{d}),
\end{aligned} \quad (8.39)
$$

where

$$
\begin{aligned}
A_{ijkl} &= \frac{1}{E_0(1-\hat{d})}\left((A_1 - A_2)\delta_{ik}\delta_{jl} + A_2\delta_{ij}\delta_{kl}\right),\\
A_1 &= 2\hat{\mu}(1-\nu_0) - 2\nu_0,\\
A_2 &= (1+\hat{\mu})(1-\nu_0) - 2\hat{\mu}\nu_0.
\end{aligned} \quad (8.40)
$$

Note that $\boldsymbol{EC} \equiv \boldsymbol{I}$, $(\mathrm{d}\boldsymbol{E})\boldsymbol{C} + \boldsymbol{E}\mathrm{d}\boldsymbol{C} = \boldsymbol{O}$, and $\mathrm{d}\boldsymbol{E} = -\boldsymbol{E}(\mathrm{d}\boldsymbol{C})\boldsymbol{E}$. This gives

$$
\begin{aligned}
\partial\boldsymbol{E}/\partial\hat{d} &= -2\boldsymbol{E}/\left(1-\hat{d}\right),\\
\partial\boldsymbol{E}/\partial\hat{\mu} &= -2\boldsymbol{EAE}/\left(1-\hat{d}\right).
\end{aligned} \quad (8.41)
$$

8.3.3 Damage Growth

The damage growth rate $\mathrm{d}D_\alpha$ is involved in the rate form relations (8.34) and (8.37). Its relation to the rate of plastic strain and/or stress is to be established by the application of damage growth rule. We introduce the damage energy release rate as the thermodynamic conjugate forces that is associated with the damage parameters, and is defined by the Helmholtz free energy equation:

$$\begin{aligned} Y_d &= -\rho \partial \Psi / \partial \hat{d} = -C_{ijkl}\sigma_{ij}\sigma_{kl}/(1-\hat{d}), \\ Y_\mu &= -\rho \partial \Psi / \partial \hat{\mu} = -A_{ijkl}\sigma_{ij}\sigma_{kl}/(1-\hat{d}). \end{aligned} \tag{8.42}$$

Here, (8.18) and (8.39) are used. Here is an analogy to the yield surface of plasticity: a plastic damage surface is postulated to characterize damage growth:

$$F_{PD} = Y_{PD} - B(w), \tag{8.43}$$

where the equivalent plastic damage energy release rate is defined as

$$\begin{aligned} Y_{PD} &= \sqrt{\Gamma_{\alpha\beta} Y_\alpha Y_\beta}, \\ \Gamma &= \begin{bmatrix} 1/2 & 0 \\ 0 & \gamma/2 \end{bmatrix}. \end{aligned} \tag{8.44}$$

Here, Γ represents the coefficient of damage evolution. Function $B(w)$ represents the plastic hardening with initial value B_0 as the initial damage threshold (similar to the yield stress in plasticity). The variable w means overall plastic damage (similar to the equivalent plastic strain).

The damage surface has a similar concept as the yield surface has. The rates of damage parameters are related to the normal of the damage surface:

$$\begin{aligned} \mathrm{d}D_\alpha &= -\mathrm{d}\lambda_{PD} \partial F_{PD}/\partial Y_\alpha = -\mathrm{d}\lambda_{PD}\Gamma_{\alpha\beta}Y_\beta/Y_{PD}, \\ \mathrm{d}w &= -\mathrm{d}\lambda_{PD}\partial F_{PD}/\partial(-B) = \mathrm{d}\lambda_{PD}. \end{aligned} \tag{8.45}$$

Using the inverse $\Gamma^{-1}_{\beta\alpha}\mathrm{d}D_\alpha = -\mathrm{d}\lambda_{PD}Y_\beta/Y_{PD}$, we obtain the rate of overall plastic damage (similar to the equivalent plastic strain rate):

$$(\mathrm{d}w)^2 = \left(\Gamma^{-1}\right)_{\alpha\beta} \mathrm{d}D_\alpha \mathrm{d}D_\beta, \tag{8.46}$$

with the notion that if $\gamma = 0$, we drop the term related to Y_μ.

The consistency condition of damage evolution gives

$$\mathrm{d}f_{PD} = \partial Y_{PD}/\partial Y_\alpha \mathrm{d}Y_\alpha - B'(w)\mathrm{d}w = 0. \tag{8.47}$$

On the other hand, we have from (8.45)

$$\mathrm{d}Y_\alpha = \frac{\partial Y_\alpha}{\partial D_\beta}\mathrm{d}D_\beta + \frac{\partial Y_\alpha}{\partial \sigma_{ij}}\mathrm{d}\sigma_{ij} = -\mathrm{d}w\frac{\partial Y_\alpha}{\partial D_\beta}\frac{\partial Y_{PD}}{\partial Y_\beta} + \frac{\partial Y_\alpha}{\partial \sigma_{ij}}\mathrm{d}\sigma_{ij}, \tag{8.48}$$

then we are able to obtain

$$B'(w)\mathrm{d}w = \frac{\partial Y_{PD}}{\partial Y_\alpha}\mathrm{d}Y_\alpha = -\mathrm{d}w\frac{\partial Y_{PD}}{\partial Y_\alpha}\frac{\partial Y_\alpha}{\partial D_\beta}\frac{\partial Y_{PD}}{\partial Y_\beta} + \frac{\partial Y_{PD}}{\partial Y_\alpha}\frac{\partial Y_\alpha}{\partial \sigma_{ij}}\mathrm{d}\sigma_{ij},$$

$$\mathrm{d}w = \frac{\dfrac{\partial Y_{PD}}{\partial Y_\alpha}\dfrac{\partial Y_\alpha}{\partial \sigma_{ij}}\mathrm{d}\sigma_{ij}}{B'(w) + \dfrac{\partial Y_{PD}}{\partial Y_\alpha}\dfrac{\partial Y_\alpha}{\partial D_\beta}\dfrac{\partial Y_{PD}}{\partial Y_\beta}}. \tag{8.49}$$

This relation provides the calculation for growth of overall plastic damage, which further leads to the integration of $B(w)$.

8.3.4 Application to Plates and Shells

This damage model can be extended to applications of plate/shell element corresponding to the generalized plane stress. A reduced damage effect tensor is defined in the five-components form:

$$\boldsymbol{M}_2 = \frac{1}{1-\hat{d}}\begin{bmatrix} 1 & \hat{\mu} & & & \\ \hat{\mu} & 1 & & & \\ & & 1-\hat{\mu} & & \\ & & & 1-\hat{\mu} & \\ & & & & 1-\hat{\mu} \end{bmatrix}, \tag{8.50a}$$

$$\boldsymbol{M}_2^{-1} = \frac{1-\hat{d}}{1-\hat{\mu}}\begin{bmatrix} \dfrac{1}{1+\hat{\mu}} & -\dfrac{\hat{\mu}}{1+\hat{\mu}} & & & \\ -\dfrac{\hat{\mu}}{1+\hat{\mu}} & \dfrac{1}{1+\hat{\mu}} & & & \\ & & 1 & & \\ & & & 1 & \\ & & & & 1 \end{bmatrix}. \tag{8.50b}$$

The associated compliance tensor $\boldsymbol{C}_2$ can also use the reduced form for the usual elasticity, with

$$E_2 = \frac{E_0\left(1-\hat{d}\right)^2}{1+\hat{\mu}^2-2\hat{\mu}\nu_0},$$
$$\nu_2 = \frac{(1+\hat{\mu}^2)\nu_0 - 2\hat{\mu}}{1+\hat{\mu}^2-2\hat{\mu}\nu_0}. \tag{8.51}$$

For calculating the damage evolution, we still have (8.41) for the derivatives of $\boldsymbol{C}_2$ with respect to the damage parameters. Now the tensor $\boldsymbol{A}$ has a reduced form too, with

$$\begin{aligned} A_1 &= \hat{\mu} - \nu_0, \\ A_2 &= 1 - \hat{\mu}\nu_0. \end{aligned} \tag{8.52}$$

8.3.5 Determination of Parameters

The challenge commonly relevant to all the damage models is that the computational results are finally to be compared to the lab tests. The damage models usually introduce one or more parameters to describe the damage state and their evolution. Developing methods to determine the parameters is almost as important as the theoretical development of the model. Robust and easy approach to conduct the test is needed for validating the model. This is also necessary for practical applications. For the isotropic damage model described above, we introduce a method to obtain the needed damage parameters by using simple tensile tests. We brief the procedure in the following (see Chow et al. (2000) for more discussions and examples):

1. Obtain the elasticity constants E_0 and ν_0 from the base material.
2. Perform a uniaxial tensile test. Pull the specimen to plastic strain ε_1, and then unload. Measure E_1 and ν_1. Reload up to ε_2, unload and measure E_2 and ν_2. Repeat this process until enough data are obtained, illustrated in Figure 8.2.
3. With E and ν as functions of ε, obtained in Step (2), use (8.23) to solve $\hat{d}$ and $\hat{\mu}$ (taking positive roots only) as functions of ε.

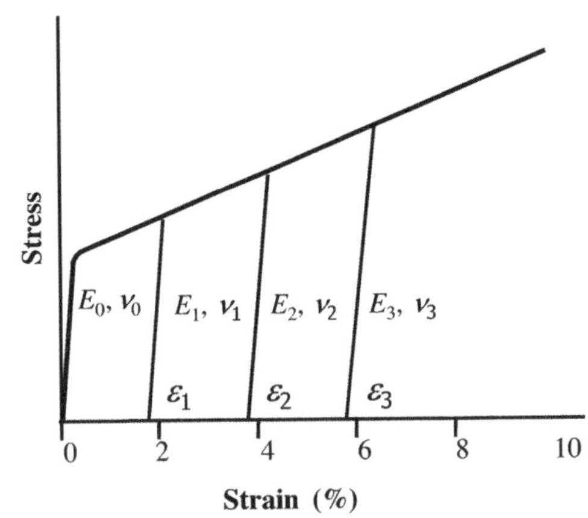

FIGURE 8.2 Damage growth determined from test. (Reprinted from Chow et al. (2000). Copyright (2000), with permission from 2000 SAE International.)

4. With stress $\sigma_1 = \sigma$ ($\sigma_2 = \sigma_3 = 0$ now) obtained in Step (2), also $\hat{d}$ and $\hat{\mu}$ from Step (3), calculate Y_d and Y_μ using (8.42):

$$Y_d = -\frac{\sigma^2}{E(1-\hat{d})},$$

$$Y_\mu = -\frac{A_1\sigma^2}{E(1-\hat{d})^2}.$$

5. Based on the plastic damage flow rule (8.45), calculate $\gamma = Y_\mu \mathrm{d}\hat{d} / Y_d \mathrm{d}\hat{\mu}$. A constant γ can be found approximately.
6. The constitutive relation $\sigma_y = \sigma_y(\varepsilon^p)$ in terms of nominal stress σ_y and nominal equivalent plastic strain ε^p can be derived from the test data.
7. Calculate $\mathrm{d}w$ by using (8.46) and obtain $B(w)$ from the consistency condition.

The constitutive relation in Step (6) can be converted to $\tilde{\sigma}_y = \tilde{\sigma}_y(\tilde{\varepsilon}^p)$ in terms of the effective stress and the effective equivalent plastic strain. Both of them can be used for software implementation. The user needs to determine which one should be provided as input data. In fact, if the later is available from other source, for example, lab test for the virgin material without damage, then this step can be used for validation. The damage model, however, assumes that the supplied material has certain type of preexisting damage, such as observed in real material. The tensile test data contains the effect of damage. In this case, in order to obtain the undamaged data we would need some iterative approaches or certain kind of approximation.

8.4 CHOW'S ANISOTROPIC MODEL OF CONTINUUM DAMAGE MECHANICS

In general, the material isotropy will no longer hold when damage initiates and is growing. Both theoretical studies and experiments have discovered that change in Poisson's ratio is induced along with the decrease of Young's modulus due to damage growth. Hence, the material anisotropy of plasticity needs to be included and the anisotropic damage model needs to be taken into account. The Gurson's model assumes no difference in orientation of voids growth; therefore, it is for the isotropic damage. The Chow's isotropic damage model has a framework relatively convenient to be extended to anisotropic case. Chow and Wei (2001) proposed several approaches with various forms of the damage effect matrix $\boldsymbol{M}$ based on the characteristics of applications. The following gives a brief description.

Hill (1950) studied the anisotropic plasticity. The equivalent stress (like the von Mises stress for isotropic material) of the effective material with anisotropy is now defined by

$$\tilde{\sigma} = \sqrt{\tilde{\boldsymbol{\sigma}}:\boldsymbol{H}_0:\tilde{\boldsymbol{\sigma}}}, \tag{8.53}$$

$$\boldsymbol{H}_0 = \begin{bmatrix} g+h & -h & -g & & & \\ -h & h+f & -f & & & \\ -g & -f & f+g & & & \\ & & & 2r & & \\ & & & & 2m & \\ & & & & & 2n \end{bmatrix}. \tag{8.54}$$

Using the damage effect tensor $\boldsymbol{M}$, with the relation (8.15), the equivalent stress in terms of the nominal stress is expressed as

$$\begin{aligned} \Phi(\boldsymbol{\sigma}) &= \sqrt{\boldsymbol{\sigma}:\boldsymbol{H}:\boldsymbol{\sigma}}, \\ \boldsymbol{H} &= \boldsymbol{M}^T:\boldsymbol{H}_0:\boldsymbol{M}. \end{aligned} \tag{8.55}$$

Examples of the damage effect tensors are (with the six-component stress representation)

$$\boldsymbol{M} = \begin{bmatrix} \begin{matrix} \frac{1}{1-d_1} & 0 & 0 \\ 0 & \frac{1}{1-d_2} & 0 \\ 0 & 0 & \frac{1}{1-d_3} \end{matrix} & \boldsymbol{O}_3 \\ \boldsymbol{O}_3 & \begin{matrix} \frac{1}{\sqrt{(1-d_2)(1-d_3)}} & 0 & 0 \\ 0 & \frac{1}{\sqrt{(1-d_1)(1-d_3)}} & 0 \\ 0 & 0 & \frac{1}{\sqrt{(1-d_1)(1-d_2)}} \end{matrix} \end{bmatrix}, \tag{8.56a}$$

$$\boldsymbol{M} = \begin{bmatrix} \frac{1}{1-d_1} & & & & & \\ & \frac{1}{1-d_2} & & & \boldsymbol{O}_3 & \\ & & \frac{1}{1-d_3} & & & \\ & & & \frac{2}{2-d_2-d_3} & & \\ & \boldsymbol{O}_3 & & & \frac{2}{2-d_1-d_3} & \\ & & & & & \frac{2}{2-d_1-d_2} \end{bmatrix}, \tag{8.56b}$$

$$
\boldsymbol{M} = \begin{bmatrix} e^{d_1} & & & & & \\ & e^{d_2} & & & \boldsymbol{0}_3 & \\ & & e^{d_3} & & & \\ & & & e^{(d_2+d_3)/2} & & \\ & \boldsymbol{0}_3 & & & e^{(d_3+d_1)/2} & \\ & & & & & e^{(d_1+d_2)/2} \end{bmatrix}. \tag{8.56c}
$$

The common feature of these examples is that the damage effect on the three normal stresses is different from each other, and the damage effect on the three shear stresses is also different from each other. The process for plasticity computation is essentially the same as discussed in Section 8.3 for the isotropic damage model.

CHAPTER 9

MODELS OF NONLINEAR MATERIALS

Besides plasticity, many other types of materials exhibit nonlinearity in the practical applications. More efforts are required to establish the theory and mechanics model for these materials. There are less experiences and thorough investigations with these materials comparing to metals.

9.1 VISCOELASTICITY

Viscous behavior is observed from many materials. The response depends not only on the deformation but also on the rate of deformation when loaded. Associated with viscous property, the material also experiences relaxation, in which the stress gradually decreases when deformation is frozen; or creep, in which the deformation gradually increases when the load is kept constant.

9.1.1 Spring–Damper Model

A linear spring usually represents the elasticity, and a damper (dashpot) can represent viscosity accordingly, for which the stress is related to strain rate instead of strain:

$$f = \eta \dot{u} \quad \text{or} \quad \sigma = \eta \dot{\varepsilon}, \tag{9.1}$$

where η is the viscous coefficient.

Introduction to the Explicit Finite Element Method for Nonlinear Transient Dynamics, First Edition.
Shen R. Wu and Lei Gu.

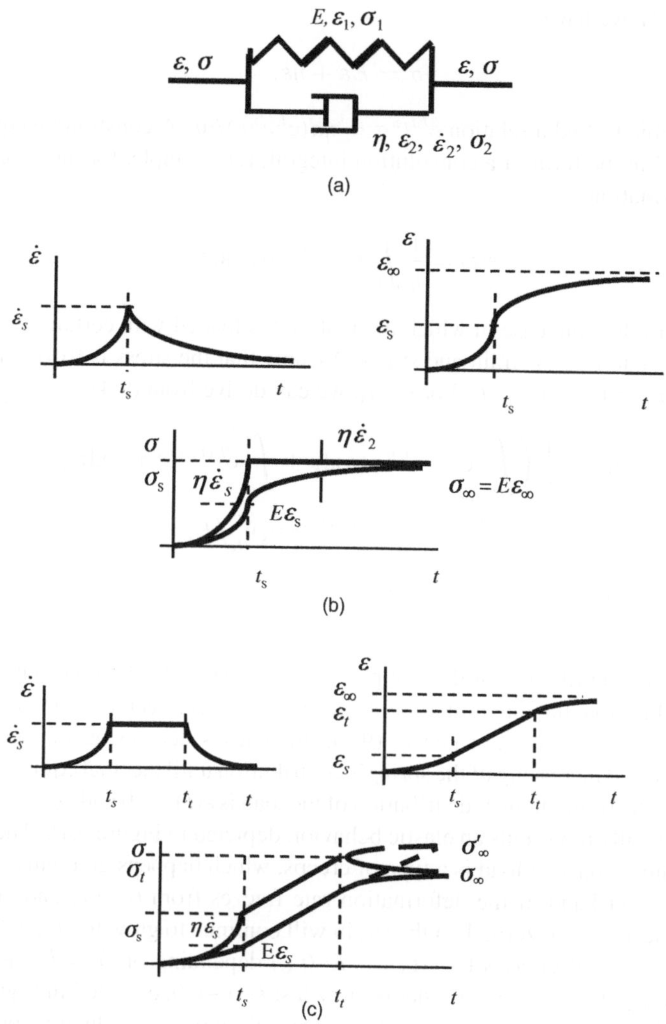

FIGURE 9.1 Kelvin model: (a) model description; (b) creep; (c) relaxation.

The viscoelasticity of materials has been simulated by using spring–damper models, cf. Fung (1965). One of the simple models consists of a spring and a damper combined in a parallel form, depicted in Figure 9.1a, named Kelvin material. The stress–strain relation is described below:

$$\begin{aligned}
\sigma &= \sigma_1 + \sigma_2, \\
\varepsilon &= \varepsilon_1 = \varepsilon_2, \\
\sigma_1 &= E\varepsilon_1, \\
\sigma_2 &= \eta\dot{\varepsilon}_2.
\end{aligned} \tag{9.2}$$

From (9.2), we have

$$\sigma = E\varepsilon + \eta\dot{\varepsilon}. \tag{9.3}$$

We are able to find a relation $e^{Et/\eta}\sigma = \eta d(e^{Et/\eta}\varepsilon)/dt$. A constitutive equation is then derived in the form of a convolution integral, for example, for the case without initial deformation:

$$\varepsilon(t) = \frac{1}{\eta}\int_0^t e^{E(\tau-t)/\eta}\sigma(\tau)d\tau. \tag{9.4}$$

Consider a loading case in which the material is loaded with certain deformation rate from a state free of strain and stress. Assume that the stress is frozen at time t_s, that is, $\sigma(t) = \sigma(t_s)$ for $t \geq t_s$. For $t > t_s$, we can derive from (9.4):

$$\begin{aligned}\varepsilon(t) &= \frac{1}{\eta}\left(\int_0^{t_s} e^{E(\tau-t)/\eta}\sigma(\tau)d\tau + \int_{t_s}^t e^{E(\tau-t)/\eta}\sigma(t_s)d\tau\right)\\ &= e^{E(t_s-t)/\eta}\varepsilon(t_s) + e^{E(\tau-t)/\eta}\sigma(t_s)\Big|_{t_s}^{t}/E,\\ \dot{\varepsilon}(t) &= (\sigma(t_s)/E - \varepsilon(t_s))e^{E(t_s-t)/\eta}/\eta.\end{aligned}$$

Due to the contribution of the damper, $\sigma(t_s) \geq E\varepsilon(t_s)$. Under constant stress, for long time duration, the magnitude of the deformation rate decreases and approaches zero when $t \to \infty$. As indicated in (9.3), this causes decreasing of the damper's contribution and increasing of the spring's contribution until the final equilibrium state is reached. The result of the redistribution of the load is $\dot{\varepsilon}(t) \to 0$ and $\varepsilon(t) \to \sigma(t_s)/E$. It asymptotically represents an elastic behavior, depicted in Figure 9.1b. The increase of strain under constant load is a type of creeps, which happens gradually.

On the other hand, if the deformation rate freezes from t_1 to t_2 and, afterward, gradually decreases to zero, then the strain will continue to grow to $\varepsilon(t_\infty)$. The stress then increases or decreases to $\sigma(t_\infty) = E\varepsilon(t_\infty)$ depending on $\dot{\sigma} = E\dot{\varepsilon} + \eta\ddot{\varepsilon}$ being positive or negative with $\ddot{\varepsilon} < 0$, that is, how fast $\dot{\varepsilon}(t) \to 0$, depicted in Figure 9.1c.

Maxwell material is another simple viscoelastic model, in which a spring and a damper are combined sequentially, shown in Figure 9.2a. The stress–strain relation is described below:

$$\begin{aligned}\sigma &= \sigma_1 = \sigma_2,\\ \varepsilon &= \varepsilon_1 + \varepsilon_2,\\ \sigma_1 &= E\varepsilon_1,\\ \sigma_2 &= \eta\dot{\varepsilon}_2.\end{aligned} \tag{9.5}$$

For this case, we have

$$\dot{\varepsilon} = \frac{\dot{\sigma}}{E} + \frac{\sigma}{\eta}. \tag{9.6}$$

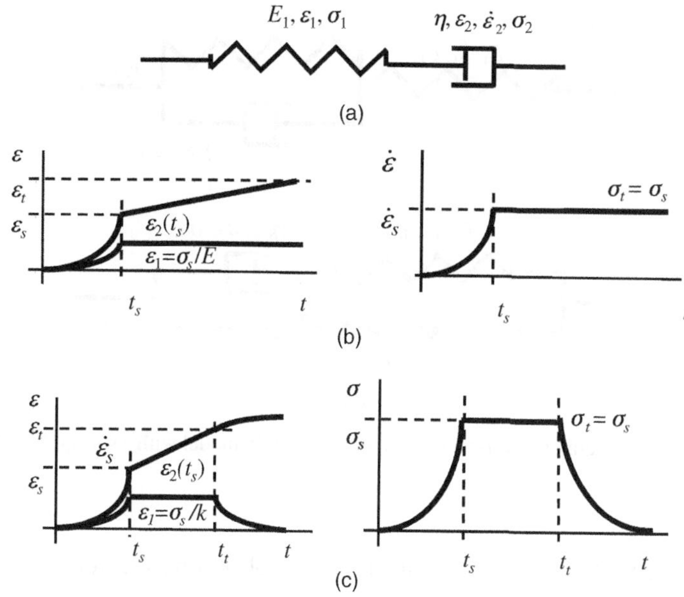

FIGURE 9.2 Maxwell model: (a) model description; (b) creep; (c) relaxation.

Similarly, we find a relation $Ee^{Et/\eta}\dot{\varepsilon} = \mathrm{d}(e^{Et/\eta}\sigma)/\mathrm{d}t$ and a constitutive equation in the form of convolution integral:

$$\sigma(t) = E\int_0^t e^{E(\tau-t)/\eta}\dot{\varepsilon}(\tau)\mathrm{d}\tau. \tag{9.7}$$

If the stress holds constant in the loading process from $t = t_s$, then the spring's deformation will keep constant $\varepsilon_1 = \varepsilon_1(t_s) = \sigma(t_s)/E$. We also have $\dot{\varepsilon}_2 \equiv \sigma(t_s)/\eta$ and $\varepsilon = \sigma(t_s)/E + (t - t_s)\sigma(t_s)/\eta$. The total deformation keeps growing. This is a type of creeps, shown in Figure 9.2b.

Let us consider another case, in which the deformation rate $\dot{\varepsilon}_1$ freezes from t_s to t_t. By (9.6), $\dot{\sigma}$ should be zero, σ and ε_1 should hold constant in this process. Meanwhile ε_2 grows linearly, as depicted in Figure 9.2c, same feature as shown in Figure 9.2b. If $\dot{\varepsilon}$ decreases to zero gradually after t_t, that is, to a state with no more deformation, then $\varepsilon_1 + \varepsilon_2 = \text{const}$. With positive stress, we have $\dot{\varepsilon}_1 = -\dot{\varepsilon}_2 < 0$. It is possible that $\dot{\varepsilon}_1 < 0$ starts directly from t_t. The spring tends to bounce back along with the decreasing stress and finally recovers completely. σ decreases to zero gradually with $\dot{\sigma} < 0$. Meanwhile, ε_2 keeps growing up to its maximum with $\dot{\varepsilon}_2 = 0$ and $\sigma = 0$. This event of gradual decrease of stress represents a type of stress relaxation, illustrated in Figure 9.2c.

Other models with complex combination of springs and dampers are also used for analyzing the viscoelastic material properties, for example, a Kelvin model or Maxwell model connecting with a spring or dashpot in a parallel or sequential way, depicted in Figure 9.3. The constitutive equations can be established similarly.

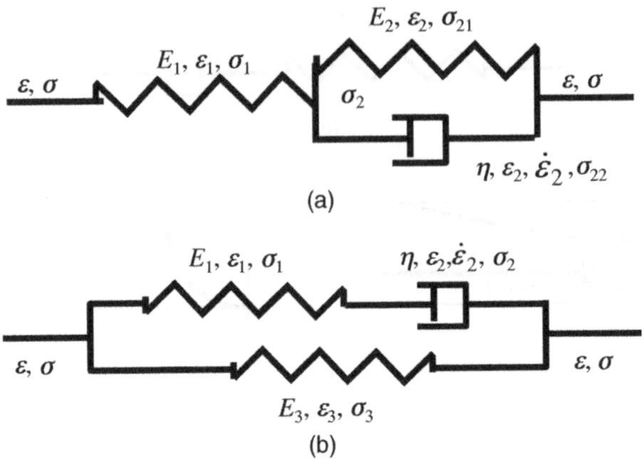

FIGURE 9.3 Generalized Maxwell models: (a) a Kelvin model with a spring; (b) a Maxwell model with a spring.

9.1.2 A General Three-dimensional Viscoelasticity Model

The discussion in Section 9.1.1 can be extended to the three-dimensional (3D) situations for the general applications.

The viscous behavior is described by the Boltzmann's superposition principal in terms of the deviatoric components:

$$S_{ij} = \int_0^t 2G(t-\tau)\dot{\varepsilon}_{ij}(\tau)\mathrm{d}\tau, \tag{9.8}$$

where $G(t)$ is the relaxation modulus. A commonly used model is given below:

$$G(t) = G_\infty + (G_0 - G_\infty)\mathrm{e}^{-\beta t}. \tag{9.9}$$

Here, G_∞ is the long-time shear modulus, G_0 is the short-time shear modulus, and β is the decay factor. For more general but complex situations, we can adopt a fourth-order tensor form $g_{ijkl}(t)$ to replace the scalar function $G(t)$.

The volumetric behavior is represented by the usual elastic model with a bulk modulus K:

$$p = -K\varepsilon_v. \tag{9.10}$$

It can also contain the viscous property, such as the expression of Ferry (1980),

$$-p = \int_0^t K(t-\tau)\dot{\varepsilon}_{jj}(\tau)\mathrm{d}\tau. \tag{9.11}$$

In some applications, Prony series are used for the relaxation functions:

$$\begin{aligned} g(t) &= \alpha_0 + \sum \alpha_m e^{-\beta_m t}, \\ k(t) &= \sum K_m e^{-\beta_{K_m} t}. \end{aligned} \tag{9.12}$$

The convolution integral (9.8) can be calculated approximately:

$$\begin{aligned} S_{ij} &= \int_0^t 2G(t-\tau)\dot{\varepsilon}_{ij}(\tau)\mathrm{d}\tau \\ &= 2(G_\infty + (G_0 - G_\infty)e^{-\beta t+\beta\tau})\varepsilon_{ij}(\tau)\big|_0^t - 2\beta e^{-\beta t}\int_0^t (G_0 - G_\infty)e^{\beta\tau}\varepsilon_{ij}(\tau)\mathrm{d}\tau. \end{aligned}$$

The computation in the explicit framework can be done with $\varepsilon_{ij}(t_n) = \varepsilon_{ij}(t_{n-1}) + \dot{\varepsilon}_{ij}\Delta t$.

9.2 POLYMER AND ENGINEERING PLASTICS

Polymer is a type of widely used nonmetal materials with many varieties. Applications of polymer materials have been growing rapidly, credit to the development of chemical engineering. The volume production of polymer reached the same level of steels in 1980s. Polymer is basically formed in a fashion of long-chain molecules, which usually consists of certain substructure(s) periodically linked together by chemical bonds. The polymer is commonly produced in a controlled thermal–chemical process. Its mechanical properties are dependent of the temperature, manufacturing process, and other related factors.

9.2.1 Fundamental Mechanical Properties of Polymer Materials

Generally, polymer material changes its properties with the change of environmental temperature due to its molecular motion, and presents different physical states. When temperature is below the glass point T_g, the polymer exhibits a glass status with moderate Young's modulus. The deformation is recoverable after load is removed. As temperature increases, the molecular activity also increases and the Young's modulus decreases. These happen simultaneously until the temperature reaches the viscous flow point T_f. When loaded below temperature T_f, large deformation may occur but is still recoverable. With further increasing temperature beyond T_f, the polymer will be at a viscous flow state. Large deformation can easily happen and is unrecoverable. The molecular chain can break when reaching the separation point T_d. At room temperature, various types of polymers can possess very different thermal states. This is true even in the case that the polymer has the same chemical constituent but is formed by different process. Hence, they exhibit various types of mechanical behaviors.

For uniaxial tension/compression test, most of the polymer materials behave elastic–plastically. They may sustain quite large plastic deformation before rupture.

This is what the word "plastics" stands for. Departing from the metal materials, the polymer can have a softening deformation after the yield point. Then, the stress can go higher than the first yield point and reach the second yield point. An explanation is then derived from the chemical structure. The long molecular chains are usually twisted and rolled in a complex manner. The long chains start to stretch out and result in a softening behavior when being loaded beyond the yield point of the integrated structure. The softening process ends when the stretching reaches certain level and the strength increases. Increasing load capacity is observed. The chains are straightened and reach the second yield point. The temperature effect is another explanation. Large plastic deformation generates heat energy. The loading process is not considered isothermal but better described as adiabatic due to the slow process of heat transfer in polymer materials. The rising temperature changes the material property to exhibit softening behavior. It is worth noting that the softening is observed on the basis of true stress and true strain, but not the engineering stress. It has been argued for metal materials, in which during the tensile test, the cross-sectional area will shrink and the true stress is higher than the engineering (nominal) stress. As reported in Mulliken and Boyce (2006), softening is observed even in compression test with lateral expansion.

Also, creep and stress relaxation are observed easily with polymer materials partially or mainly due to their long molecular chain structure. This kind of behavior also contributes to the flexibility of the polymers.

The mechanical properties depend partially on the molding and forming process, which is thermal–chemical process, as reported in Viana et al. (2004). The process results in the formation of the specific intrinsic macromolecular structure and its spatial organization, which affect the mechanical properties. One example is the injection molding creating a layer of skin in the polymer products. The skin usually has higher modulus, higher yield stress, and lower ductility than the core has. For more information and general discussions, we refer to Ward and Sweeney (2004).

9.2.2 A Temperature, Strain Rate, and Pressure Dependent Constitutive Relation

To study the constitutive relations of polymer materials, Mulliken and Boyce (2006) conducted comprehensive tests on polycarbonate (PC) and poly(methyl methacrylate) (PMMA). The TA Instruments Q800 Dynamic Mechanical Analyzer (DMA) was used to find the rate dependent material property at a given temperature ranging from $-140°C$ to $180°C$. The oscillator with a specified frequency created an environment with a given strain rate. Uniaxial compression test was performed using Instron servo-hydraulic test machine for strain rate ranging from $10^{-4}/s$ to $1/s$ and the split Hopkinson pressure bar for high strain rate. It has been observed that the material properties depend on the temperature and the strain rate. The rate-dependence was discussed in Section 7.1.6 for plasticity. Several empirical formulas were described with multiplier or additive component of strain rate. Many of these contain a linear form of $\log \dot{\varepsilon}$. It was found in Mulliken and Boyce (2006) that the relation between the yield stress of polymer and $\log \dot{\varepsilon}$ was not simply linear, but better characterized by a bilinear form.

The test of PMMA material with strain rate below 1,400/s exhibited plastic deformation before rupture. With rate higher than 1,400/s, it presented a brittle failure mode without evident plastic deformation. On the other hand, for PC material, plastic deformation occurred before rupture even with strain rate as high as 5,050/s.

It was observed that the storage modulus and loss modulus were affected by the α-transition and β-transition, which are two temperature characteristics; see Meyers and Chawla (1999) for the concepts and definitions. Mulliken and Boyce (2006) proposed a constitutive relation and postulated a decomposition of contributions characterized by α- and β-transition. The material model contained two material characteristics in parallel combination. Part A consisted of two Maxwell type of elastic–viscoplastic elements to represent the intermolecular contributions with α- and β-components, respectively. The elasticity constants, for example, shear modulii μ_δ and bulk modulii K_δ, are both temperature dependent. Part B, as nonlinear hardening element, accounted for entropic resistance of molecular alignment.

This constitutive relation represents the dependence of yield on strain rate, temperature, and pressure. The softening behavior is also included. It contains 16 parameters, among them 7 are for each of the α- and β-components. It is an extension of Ree and Eyring's (1955) model with α-component, and Arruda and Boyce's (1993) model with dependence on strain rate, temperature, and pressure derived under low strain rate.

9.2.3 A Nonlinear Viscoelastic Model of Polymer Materials

Kobayashi and Wang (2001) reported the results of a series of tests on several polymer materials such as epoxy resin, PMMA, and PC, with a wide range of strain rates and strain up to 7–8%. It was observed that for a given range of stain from ε_1 to ε_2, the stress increment $\Delta\sigma = \sigma_2 - \sigma_1$ was basically independent of strain rate. For a constant strain increment $\Delta\varepsilon$, however, the tests presented nonconstant $\Delta\sigma$ which decreased with increasing strain level; hence, there is a decreasing modulus.

Similar to the observation of Mulliken and Boyce (2006), it was found that the relation between σ and $\log \dot{\varepsilon}$ was basically bilinear with a sudden increase at $\dot{\varepsilon}$ near 100/s. This indicated two dominant relaxation times, corresponding to low and high strain rate, respectively. A nonlinear viscoelastic model, named ZWT (Zhu–Wang–Tang) material was then proposed.

In ZWT model, a nonlinear spring and two Maxwell elements were connected in a parallel fashion to represent the nonlinear elastic and viscoelastic property. The two Maxwell elements corresponded to the bilinear relation between σ and $\log \dot{\varepsilon}$. The constitutive relation was written as

$$\begin{aligned} \sigma &= f_e(\varepsilon) + E_1 \int_0^t e^{E_1(\tau - t)/\eta_1} \dot{\varepsilon}(\tau) d\tau + E_2 \int_0^t e^{E_2(\tau - t)/\eta_2} \dot{\varepsilon}(\tau) d\tau, \\ f_e(\varepsilon) &= E_0 \varepsilon + \alpha \varepsilon^2 + \beta \varepsilon^3. \end{aligned} \tag{9.13}$$

This can be viewed as a special case of Green–Rivlin (1957) model in a multi-integral form. The nonlinearity comes from the nonlinear spring, which is rate independent. The model is an extension from linear rate dependent viscoelastic model, and viewed as rate independent or weak nonlinear.

The test data for epoxy, PC, and several PMMA materials showed that the relaxation time $\theta_1 = \eta_1/E_1$ was of order of 10–100 s and $\theta_2 = \eta_2/E_2$ was of the order of 10^{-4} to 10^{-6} s. Therefore, the first Maxwell element represented low strain rate behavior when the second Maxwell element quickly relaxed. The second Maxwell element represented high rate behavior. Under high rate loading, the first element did not have enough time to relax and still contributed to the material strength. The method described the mechanical property for thermosetting plastics (cross-linked polymers) very well. An additional viscous term $\eta\dot{\varepsilon}$ was also suggested for the thermoplastic polymers.

9.3 RUBBER

Rubber is widely used in various industries. It can be made from the natural rubber trees or by chemical synthesis. There are many varieties of rubber materials. Most of them can experience very large recoverable deformation when subjected to loading. Its Young's modulus is around 1 MPa, which is relatively lower than that of the other solid materials. On the other hand, rubber material's viscosity is also important for applications in reducing and absorbing vibration. In addition, rubber material is often considered incompressible with Poisson ratio close to 0.5. Numerically, a constraint condition $\boldsymbol{J} = 1$ can be used with Lagrange multiplier method or penalty method, where $J = \det \boldsymbol{F}$ is the determinant of the deformation gradient, representing volume ratio. In this chapter, we introduce several material models for large elastic deformation, which are generally available in the commercial software.

9.3.1 Mooney–Rivlin Model of Rubber Material

When Poisson ratio $\nu = 0.5$, the material's bulk modulus $K = \infty$, therefore is incompressible. For nearly incompressible rubber (ν near 0.5), Mooney (1940) and Rivlin (1948) developed a hyperelastic material model, using the strain invariants instead of the usual elasticity parameters. A strain energy density function was introduced for the hyperelastic behavior:

$$W = A(I_1 - 3) + B(I_2 - 3) + C\left(I_3^{-2} - 1\right) + D(I_3 - 1)^2, \tag{9.14}$$

where A and B were material parameters. $I_j, j = 1, 2, 3$, were the three invariants of the right Cauchy–Green strain tensor $\boldsymbol{C} = \boldsymbol{F}^T\boldsymbol{F}$. Expressed in terms of the principal stretches λ_j,

$$\begin{aligned} I_1 &= \lambda_1^2 + \lambda_2^2 + \lambda_3^2, \\ I_2 &= \lambda_1^2\lambda_2^2 + \lambda_2^2\lambda_3^2 + \lambda_3^2\lambda_1^2, \\ I_3 &= \det C = J^2 = \lambda_1^2\lambda_2^2\lambda_3^2. \end{aligned} \tag{9.15}$$

According to Ogden (1984), the principal stresses are given by

$$J\sigma_i = \lambda_i \partial W/\partial \lambda_i, \quad i = 1, 2, 3. \tag{9.16}$$

For example,

$$J\sigma_1 = 2\lambda_1^2 \left(\frac{\partial W}{\partial I_1} + \frac{\partial W}{\partial I_2} (\lambda_2^2 + \lambda_3^2) + \frac{\partial W}{\partial I_3} \lambda_2^2 \lambda_3^2 \right).$$

For uniform compression test,

$$\begin{aligned} &\lambda_1 = \lambda_2 = \lambda_3 = \lambda, \\ &\sigma_1 = \sigma_2 = \sigma_3 = -p = \sigma_{jj}/3, \\ &I_1 = 3\lambda^2,\ I_2 = 3\lambda^4,\ I_3 = \lambda^6,\ J = \lambda^3. \end{aligned} \tag{9.17}$$

We have

$$\sigma_i = -p = \frac{2}{\lambda} \left(\frac{\partial W}{\partial I_1} + 2\lambda^2 \frac{\partial W}{\partial I_2} + \lambda^4 \frac{\partial W}{\partial I_3} \right).$$

The bulk modulus is defined as

$$K = \frac{2(1+\nu)G}{3(1-2\nu)} = -\frac{p}{\varepsilon_v} = \frac{-p}{J-1}. \tag{9.18}$$

From (9.14) to (9.17), we have

$$\begin{aligned} &\partial W/\partial I_1 = A, \\ &\partial W/\partial I_2 = B, \\ &\partial W/\partial I_3 = -2C/I_3^3 + 2D(I_3 - 1) = -2C\lambda^{-18} + 2D(\lambda^6 - 1). \end{aligned}$$

For nearly incompressible situation, $\lambda - 1 = o(1)$. The pressure is finite, which leads to

$$C = 0.5A + B. \tag{9.19}$$

Furthermore,

$$K = \frac{2(1+\nu)G}{3(1-2\nu)} = \frac{-p}{J-1} = \frac{2}{\lambda^4 - \lambda}(2B(\lambda^2 - 1) - 2C(\lambda^{-14} - 1) + 2D\lambda^4(\lambda^6 - 1)).$$

For nearly incompressible situation, $K = \lim_{\lambda \to 1} K = \frac{2}{3}(14A + 32B + 12D)$.

For small pure shear test, for example, $\lambda_1 = 1 + \varepsilon$, $\lambda_2 = 1 - \varepsilon$, $\lambda_3 = 1$, we have $J = 1 - \varepsilon^2$, $I_1 = 3 + 2\varepsilon^2$, $I_2 = 3 + \varepsilon^4$, and $I_3 = (1 - \varepsilon^2)^2$. Using shear stress $\tau_{12} = \sigma_1 - \sigma_2$ and shear strain $\varepsilon_{12} = \varepsilon_1 - \varepsilon_2 = \lambda_1 - \lambda_2$, we can obtain shear modulus $G = 2(A + B)$. The calculation of bulk modulus K then leads to

$$K = \frac{2(1+\nu)G}{3(1-2\nu)} = \frac{4(1+\nu)(A+B)}{3(1-2\nu)},$$
$$D = \frac{A(5\nu - 2) + B(11\nu - 5)}{2(1-2\nu)}. \tag{9.20}$$

For software implementation, A and B are material parameters inputted by the user along with Poisson ratio ν, which is close to 0.5. In each time step, the deformation gradient $\boldsymbol{F}$, the right Cauchy-Green strain tensor $\boldsymbol{C}$, and the principle stretches λ_i all need to be calculated.

9.3.2 Blatz–Ko Model

A hyperelastic model was proposed by Blatz and Ko (1962) for rubber material. The strain energy density function was defined below with the invariants used in Section 9.3.1,

$$W = \frac{\mu}{2}\left(I_1 + \frac{I_3^{-\alpha} - 1}{\alpha} - 3\right) + \frac{\mu(1-\beta)}{2}\left(\frac{I_2}{I_3} + \frac{I_3^{\alpha} - 1}{\alpha} - 3\right),$$
$$\alpha = \frac{1}{1-2\nu}. \tag{9.21}$$

A simplified version used $\beta = 1$, which resulted in

$$W = \frac{\mu}{2}\left(I_1 + \frac{I_3^{-\alpha} - 1}{\alpha} - 3\right). \tag{9.22}$$

The second Piola–Kirchhoff stress was calculated from the strain energy density function. The Cauchy stress was obtained by transformation: $\sigma = J^{-1}\mathbf{F}\tilde{\sigma}\mathbf{F}^T$,

$$\tilde{\sigma}_{ij} = 2\frac{\partial W}{\partial C_{ij}} = \mu\left(\delta_{ij} - J^{-\frac{1}{1-2\nu}} J C_{ij}^{-1}\right),$$
$$\sigma_{ij} = \frac{\mu}{J}\left(\frac{\partial x_i}{\partial X_k}\frac{\partial x_j}{\partial X_k} - J^{-\frac{1}{1-2\nu}} J \delta_{ij}\right). \tag{9.23}$$

9.3.3 Ogden Model

Another commonly used rubber material model was introduced by Ogden (1972). The strain energy density function was organized in a general form plus a penalty term $f(J)$ for incompressibility condition $J = 1$, for example, $f(J) = K(J - 1)^2/2$, $f(J) = K(J - 1 - \ln J)$, with a large bulk modulus K,

$$W = \sum_{k=1}^{3}\sum_{j=1}^{n} \frac{\mu_j}{\alpha_j}((\lambda_k^*)^{\alpha_j} - 1) + f(J), \tag{9.24}$$

where $\lambda_k^* = \lambda_k J^{-1/3}$. With $J = \lambda_1\lambda_2\lambda_3$, we have $\partial J/\partial\lambda_k = J/\lambda_k$ and

$$\partial\lambda_k^*/\partial\lambda_j = \delta_{kj}J^{-1/3} - \lambda_k J^{-1/3}/3\lambda_j = (\delta_{kj}\lambda_j^* - \lambda_k^*/3)/\lambda_j.$$

In view of (9.16), we derive the principle Cauchy stress

$$\begin{aligned}\sigma_i &= \frac{\lambda_i}{J}\frac{\partial W}{\partial \lambda_i} \\ &= \frac{\lambda_i}{J}\left(\sum_{k=1}^{3}\sum_{j=1}^{n} \mu_j(\lambda_k^*)^{\alpha_j-1}\frac{1}{\lambda_i}(\delta_{ki}\lambda_i^* - \lambda_k^*/3) + f'(J)\frac{J}{\lambda_j}\right) \\ &= \sum_{j=1}^{n} \frac{\mu_j}{J}\left((\lambda_i^*)^{\alpha_j} - \frac{1}{3}\sum_{k=1}^{3}(\lambda_k^*)^{\alpha_j}\right) + f'(J).\end{aligned} \tag{9.25}$$

This model in fact is a general formulation, which includes Mooney–Rivlin model as a special case by choosing certain terms and parameters. For more discussions on the constitutive models of rubber materials, we refer to Ogden (1984), and Boyce and Arruda (2000).

9.4 FOAM

Besides the metal materials, there are other materials used for energy absorption with large deformations in impact engineering. Polyurethane foam is one of those. In fact, foam is a kind of porous materials, usually made of polymer materials such as polyurethane or polypropylene. There are also metal foams. Here, we mainly discuss the polymer foams. In the blowing and molding process, the polymer forms thin-walled cells (closed or open), which contain air. The solid polymer occupies small fractional volume. Hence, the bulk density is very low. The foam can sustain very large compressive deformation without losing load capacity. Its tensile strength is relatively low. The high compressive load capacity and the low density turn the foam into a useful lightweight material with good capacity of energy absorption per unit

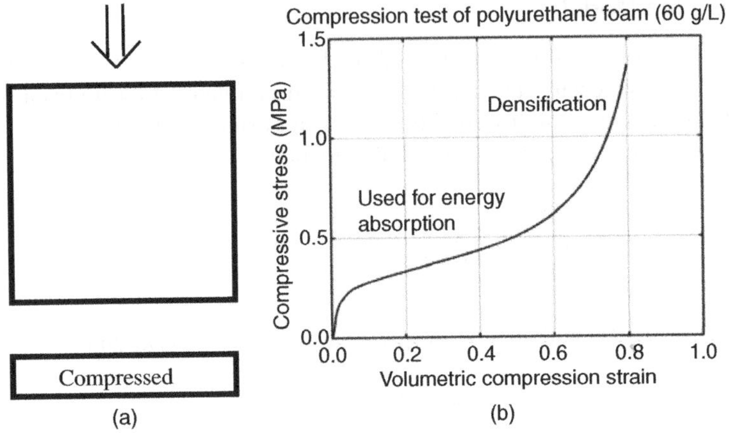

FIGURE 9.4 Uniaxial compression test of foam material: (a) compression test of foam; (b) force–deformation curve of foam under compression test.

mass. Foam is often used in the form of blocks and modeled by solid elements. We can consider foam as a pseudo isotropic continuum.

The uniaxial compression test for a cubic specimen of foam material, with a side length of 100 mm, is depicted in Figure 9.4a. The force–deformation curve shown in Figure 9.4b presents a highly nonlinear behavior, very different from the plasticity of metals and the polymers. The foam is full of air bubbles. After yielding, the block specimen of foam undergoes very large compressive deformation with some hardening. The compression can easily reach 70% or more of the original dimension. The material is then densified with fast growing stiffness. Meanwhile the nominal density of the block material increases with reduced volume. In fact, the air bubbles contained in the specimen are gradually compressed and possibly partially squeezed out. The cells of polymer material are gradually collapsed and compressed significantly to provide much stronger strength, with higher volume fraction. Such deformation process is named densification, which is a characteristic behavior of foam. This is a response of the porous material as a whole, but not of the base material. Note that the hardening part in the commonly used metal materials usually has decreasing tangent modulus. Another interesting behavior of the foam is that the effective Poisson ratio is almost zero. The specimen shows almost no lateral deformation with such large compression, as illustrated in Figure 9.4a.

The regular plasticity theory assumes incompressibility in plastic deformation with a convex yield surface. Such type of theory is not suitable for describing foam's large capacity of compressibility and hardening. We notice that the foam as a physical body is not a continuum. Instead it is a type of porous materials. It does not necessarily follow the mechanical rule developed for metals as continuum. Due to the inclusion of air bubbles, the constitutive equation for the foam material could be quite complex and remains to be a challenging task.

9.4.1 A Cap Model Combining Volumetric Plasticity and Pressure Dependent Deviatoric Plasticity

As to porosity, the constitutive model developed by Krieg (1972) for soil and concrete seems to be a candidate for foam. The model combined the volumetric behavior and the pressure dependent deviatoric behavior. A yield function was introduced as a combination of volumetric and deviatoric effects:

$$F_V = p - f(\gamma),$$
$$F_S = J_2 - (a_0 + a_1 p + a_2 p^2), \tag{9.26}$$

where $p = -\sigma_{kk}/3$ was the pressure (positive for compression), γ was the volumetric strain $(V - V_0)/V_0$, f was a function representing the basic compressive behavior, and J_2 was the second invariant of the deviatoric stress as used in the classical plasticity. The coefficients a_j were considered material constants to be determined by fitting the test data. A cutoff of tensile stress was also introduced in the model to simulate the low tensile strength.

In computation, the pressure was determined from the first equation of (9.26) by the calculated volumetric strain. The second equation of (9.26) then played the role of a yield function with the calculated pressure. This model, named cap model, is available in some commercial explicit software.

Pressure can be developed under uniaxial compression. With this model, however, the pressure activates negative stress in the other directions and causes lateral expansions, which fails to correlate to the tests.

9.4.2 A Model Consisting of Polymer Skeleton and Air

To investigate foam mechanics, Neilsen et al. (1987) conducted a series of tests at Sandia National Laboratories, focusing on hydrostatic compression and triaxial compression. In the triaxial compression test, an additional uniaxial compression was exercised following the hydrostatic compression. The tests used a closed device and sealed cylindrical specimen with no air leaking from the foam. The results showed that the mean stress at volumetric yield was independent of the deviatoric stress. In addition, under certain level of hydrostatic pressure, the subsequent triaxial test showed a second yield point with a second piece of hardening stress–strain curve, illustrated in Figure 9.5. On the other hand, the unsealed specimen under uniaxial compression did not have lateral expansion, that is, the Poisson ratio of the bulk material seemed to be zero. Hence, the deformation in one principal direction did not cause stress in the other principal directions.

The foam has higher porosity than soil or concrete. For instance, the polyurethane-closed cell foam, used as impact energy absorber, is full of air bubbles trapped in the thin-walled polymer cells. The volume fraction of the solid polyurethane material is only a small percentage. The bulk density is very low.

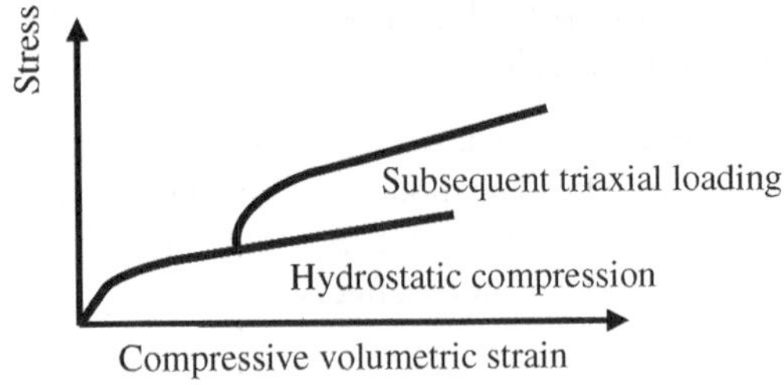

FIGURE 9.5 A second hardening behavior of foam material under complex 3D loading.

The air inside the cell can support certain compressive load as long as the cell wall does not break and air does not escape substantially. The pressure contributes to the load capacity of foam. Based on the observations from the tests, Neilsen et al. (1987) proposed a phenomenological constitutive equation to take into account the contribution by the compressed air, briefly discussed below. The foam consists of two parts, the cell and the air. The polymer structure is named skeleton, forming the thin-walled cells with air inside. The cells are assumed to be evenly distributed and the polymer material is homogeneous. The foam response is decomposed into the skeleton response and the air response. The nominal foam stress is assumed to be

$$\sigma_{ij} = \sigma_{ij}^{sk} + \sigma^{\text{air}}\delta_{ij}. \tag{9.27}$$

The ideal gas model is assumed for the air response:

$$pV/T \equiv \mu R. \tag{9.28}$$

When the foam is compressed from initial volume V_0 to V_1, the nominal volumetric strain is given by

$$\gamma = (V_1 - V_0)/V_0. \tag{9.29}$$

The volume consists of two parts, the polymer with volume fraction η and the air. During deformation of the foam, the polymer volume is assumed to be constant, but the air is compressed from $\tilde{V}_0 = V_0(1-\eta) = V_0 - \eta V_0$ to $\tilde{V}_1 = V_1 - \eta V_0$. From (9.28) to (9.29), it results in

$$p_1 = p_0 \frac{1-\eta}{\gamma + 1 - \eta} \frac{T_1}{T_0}, \tag{9.30}$$

where p_0 is the initial air pressure, assumed to be equal to the ambient pressure. Hence,

$$\sigma^{\text{air}} = -p_1 + p_0 = -p_0 \frac{\gamma + (1-\eta)(1 - T_1/\mathrm{T}_0)}{\gamma + 1 - \eta}. \tag{9.31}$$

The next step is to define the yield function. According to the observation from test that the lateral strains are zeroes under uniaxial loading, the skeleton response in one principal stress direction is not affected by the other principal skeleton stresses. To represent hydrostatic behavior, the yield function is then defined for the principal stress, component wisely:

$$F = A\,\text{Heaviside}(II') + B(1 + C\gamma), \tag{9.32}$$

where *II'* is the second invariant of deviatoric strain. Constant *B* is the yield stress of the skeleton under purely hydrostatic loading. *B***C* represents the volumetric response after yielding due to purely hydrostatic loading. Constant *A* with the Heaviside function represents the coupling effect and is equal to the difference between the axial yield stress under hydrostatic loading and the axial yield stress under deviatoric loading. The constitutive relation is basically isotropic.

Based on the test data for a set of foams with various densities and volume fractions, Neilson et al. (1987) obtained empirical formulas which were dependent of the volume fraction. The parameters and the Young's modulus for the linear elastic behavior were derived:

$$\begin{aligned} A &= 3440\eta^{1.676}(psi), \\ B &= 2780\eta^{1.645}(psi), \\ C &= 2.21 - 21.1\eta, \\ E &= 45400\eta^{2.20}(psi). \end{aligned} \tag{9.33}$$

This model was implemented in PRONTO, a transient dynamics software developed in Sandia National Laboratories, also available in some commercial software. For more discussion about this model, see Neilsen et al. (1989).

9.4.3 A Phenomenological Uniaxial Model

The foam material models described in Sections 9.4.1 and 9.4.2 involve pressure and other parameters, which need to be determined by well-conducted laboratory test. Only simple test facility, however, is available in many industrial applications. On the other hand, the pressure built up from uniaxial compression takes action in all directions and causes lateral expansion, which is not observed in the experiments. This effect is considered for the yield function of the cap model described in Section 9.4.2, but not the stress. For certain class of applications, the foam is mainly loaded in one direction and the hydrostatic behavior is not critically important. All of these have motivated a simple uniaxial model without the pressure. The yield function is assigned to the principal directions:

$$\begin{aligned} \sigma_j &= E\varepsilon_j \quad \text{if } \sigma_j < \sigma_y, \\ \sigma_j &= f(\varepsilon_j) \quad \text{if } \sigma_j \geq \sigma_y, \end{aligned} \qquad j = 1, 2, 3. \tag{9.34}$$

Here, σ_y is the initial yield stress. Function f can use tabulated test data as well as the traditional empirical functional form with several parameters needed to be determined. The yield condition not only acts individually but also acts the same for all the principal directions due to isotropy. Under uniaxial loading, for example, $\varepsilon_1 \neq 0$ and $\varepsilon_2 = \varepsilon_3 = 0$, there is no lateral action at all. Equation (9.34) can also be used with element local $x/y/z$ directions defined by node connectivity if the mesh is quite regular.

Improvement with the effect of interaction with two or three directions has been implemented in LS-DYNA. This is based on the concept of stress objectivity when the area change due to compression in other directions is considered. The model as modified from (9.34) was presented by Chou et al. (1994):

$$\begin{aligned} \hat{\sigma}_i &= f(\varepsilon_i), \\ \sigma_i &= \hat{\sigma}_i / \lambda_j \lambda_k, \quad j \neq k \neq i, \end{aligned} \tag{9.35}$$

where λ_j, $j = 1, 2, 3$, were the principal stretches calculated from the left stretch tensor V_{ij}

$$\det(V_{ij} - \lambda \delta_{ij}) = 0. \tag{9.36}$$

Theory about the stretch tensor can be found in textbooks of continuum mechanics, for example, Truesdell and Toupin (1960).

Some of the foam elements are not under uniaxial loading in real applications. While some of the others do not have the loading constantly aligned with element local system directions. A set of nonstandard specimens with various geometries using the usual compression test has been proposed for evaluating the robustness of the simple material model. As reported in Chang (1995), a cubic block with side length of 76.2 mm was used as the standard test specimen. Other specimens in the shapes of triangle, trapezoid, inverted U with a rectangular groove, etc., were cut from the cube. The cube was graded with $6 \times 6 \times 6$ brick elements. Different meshes were generated for triangular and trapezoidal specimens, using wedges, bricks, and parallelepipeds. Material property obtained from the uniaxial quasistatic test with the standard cubic specimen was used as input data to simulate tests with these specimens. The numerical results agreed with the test data fairly well in general. Simulation of dynamic test with some of these specimens (not necessarily with the same dimensions) has also been conducted, cf. Chou et al. (1994). The results were reported to concur with the test reasonably.

9.4.4 Hysteresis Behavior

Study found that some foam belongs to a kind of highly hysteretic material, which can go back to its original shape quickly when unloaded. But other foam takes long time to recover, or even takes too much time to return to its original shape upon unloading. Unlike the traditional plasticity, the unloading of foam material after yielding is not

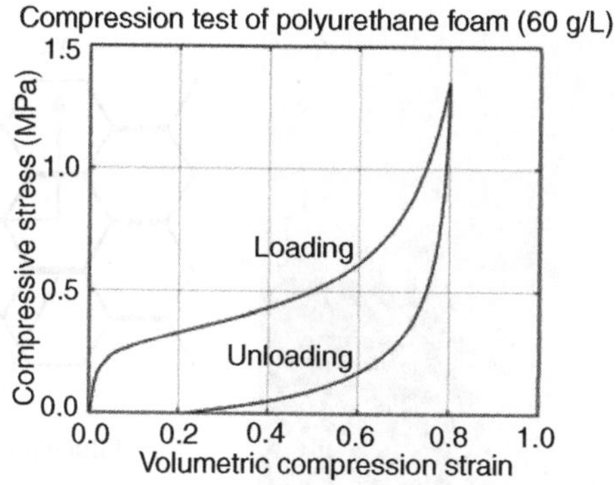

FIGURE 9.6 Loading and unloading of foam under uniaxial compression test.

a linear elastic behavior, as depicted in Figure 9.6. In addition, as presented in Chang et al. (1994), the foam might lose 50%–80% of its energy absorption capability when reloading after unloading. An enhanced uniaxial constitutive model was proposed, modified from (9.35), allowing unloading to follow another function:

$$\hat{\sigma}_i = g(\alpha_j, \varepsilon_i) f(\varepsilon_i), \tag{9.37}$$

where $g(\alpha_j, \varepsilon_i)$ was a scaling function with state variables α_j. As reported in Chang et al. (1994), when using two parameters, α_1 for hysteresis unloading factor and α_2 for unloading shape factor, respectively, the unloading behavior could be modeled with sufficient accuracy.

It deserves further study to derive function g in a systematic way.

9.4.5 Dynamic Behavior

It is found that the foam material exhibits strong sensitivity on strain rate. Under dynamic loading, the foam has higher resistance and absorbs more energy than the case it is under static test. Strain rate effect is an important feature of foam, like the case of steels. The constitutive relation can be extended to include the strain rate effect similar to the case discussed for the metal plasticity. For example, a comprehensive constitutive model with dependence on strain rate was developed by Chang (1995).

9.5 HONEYCOMB

Aluminum honeycomb is another type of material often used as energy absorber in impact engineering. Like foam material, the honeycomb is also mainly used in the form of blocks. On the other hand, the honeycomb is anisotropic. The honeycomb block is much stronger in the hexagonal wall direction than in the lateral directions. In

(a)

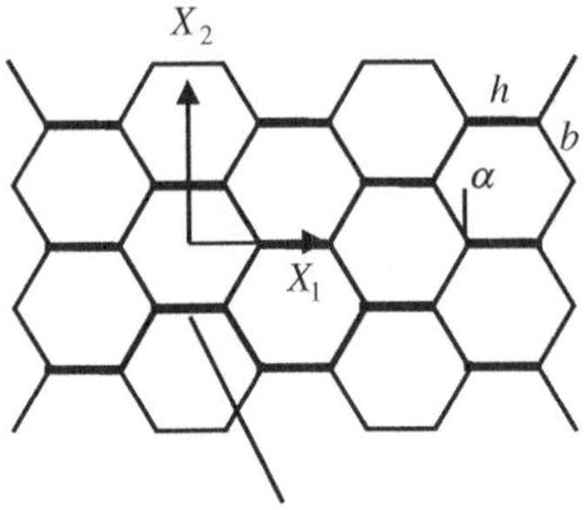

(b)

FIGURE 9.7 Honeycomb: (a) honeycomb specimen; (b) local system for honeycomb.

many cases, the honeycomb block in engineering applications is placed in a position aligned with the loading direction.

9.5.1 Structure of Hexagonal Honeycomb

Hexagonal honeycomb made of aluminum foils has important applications. The honeycomb is generally made by gluing the thin foils together, layer-by-layer, and folding in a periodic fashion to form the hexagonal-celled structure, depicted in Figure 9.7a. The cells form a thin-walled structure. The thickness is small comparing to the cell size, usually ranging from 1% to 10%. The thickness of the foil becomes the wall thickness of the honeycomb. However, two sides of the hexagons have double thickness due to the glued structure.

Usually a Cartesian coordinate system is defined for the honeycomb. For example, as shown in Figure 9.7b, x_3 is in the wall direction, or named as the axial direction or the out of plane direction. In the cross-section plane, x_1 is along the glue line, that is, the wall with double thickness. x_2 is normal to x_1. The honeycomb is basically orthotropic. When loaded in x_1 or x_2 direction, the foils are mainly subjected to bending. However, the foils are subjected to membrane compression or tension when loaded in x_3 direction. The x_3-direction is stronger than the other directions. The out-of-plane shear is also stronger than the in-plane shear.

9.5.2 Critical Buckling Load

A typical deformation mode under x_3-compression with a crushed test specimen is presented in Figure 9.8. The walls of foil cell experience a progressive collapse mode, which has been studied by several researchers, cf. McFarland (1963). The macroscopic displacement consists of the progressive folds developed in the microstructure. Such folding structure trigged by the first fold as initial buckling is typical of thin-walled structure. Wierzbicki (1983) developed a folding mechanism using

FIGURE 9.8 Progressive collapse of honeycomb under compression in X_3-direction.

rigid-plasticity model to study the crushing strength of honeycomb. An empirical formula for nominal collapse stress was proposed:

$$\sigma_3^* \approx 6.6\sigma_y(t/b)^{5/3}. \tag{9.38}$$

Gibson and Ashby (1988) analyzed the linear elastic properties comprehensively. Zhang and Ashby (1992a, 1992b) stated that the elastic buckling load was essential, due to the small thickness comparing to the cell size. The cell walls were treated as long panels. The upper (or lower respectively) bound of the buckling load was proposed with the assumption of clamped (or simply supported respectively) neighboring cell walls:

$$\begin{aligned} \sigma_{3-\text{upper}}^* &\approx 6E(\bar{\rho}/\rho)^3, \\ \sigma_{3-\text{lower}}^* &\approx 3.8E(\bar{\rho}/\rho)^3, \end{aligned} \tag{9.39}$$

where $\bar{\rho}$ and ρ were the nominal density and the density of the wall material, respectively.

It is observed from the uniaxial compression test, that the compressive force (or the nominal stress) remains in an oscillatory band with nearly constant mean value after the initial buckling, depicted in Figure 9.9; see more data in Gibson and Ashby (1988) and Hinnerichs et al. (2006). Therefore, the above analysis of critical values of the nominal stress serves the purpose of describing an important part of the mechanical behavior of honeycomb.

Zhang and Ashby (1992a, 1992b) also analyzed the buckling phenomena under transverse shear load σ_{31} and σ_{32}, and the in-plane biaxial load.

9.5.3 A Phenomenological Material Model of Honeycomb

As discussed in Section 9.5.2, when the honeycomb is loaded in the wall direction, the thin foils will buckle and the honeycomb block will deform in a progressive collapse mode shown in Figure 9.8. The foils form a series of folds with a wavy buckling mode. The stress–strain diagram from a uniaxial compression test of a block

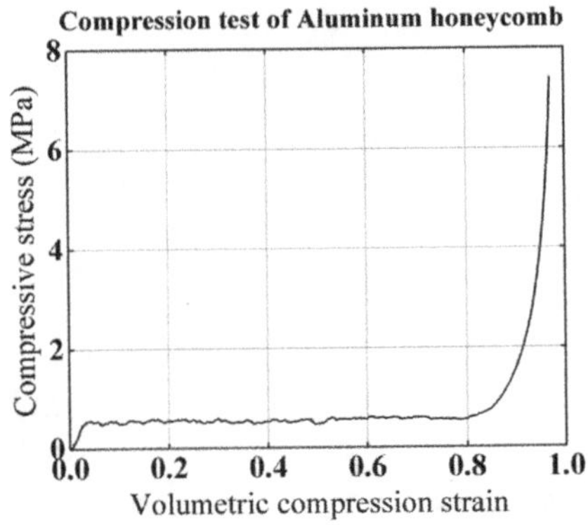

FIGURE 9.9 Response of honeycomb in X_3—compression test.

specimen, shown in Figure 9.9, presents an almost flat plateau. The honeycomb block is empty inside the hexagonal cells and does not have material densification from the compression of the open cells. The plateau can extend to withstand compression more than 80%. The progressive collapse mode ends with stack-up of the foil material, resembling the densification of foam. The honeycomb behaves like a solid block at this moment.

Observed from the laboratory tests, the approach of buckling analysis might provide some fundamental mechanical property of honeycomb. The estimation becomes critical for the nearly constant plateau regarding representing the honeycomb material accurately. Similar to the phenomenological material model of foam, a simple model of honeycomb can also be developed for impact engineering analysis. In any lateral direction, the honeycomb has a periodic-layered structure. Under compression load, it can collapse layer by layer, which is also a progressive collapse mode. The compressive force–deflection curve also shows a plateau. The overall strength is much weaker than that in the wall direction. Note that the honeycomb is most likely used under compression condition. Its tensile behavior is basically elastic; however, its failure behavior is by no means a simple characteristic. Due to the nature of anisotropy, the force–deflection curves from tests in the wall and lateral directions can be used as stress–strain relations independently. In addition, the shear stress–strain relation can be obtained and implemented in the constitutive equations similarly. In this case, six-component wise stress–strain relations are considered independent properties.

One of the main applications of the honeycomb material model is to simulate the impactor. Although the honeycomb behavior is very complex in the applications under complex loading, the simple phenomenological constitutive equation provides a valuable modeling approach.

Another type of usages is for the impact test facility. Honeycomb can be used as the energy absorber to control the impact condition when excessive impact energy is expected. The impactor can be brought to rest by using honeycomb energy absorber. In this way, the whole impact test can be controlled under certain level of speed. Due to the plateau of the compressive force–deflection curve, its capacity of energy

FIGURE 9.10 Large deformation-induced fracture in vehicle offset impact test.

absorption is easy to be estimated and certain size of the honeycomb block can be designed for the test.

9.5.4 Behavior of Honeycomb under Complex Loading Conditions

The simple model described here may not be accurate enough under general loading condition. Continuous efforts have been devoted to the development of both material model and analysis method.

For instance, in the front offset vehicle impact, very large deformation may occur in part of the honeycomb block because of the complex loading conditions, whereas much less deformation happens in the other part, as shown in Figure 9.10. Severe shear is presented in the transition area, with rupture of the honeycomb block, debonding, and tearing off of the foils. In addition to a more general constitutive relation to represent the honeycomb response to the complex loading, appropriate rupture criterion is needed. However, it is a challenging task.

To investigate the characteristics of honeycomb under complex loading conditions, Hong et al. (2003) and Tran et al. (2006) conducted a mixed load test with compression and shear load in the lateral directions. The ratio of shear load S over the normal load P was denoted by

$$\eta = S/P = \tan\gamma, \tag{9.40}$$

where γ represented the angle formed by the shear direction and the normal direction. As reported, the collapse modes presented a type of microscopic folding similar to the case of pure normal compression. To estimate the load capacity, a yield function based on Hill's (1950) orthotropic yield function was proposed:

$$\sigma_{cr}^2 = \sigma_3^2 + A\tau_{31}^2 + B\tau_{32}^2. \tag{9.41}$$

Reasonably good correlation to the test data was reported in the quoted reference. The critical load could also include the effect of loading speed.

Among others, Hinnerichs et al. (2006) conducted comprehensive biaxial tests with normal load and one lateral compression and also two lateral compressions in addition. The tests also included the orientation angle with respect to the in-plane principal directions.

9.6 LAMINATED GLAZING

Glass is an important material used in architecture, automobile, etc. Glass is generally considered a type of brittle material, which breaks easily when loaded, particularly under impact loading. The fragments of the broken glass can result in personnel injuries. To mitigate the potential damage caused by the broken glass, laminated glass is widely adopted in many applications. For example, two layers of soda lime glass adhered with a thin layer of polymeric material can form a type of sandwich composite. Fragments will stay with the adhesive layer without going airborne in case if the glass breaks. Polyvinyl butyryl (PVB) is often used for the adhesive. Application of this type of laminated glass for automobile windshield traces back to 1910s. It has gained important applications in building constructions too.

The key questions to be answered for the application of glass are: under what load conditions and how the glass breaks. Among many researchers, Dharani et al. (2003) and Zhao et al. (2006) studied the simulated human head impact into the laminated glazing. Du Bois et al. (2003) investigated the head impact and the role of windshield glass in vehicle's roof strength.

The glass behaves differently under different loading conditions and in different applications. Whereas breakage occurs most likely under tensile loading, many factors such as surface condition, manufacturing process, and relative humidity, affect the rupture strength of the glass. Bansal and Doremus (1986) published a rich collection of glass data obtained from tests, and considered strength not an intrinsic property of glass. The theoretical strength of glass was considered high, in the level of 10 GPa (giga Pascals). However, a tiny flaw could significantly reduce the strength, for example, to the level of 0.1 GPa.

9.6.1 Application of J-integral

Dharani et al. (2003) used maximum principal tensile stress and the technique of J-integral to investigate the condition for crack initiation in the laminated glazing. A substructuring method along with the J-integral was proposed. The head impact problem was simulated by a spherical head form impacting into the laminated glazing. Under impact conditions, the two layers of glass might fail due to compression stress or stress induced by bending load. The failure due to compressive stress σ_z presented a Hertzian cone, which originated from the impact surface just outside the contact zone. The tensile stress reached maximum in this area. On the other hand, the bending stress σ_r depended on the impact energy. The crack due to bending stress initiated at the location where the maximum principal tensile stress occurs, but not at the impact point.

Their finite element model utilized axisymmetric two-dimensional (2D) elements (physically representing solids). The glass plies were meshed in several layers of four-node bilinear elements. The skin of the head form for the test was modeled as viscoelastic and the aluminum skull as elastic. The glass was considered elastic and brittle, which meant that the rupture or crack could happen during the elastic deformation. The PVB interlayer was considered viscoelastic.

The large tensile stress might occur in both glass plies. In the numerical procedure, the location with maximum principal tensile stress was identified at every time step. To detect the possibility of crack initiation, a small crack normal to the glass surface was introduced in the surrounding area. The J-integral was then applied in the neighborhood area. If the J-integral reached the critical value of the glass, crack was then considered to set off. To perform the J-integral, a substructuring analysis was proposed and a refined mesh was used to achieve good accuracy.

Hertzian cone crack formation was not found in their study. The maximum principal tensile stress due to bending effect occurred at the centerline, for both the impact ply and the nonimpact ply, and on the nonimpact side. Furthermore, it was higher on the surface of nonimpact ply than that on the impact ply. However, the J-integral showed that the critical value was first reached at the impact ply at the inter surface with the PVB adhesive where the crack initiated. The crack in the nonimpact side occurred later. This pattern of crack was observed consistently for a range of PVB thickness.

9.6.2 Application of Anisotropic Damage Model

Zhao et al. (2006) reported a study of damage in laminated glazing due to head impact, analyzed by using continuum damage mechanics, described below. The problem was the same as discussed in Section 9.6.1.

An anisotropic damage mechanics model is introduced to describe the mechanical behavior of the glass:

$$\sigma_{ij} = \left(K^e_{ijkl} + K^d_{ijkl}\right) \varepsilon_{kl}, \tag{9.42}$$

where K^e is the elasticity tensor and K^d is for damage parameters defined below:

$$\begin{aligned} K^e_{ijkl} &= \lambda \delta_{ij}\delta_{kl} + \mu(\delta_{ik}\delta_{jl} + \delta_{il}\delta_{jk}), \\ K^d_{ijkl} &= C_1(\delta_{ij}D_{kl} + \delta_{kl}D_{ij}) + C_2(\delta_{jk}D_{il} + \delta_{il}D_{jk}). \end{aligned} \tag{9.43}$$

The case with all components $D_{ij} = 0$ is for the undamaged material, which is the usual undamaged elasticity. The parameters C_1 and C_2 are determined in the following way: for uniaxial tension test, when $D_{11} = 1$, crack happens to the material and results in $\sigma_{11} = 0$.

The damage growth follows a simple rule: the damages corresponding to tensile principal stress σ_i and shear stress τ_{ij} act independently,

$$D_{ii} = \begin{cases} 0 & \text{if } \sigma_i \leq \tilde{\sigma} \\ \dfrac{\sigma_i - \tilde{\sigma}}{\sigma_{crit} - \tilde{\sigma}} & \text{if } \tilde{\sigma} < \sigma_i < \sigma_{crit}, \quad i = 1, 2, 3, \\ 1 & \text{if } \sigma_{crit} \leq \sigma_i \end{cases} \tag{9.44}$$

$$D_{ij} = \begin{cases} 0 & \text{if } \tau_{ij} \leq \tilde{\tau} \\ \dfrac{\tau_{ij} - \tilde{\tau}}{\tau_{crit} - \tilde{\tau}} & \text{if } \tilde{\tau} < \tau_{ij} < \tau_{crit}, \quad ij = 12, 23, 31 \\ 1 & \text{if } \tau_{crit} \leq \tau_{ij} \end{cases} \tag{9.45}$$

The damage parameters are nondecreasing and ensure the irreversible process. For dynamic problem, at each time step n, simply take

$$D_{ij} = \max\left(D_{ij(n)}, D_{ij(n-1)}\right). \tag{9.46}$$

In the study, Zhao et al. (2006) used linear elasticity to model the PVB adhesive and found that only 2% difference was observed in stress obtained by modeling the PVB by viscoelasticity or elasticity.

The simulation discovered that the damaged zone size of the nonimpacted side of glass ply was much larger than that of the impact side. Most of the web-shaped damage (caused by radial stress σ_r) was concentrated on the nonimpact surface. Most of the star-shaped damage (caused by circumferential stress σ_θ) was also concentrated on the nonimpact surface. The later had a larger zone than the former. This observation concurred with many laboratory tests. The study showed that the glass ply thickness of the nonimpact side had significant effect, whereas the thickness of the impact side did not. The thickness of the PVB interlayer had no significant effect on the impact resistance. On the other hand, larger glass area under loading resulted in less damage under the same impact condition.

9.6.3 A Simplified Model with Shell Element for the Laminated Glass

The study on laminated glass' behavior described in Sections 9.6.1 and 9.6.2 utilized solid modeling. Du Bois et al. (2003) developed a shell element model to simulate the collision of human head and the windshield glass of the vehicle. The model was also used to study the effect of windshield glass on the vehicle roof strength, to meet the regulatory requirements. In the numerical model, two coincidental elements represented the laminated windshield glass. One of them was the shell element for the glass, in which two plies were integrated into one. The other one was a membrane element for the PVB adhesive. Both types of elements used full integration for in-plane terms. The glass was assumed to be plastic and allowed to fail when the strain reached its limit for the applications. This model assumed that both layers of glass would fail at the same time if the glass was going to break. The PVB adhesive was considered nearly incompressible. Hence, several material models, usually used for rubber, were employed to evaluate the performance of the laminated glass.

In Du Bois et al. (2003), the glass was allowed to fail by strain at 0.1%. Several modeling approaches were evaluated. It was reported that for the method described here, the PVB adhesive modeled by using Mooney–Rivlin law achieved best results in comparison to the experiment.

PART IV

CONTACT AND CONSTRAINT CONDITIONS

CHAPTER 10

THREE-DIMENSIONAL SURFACE CONTACT

During large deformation, different parts of the structure can come into contact with one another. Even different portions of the same component can come into contact with one another. Thus, in numerical simulation, the contact algorithm is essential to prevent structural penetration so that the analysis can represent the true physical events. The development of contact algorithm in explicit finite element is a major contribution to computational mechanics and numerical methods.

As a matter of fact, analyzing contact problems has become critical in today's engineering applications as well as in theoretical studies. Contact problem has been an active research field in applied mechanics, numerical methods, and applied mathematics. It is also an active field related to development of software and computer architecture. The readers are referred to several well-written books with concentration on contact problems and to the comprehensive references quoted within. For example, see Kikuchi and Oden (1988), Han and Sofonea (2002), Shillor et al. (2004), and Eck et al. (2005) for mathematical theories; Johnson (1985), Zhong (1993), and Wriggers (2002) for the mechanics aspects and numerical approaches. In addition, the survey articles provide valuable information regarding the history of development, cf. Zhong and Mackerie (1994), Wriggers (1995), and Bourago (2002).

10.1 EXAMPLES OF CONTACT PROBLEMS

We will use several examples of contact problems to illustrate some features of contact mechanics. These examples are solved analytically. They serve the purpose

Introduction to the Explicit Finite Element Method for Nonlinear Transient Dynamics, First Edition.
Shen R. Wu and Lei Gu.

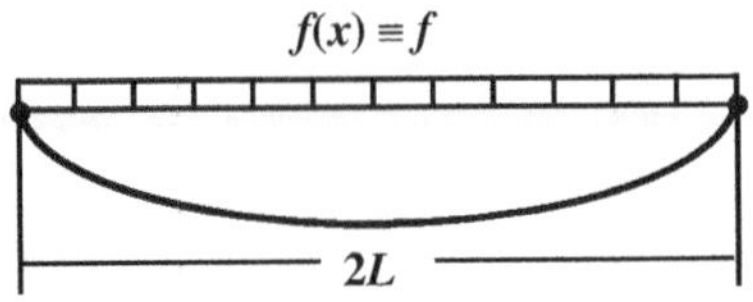

FIGURE 10.1 String under uniform loading. (Reproduced with the permission of WIT Press, Southampton and Boston.)

to reveal the features of contact problems. They also present the challenges that the finite element methods are facing.

10.1.1 Uniformly Loaded String with a Flat Rigid Obstacle

As a classical example commonly seen in textbooks of differential equations and elasticity, a string is fixed at both ends with uniform cross section area and material property. The string has a length $2L$. It is subjected to a uniformly distributed load f (per unit length, downward as positive) and is balanced by the internal tension force T. For small loading and small deformation, the deflection of the string (upward as positive) is described by the following linear equation and boundary conditions:

$$\begin{aligned} -Tu'' + f &= 0, \\ u(0) = u(2L) &= 0. \end{aligned} \tag{10.1}$$

The solution as a parabolic function is obtained by direct integration, shown in Figure 10.1:

$$u = -fx(2L - x)/2T. \tag{10.2}$$

Suppose there is a flat rigid obstacle at a small distance d below, as shown in Figure 10.2. Under the same load f, the string deflects downward and contacts the obstacle somewhere but is stopped without penetration. Upon contact with the obstacle, the deflection stops to increase. The obstacle plays the role of a constraint. This is a simple example of Signorini (1933) problem. By symmetry, we only consider the left half portion $0 \leq x \leq L$. Assume that the string touches ground at a priori unknown point $x_1 \leq L$ as the first contact point. Once the contact at x_1 happens, we

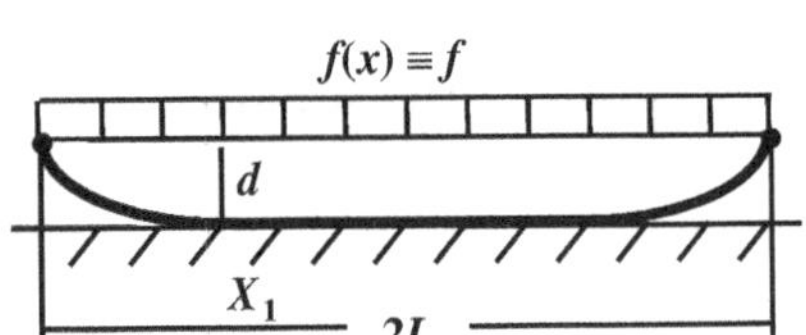

FIGURE 10.2 String under uniform loading with a rigid obstacle. (Reproduced with the permission of WIT Press, Southampton and Boston.)

can find one solution satisfying the constrained system. In this case, we assume that the string keeps lying on the obstacle for $x_1 \le x \le L$ and obtain

$$\begin{aligned} -Tu'' + f &= 0, \quad & 0 \le x \le x_1, \\ u &= -d, & x_1 \le x \le L. \\ u(0) &= 0, \\ u'(L) &= 0, \end{aligned} \tag{10.3}$$

After direct integration, we have the solution of (10.3) with two unknowns c_1 and x_1 to be determined:

$$\begin{aligned} u &= fx^2/2T + c_1 x, \quad & 0 \le x \le x_1, \\ u &= -d, & x_1 \le x \le L. \end{aligned} \tag{10.4}$$

By enforcing C^1 continuity at x_1 (reasonable for the flexible string),

$$\begin{aligned} u(x_1 - 0) &= u(x_1 + 0), \quad & fx_1^2/2T + c_1 x_1 = -d, \\ u'(x_1 - 0) &= u'(x_1 + 0), & fx_1/T + c_1 = 0. \end{aligned} \tag{10.5}$$

We obtain c_1 and the contact point x_1, as part of the solution of (10.3):

$$\begin{aligned} c_1 &= -fx_1/T = -\sqrt{2fd/T}, \\ x_1 &= \sqrt{2Td/f}. \end{aligned} \tag{10.6}$$

We further investigate if the solution obtained is unique. Assume that the string bounces off (separates) from the obstacle at $x_2 > x_1$, $x_2 < L$, without contact to the obstacle. For the whole string, by symmetry, it has another contact point $x_4 = 2L - x_1$ in the right half portion. Therefore, the string has to bend down at another point, such as $x_3 = 2L - x_2 \le x_4$, as a second contact, depicted in Figure 10.3a. This yields a portion of the loaded string without contact in $[x_2, x_3]$, with the equations $-Tu'' + f = 0$ and $u(x_2) = u(x_3) = -d$. The solution is

$$u = -d + f(x - x_2)(x - x_3)/2T.$$

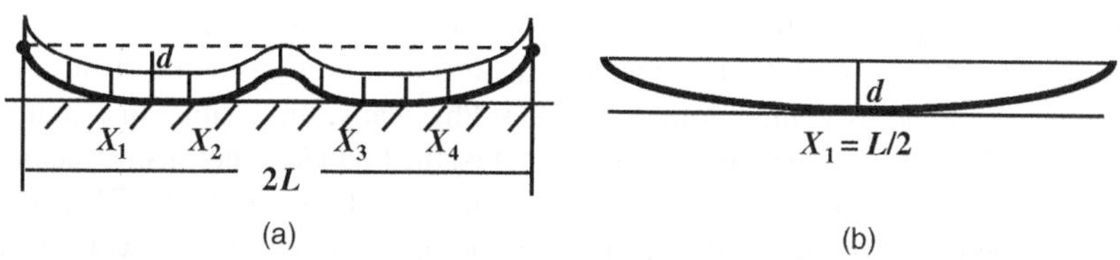

FIGURE 10.3 Uniqueness of contact point: (a) bounce off and a second contact point; (b) a special case—contact at the mid-span.

Obviously, $u < -d$ in $[x_2, x_3]$, a contradiction to the contact condition. Note that there is an exception, when $x_1 = L$, shown in Figure 10.3b. There is no issue of bouncing off the obstacle, but the solution is still unique with only one contact point.

For more discussions about the proof of uniqueness of solution for this type of contact problems, the readers are referred to Oden and Kikuchi (1980). For discussion about the tension force, see Wu (2001), where the tension was initially zero and built up by the stretch of the string at equilibrium due to loading and deformation. Here, we assume the additional tension force due to stretch is negligible in comparison to T.

Remark 10.1 Compare the two cases, with and without contact, represented by (10.1) and (10.3), respectively. We observe the following:

1. The solution (10.6) of the constrained system (10.3) is a C^1 function, which has discontinuity in the second derivative at the contact point x_1: $u''(x_1^-) = f/T$, $u''(x_1^+) = 0$. This is the only one discontinuity in this case. On the other hand, the solution (10.2) of the unconstrained system (10.1) is a parabolic function $\in C^2$, in fact C^∞ in this case. The reduction of smoothness is due to contact.
2. The solution inside the contact zone and noncontact zone still has high smoothness, even though the smoothness in the whole domain is reduced.
3. System (10.1) is a linear problem, whose solution (10.2) is proportional to the load. But the solution (10.6) of the constrained system (10.3) is no longer proportional to the load. The contact point x_1 depends on f nonlinearly. The constrained system is a nonlinear problem, even the governing equation does not contain a nonlinear term and deflection is still considered small. Here, the nonlinearity is introduced by the contact condition, unlike the traditional geometric nonlinearity such as large deformation and material nonlinearity such as plasticity in applied mechanics.
4. The contact zone can be a continuous region or a single point. The noncontact zone, however, does not seem to contain discrete points.

Since the contact point x_1 is a priori unknown and acts as a boundary between the contact zone and noncontact zone, this type of contact problem is mathematically classified as a free boundary problem.

In the case that the string initially lies on the obstacle, in contact, with $d = 0$, the string can only lift off somewhere but not penetrate the obstacle. The deformation or motion is possible only in one direction. The restriction on motion is also in one direction. This type of problem is hence also named unilateral contact.

Note that in this example, the contact happens at a portion of the string body. This type of contact is named domain contact. Physically, the contact of a material body is considered to happen at its surface, that is, the boundary, not inside the body. In our example, the contact happens at somewhere of the string body. This is due to the mathematical model of the string as a one-dimensional (1D) object. In later discussions, we will see that the contact at surface of membrane, plate, and shell is also considered as domain contact.

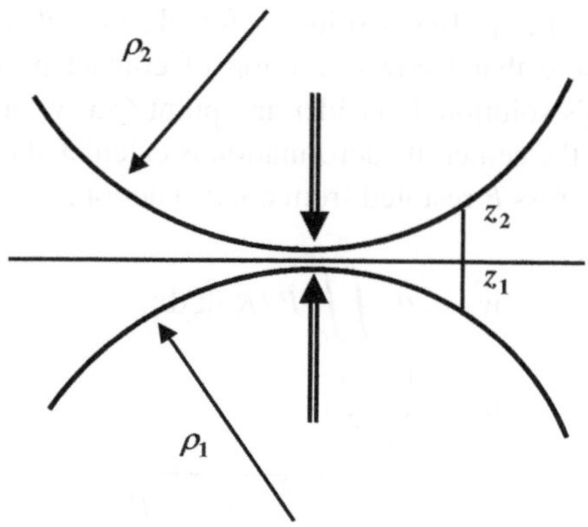

FIGURE 10.4 Contact of two balls.

10.1.2 Hertz Contact Problem

Contact between two elastic bodies has been an active research field. Hertz (1881) was among the earliest to provide an analytical solution. We present here a simple Hertz contact problem between two elastic balls. Detailed discussions can be found in many classical textbooks, for example, Love (1927), Chien and Ye (1956), Timoshenko and Goodier (1970), and Johnson (1985).

Assume that the radii of the balls are ρ_1 and ρ_2, respectively. They start at the position with contact at a point O without deformation. A pair of forces F is applied to the balls and presses them to contact each other. The forces pass through the center of balls and the contact point O, depicted in Figure 10.4. At equilibrium, due to symmetry, the contact areas form a circle C centered at O with radius r_c. The deformation of the material at contact points inside this circle, along the lines parallel to the axis, is w_1 and w_2, respectively. The initial clearances with respect to the neutral plane are z_1 and z_2, respectively. The total compression deformation is

$$\alpha = z_1 + z_2 + w_1 + w_2 = \Delta O_1 + \Delta O_2. \tag{10.7}$$

Assume that the contact area is planar and small, and that the deformation due to contact compression is small: r_c, w_1, and $w_2 << \min(\rho_1, \rho_2)$. For the points with a distance r from the centerline, we can make an approximation

$$z_1 = r^2/2\rho_1, \quad z_2 = r^2/2\rho_2, \tag{10.8}$$

$$\begin{aligned} z_1 + z_2 &= \gamma r^2, \\ \gamma &= (\rho_1 + \rho_2)/2\rho_1\rho_2. \end{aligned} \tag{10.9}$$

At the boundary of the contact zone, $w_1 = w_2 = 0$, $r = r_c$. At the centerline, $z_1 = z_2 = 0$. From (10.7) and (10.9), we have

$$\alpha = \Delta O_1 + \Delta O_2 = w_{10} + w_{20} = \gamma r_c^2. \tag{10.10}$$

Now, we use Boussinesq (1885) solution for elastic half space subjected to a concentrated force. Note that Hertz's solution of contact problem was published earlier than Bossinesq's solution. Consider any point $Q(x, y)$ inside the contact area with a distance r from the center. Its deformation is calculated by surface integration with the compressive stress P resulted from contact action:

$$\begin{aligned} w_j &= \theta_j \iint_C P/R \, \mathrm{d}\xi \mathrm{d}\eta, \\ \theta_j &= \frac{1-\nu_j^2}{\pi E_j}, \\ R &= \sqrt{(x-\xi)^2 + (y-\eta)^2}. \end{aligned} \tag{10.11}$$

Here, E_j and $\nu_j, j = 1, 2$, are the Young's modulus and Poisson ratios of the two balls.

Assume that the pressure distribution is in a shape of hemisphere with base on the contact circle (radius $= r_c$). This means a hemispherical distribution of contact force. It results in the following solution, after some manipulations:

$$\begin{aligned} F &= \iint_C p \mathrm{d}\xi \mathrm{d}\eta = 2\pi P_0 r_c^2/3, \\ r_c &= \left(\frac{3\pi F}{4} \frac{(\theta_1+\theta_2)\rho_1\rho_2}{\rho_1+\rho_2} \right)^{1/3}, \\ P_0 &= \frac{3F}{2\pi r_c^2} = \left(\frac{6F}{\pi^5} \left(\frac{\rho_1+\rho_2}{(\theta_1+\theta_2)\rho_1\rho_2} \right)^2 \right)^{1/3}, \\ \alpha &= \frac{3\pi F}{4r_c}(\theta_1+\theta_2) = \left(\frac{9\pi^2 F^2}{16} \frac{(\theta_1+\theta_2)^2(\rho_1+\rho_2)}{\rho_1\rho_2} \right)^{1/3}. \end{aligned} \tag{10.12}$$

Here, we observe nonlinearity again. The solution is not proportional to the total force F, and even not to the material constants. The contact zone is not given but solved with the system.

The stress of balls in contact was studied by Huber (1904), among others. The solutions were validated by experiments. More references can be found in Andress et al. (1922). Hertz also studied small area contact of two elastic bodies with arbitrary surfaces. By assuming the contact area to be approximated by quadratic surfaces, the solution was derived in terms of elliptic integrals. The study of stress distribution in the ellipsoidal contact can be found, for example, in Беляев (1924).

10.1.3 Elastic Impact of Two Balls

Hertz (1881) extended the equilibrium contact problem to a quasistatic impact problem. Two balls with masses m_1 and m_2 impact each other with initial velocities v_{10} and v_{20}, respectively. The velocities align with the centers of the balls, depicted

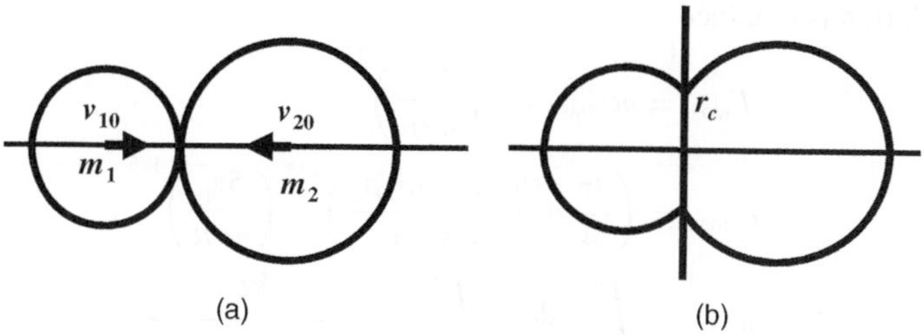

FIGURE 10.5 Impact of two balls: (a) before deformation; (b) after deformation.

in Figure 10.5. When the impact starts, the balls contact at one point. Assume the impact to be purely elastic so that the energy is conserved. The impact is indeed a transient process. The initial kinetic energy is determined by the initial velocity. During the impact process, pressure builds up in the contact area. This reaction results in increasing deformation of the balls and the decelerating motion. Eventually, the balls' relative velocity drops to zero. Rebounding follows the relative "rest."

At any time during the impact, the motion equations of the balls are

$$m_1\dot{v}_1 = -F = m_2\dot{v}_2. \tag{10.13}$$

For the total deformation of (10.7), the rate of closing distance of the centers follows the relation $\dot{\alpha} = v_1 + v_2$. Hence,

$$\begin{aligned} \ddot{\alpha} = \dot{v}_1 + \dot{v}_2 &= -F(m_1 + m_2)/m_1 m_2 \\ &= -n\mu\alpha^{3/2}, \end{aligned} \tag{10.14}$$

where (10.12) is used, with

$$\begin{aligned} n &= \sqrt{\frac{16}{9\pi^2}\frac{\rho_1\rho_2}{(\theta_1+\theta_2)^2(\rho_1+\rho_2)}}, \\ \mu &= \frac{m_1+m_2}{m_1 m_2}. \end{aligned} \tag{10.15}$$

The first integration from (10.14) is

$$\begin{aligned} &\tfrac{1}{2}\left(\dot{\alpha}^2 - v_0^2\right) = -\tfrac{2}{5}n\mu\alpha^{5/2}, \\ &v_0 = v_{10} + v_{20}. \end{aligned} \tag{10.16}$$

When the balls reach the maximum compression, $\dot{\alpha} = 0$. Then

$$\alpha_{\max} = \left(\frac{5v_0^2}{4n\mu}\right)^{2/5}. \tag{10.17}$$

The solution is obtained

$$
\begin{aligned}
F_{\max} &= n\alpha_{\max}^{3/2} = n\left(\frac{5v_0^2}{4n\mu}\right)^{3/5}, \\
r_{c\max} &= \left(\frac{3\pi}{4}n\frac{(\theta_1+\theta_2)\rho_1\rho_2}{\rho_1+\rho_2}\right)^{1/3}\left(\frac{5v_0^2}{4n\mu}\right)^{1/5}, \\
t_{\max\,(\text{impact})} &= \int_0^{t_{\max}} \mathrm{d}t = \int_0^{\alpha_{\max}} \frac{\mathrm{d}\alpha}{\sqrt{v_0^2 - 4n\mu\alpha^{5/2}/5}}, \\
&= \frac{\alpha_{\max}}{v_0}\int_0^1 \frac{\mathrm{d}x}{\sqrt{1-x^{5/2}}} = 2.94\frac{\alpha_{\max}}{v_0}.
\end{aligned}
\tag{10.18}
$$

Remark 10.2 This example gives the solution only at $t_{\max}$. The contact zone varies in time. Solving this problem for an arbitrary time point is a challenging task.

10.1.4 Impact of an Elastic Rod on the Flat Rigid Obstacle

A uniform rod with Young's modulus E, mass density ρ, and length L, moves in the axial direction with uniform initial velocity v_0 and constant acceleration a toward a rigid barrier perpendicularly. The rigid barrier is fixed and placed at a distance h ahead of the rod, as shown in Figure 10.6a. The sound speed $C = \sqrt{E/\rho}$ is introduced. The governing equations of the system are described below:

$$
\begin{aligned}
u_{t^2} - C^2 u_{x^2} &= -a, \\
u_x(t, L) &= 0, \\
u(t, 0) &\geq -h, \\
u_x(t, 0) &\leq 0, \\
(u(t, 0) + h)u_x(t, 0) &= 0, \\
u(0, x) &= 0, \\
u_t(0, x) &= 0,
\end{aligned}
\tag{10.19}
$$

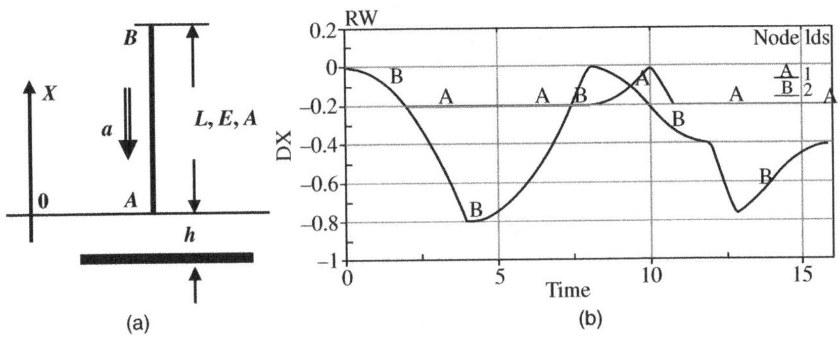

FIGURE 10.6 Elastic rod impacting the rigid obstacle: (a) initial condition; (b) solution of an example: displacement of the end points of the rod. (S.R. Wu, A variational principle for dynamic contact with large deformation. Elsevier 2009.)

$u(t, x)$ is the displacement of the rod. At the end $x = L$, it is free of load. The boundary conditions at the end point $x = 0$ is typical for contact problems. The third equation states that the end point has to be above the rigid barrier. This is a type of unilateral contact condition expressed by an inequality. The fourth equation is for the strain and stress at the end point, which is always of compressive nature due to contact, also in an inequality. The fifth equation is a combined constraint condition, a Kuhn–Tucker type of complementary condition. The meaning is, when $u(t, 0) > -h$, there is no contact and the stress is zero. When contact happens, $u(t, 0) = -h$, but the contact stress is not determined. The complementary condition always holds.

After a certain time, t_1 of free traveling, the rod closes the gap of distance h and hits the barrier. It will then rebound from the barrier (by conservation of energy). This impact problem was studied by Shi (1998a). For this type of impact problems, for any time $T > 0$, the existence and uniqueness of solution $u \in H^1((0, L) \times (0, T))$ with $\sigma = Eu_x \in L^2((0, L) \times (0, T))$ were proved by Lebeau and Schatzman (1984), and Schatzman and Bercovier (1989). The free travel time t_1 and the velocity v_1 at which the rod hits the barrier are obtained from classical dynamics:

$$t_1 = \left(\sqrt{v_0^2 + 2ah} - v_0\right) \Big/ a,$$

$$v_1 = \sqrt{v_0^2 + 2ah}.$$

After transforming to a homogeneous equation, Shi (1998a) used D'Alembert solution for the wave equation and constructed the solution for this problem. We quote the results here without the technical details:

$$u = \varphi(Ct + x) + \psi(Ct - x) - v_0 t - at^2/2. \tag{10.20}$$

Let $z = Ct$. The functions ϕ and ψ are expressed below:

$$\begin{aligned}
&\varphi(z) = 0, \quad \text{if } z < 2L, \\
&\varphi(z + 2L) = \varphi(z) + \sup_{0<s<z} \left\{ \left(2\varphi(s) + h - \frac{v_0 s}{C} - \frac{as^2}{2C^2} \right)^- \right\}, \quad \text{if } z > 0, \\
&\psi(z) = 0, \quad \text{if } z < 0, \\
&\psi(z) = \varphi(z + 2L), \quad \text{if } z > 0,
\end{aligned} \tag{10.21}$$

where we denote $(f)^- = (|f| - f)/2$. For more details, spreading out for $z < 4L$, we have

$$\varphi(z) = \begin{cases} 0, & \text{if } z < Ct_1 + 2L, \\ \dfrac{a(z-2L)^2}{2C^2} + \dfrac{v_0(z-2L)}{C} - h, & \text{if } Ct_1 + 2L < z < 4L, \end{cases} \tag{10.22a}$$

$$\psi(z) = \begin{cases} 0, & \text{if } z < Ct_1; \\ \dfrac{az^2}{2C^2} + \dfrac{v_0 z}{C} - h, & \text{if } Ct_1 < z < Ct_1 + 2L; \\ -f(Ct_1 + 2L) + \dfrac{a(z-2L)^2}{2C^2} + \dfrac{v_0(z-2L)}{C} - h, & \\ \qquad \text{if } s_0 < Ct_1 + 2L,\ Ct_1 + 2L < z < 4L; & \\ -\dfrac{a(z-2L)^2}{2C^2} - \dfrac{v_0(z-2L)}{C} + \dfrac{az^2}{2C^2} + \dfrac{v_0 z}{C}, & \\ \qquad \text{if } s_0 \geq Ct_1 + 2L,\ Ct_1 + 2L < z < s_0; & \\ -f(s_0) + \dfrac{a(z-2L)^2}{2C^2} + \dfrac{v_0(z-2L)}{C} - h, & \\ \qquad \text{if } s_0 \geq Ct_1 + 2L,\ s_0 < z < 4L. & \end{cases} \tag{10.22b}$$

Here, we denote

$$f(s) = 2\left(\frac{a(s-2L)^2}{2C^2} + \frac{v_0(s-2L)}{C} - h\right) + h - \frac{v_0 s}{C} - \frac{as^2}{2C^2}, \qquad s_0 = 4L - v_0 C/a. \tag{10.22c}$$

In fact, s_0 minimizes f with $f(s_0) = (8a^2L^2 + v_0^2C^2 + 2ahC^2)/2aC^2$.

By analyzing $u(t, 0)$ from the above result, we can find time t_2 when the end point $x = 0$ rebounds off the barrier. What follow are the time duration $t_2 - t_1$ when the contact keeps intake and the rebounding velocity $v_2 = \dot{u}(t_2, 0)$. Recall $v_1 = \sqrt{v_0^2 + 2ah}$, we have according to Shi (1998a)

$$\begin{aligned} &\text{If } 2aL \leq Cv_1, \quad t_2 = t_1 + \frac{2L}{C}, \quad v_2 = a\frac{2L}{C} - v_1, \\ &\text{If } 2aL > Cv_1, \quad t_2 = \frac{4L}{C} - \frac{v_0}{a}\left(> t_1 + \frac{2L}{C}\right), \quad v_2 = 0, \\ &\gamma = \min\left(0, a\frac{2L}{Cv_1} - 1\right), \end{aligned} \tag{10.23}$$

where γ is the restitution coefficient defined for this problem by

$$\gamma = \frac{\lim_{t \to t_2 - 0} u_t(t, 0)}{\lim_{t \to t_1 + 0} u_t(t, 0)}. \tag{10.24}$$

In some cases, $2aL \leq Cv_1$, the rebounding is sharp; whereas in other cases, $2aL > Cv_1$, the rebounding happens smoothly with the velocity gradually increasing from zero. Note that $2L/C$ is the time interval for the stress wave to travel back and forth in one cycle inside the rod. This relates to the driving force to lift the rod. Numerical examples for the calculated motion and impact force were presented in Shi (1998b).

For a special case, let the acceleration $a = 0$. Then, we have only the first situation:

$$\begin{aligned} t_1 &= h/v_0, \quad v_1 = v_0, \\ t_2 &= t_1 + 2L/C, \\ \gamma &= -1. \end{aligned}$$

Another special case is $v_0 = 0$, presented in Wu (2009). Take an example with parameters $L = 2{,}000$ (mm), $E = 2$ (kN/mm^2), $\rho = 2.0 \times 10^{-6}$ (kg/mm^3), $a = 0.1$ (mm/ms^2), $h = 0.2$ (mm). The gap is closed at $t_1 = 2$ (ms) after free falling with $v_1 = 0.02$. With these parameters, $C = 1{,}000$ and $2aL > Cv_1$. In this case, the impacting end bounces from the obstacle with a smooth velocity. The numerical solution of displacement at the two end points by using 1,000 truss elements is illustrated in Figure 10.6b. The error of the displacement at the end points comparing to the analytical one is within 1% at some key time points.

Remark 10.3 It is observed that in time domain, the displacement has only C^0 continuity at some time points when the impact and rebounding occur.

10.1.5 Impact of a Vibrating String to the Flat Rigid Obstacle

When considering the dynamic problem of the string as discussed in Section 10.1.1, we encounter a much more complicated situation. The contact to the rigid obstacle involves an impact action, which is followed by rebounding. The location of contact changes with time. Different points may come into contact with the obstacle at different times. Amerio and Prouse (1975) studied a perfectly elastic impact case, where the impact condition was described by

$$u_t(t_0 + 0, x_0) = -u_t(t_0 - 0, x_0). \tag{10.25}$$

As defined in (10.24), the restitution coefficient $\gamma = -1$. The rebounding was instant at the contact point with the velocity of same magnitude but reversed direction. The energy was conservative.

The authors also used D'Alembert solution and the approach of characteristic lines to analyze the behavior of the solution. They constructively proved the existence and uniqueness of the solution for this type of initial-boundary value problems with unilateral contact constraint. In light of this research, Cabannes (1985), among others, presented an analytical solution to the vibrating string impacting a rigid barrier, as described below. The string is fixed at both ends, $x = 0$ and $x = 1$. The distance to the obstacle is zero. The string is then stretched with pretension into a prescribed shape. Then, the motion starts. The system of equations for the position function $u(t, x)$ of the string is as follows:

$$\begin{aligned} &u_{t^2}(t,x) - u_{x^2}(t,x) = 0, \quad \text{if } u(t,x) > 0, \\ &u_t(t+0,x) = -u_t(t-0,x), \quad \text{if } u(t,x) = 0, \\ &u(t,0) = u(t,1) = 0, \quad t \geq 0, \\ &u(0,x) = u_0(x) \geq 0, \quad 0 \leq x \leq 1, \\ &u_t(0,x) = 0, \quad 0 \leq x \leq 1. \end{aligned} \tag{10.26}$$

The contact is represented by the condition $u = 0$, while $u > 0$ means no contact. The second equation of (10.26) is the same as (10.25), for the perfectly elastic impact. The contact point is not given a priori. The initial position $u_0(x)$ is strictly positive except at the end points:

$$\begin{aligned} &u_0(0) = u_0(1) = 0, \\ &u_0(x) > 0, \quad 0 < x < 1. \end{aligned} \tag{10.27}$$

Let $u_0(x)$ be smooth. Denote by $a(x)$ for the periodic and odd continuation of $u_0(x)$ with period 2. For an unconstrained problem with the same initial-boundary values, we can have a solution of D'Alembert type

$$w(t,x) = (a(t+x) - a(t-x))/2. \tag{10.28}$$

Cabannes (1985) constructed and verified the solution in the domain $0 \leq t \leq 1$, $0 \leq x \leq 1$, consisting of four triangular subregions:

In region (1): $0 \leq x \pm t \leq 1, t \geq 0, w \geq 0$. The solution was obtained

$$u_1(t,x) = w(t,x). \tag{10.29a}$$

In region (2): $0 \leq t \pm x \leq 1, x \geq 0$, the solution was found after certain manipulation:

$$\begin{aligned} &u_2(t,x) = \frac{a(t+x) - a(t-x)}{2} - \inf_{\lambda \in R} a(\lambda), \\ &R = [t-x, t+x]. \end{aligned} \tag{10.29b}$$

In region (3): $0 \le x - t \le 1 \le x + t \le 2, x \le 1$, symmetry to region (2) led to

$$u_3(t, x) = u_2(1 - t, 1 - x). \tag{10.29c}$$

In region (4): $0 \le t - x \le 1 \le t + x \le 2,\ t \le 1$, symmetry to region (1) resulted in

$$u_4(t, x) = -u_1(t, x) = -w(t, x). \tag{10.29d}$$

The solution is then extended to all values of t and x.

Remark 10.4 This example presents a case of dynamic domain contact. It is more complicated than the static case discussed in Section 10.1.1. Here, the contact zone varies with time.

10.2 DESCRIPTION OF CONTACT CONDITIONS

In Section 10.1, we demonstrated several examples of contact problems. They serve the purpose of showing the main features of contact problems and are summarized below:

- Contact is a constraint condition of impenetrability and is usually described by inequalities.
- The contact zone usually is not given a priori, but determined during motion and deformation.
- Nonlinearity is embedded due to contact condition.
- Discontinuity in derivatives of displacement solution may happen at the boundary of contact zone, as well as at certain key time points when impacting or rebounding occurs.
- The contact zone varies in time.

With all these in mind, we realize that the contact problem is by no means an easy one. Even for small deformation of linear elastic material with general geometry and loading condition, seeking an analytical solution is very challenging. We will have to turn to the numerical approach. Nonetheless, to establish a robust numerical procedure, we need to investigate the situation upon the sound foundation supported by theories of mathematics, mechanics, and numerical methods. In this section, we discuss the contact conditions without friction first, then later with frictional contact.

10.2.1 Contact with a Smooth Rigid Obstacle—Signorini's Problem

Consider a material body occupying domain Ω in a three-dimensional (3D) space (or two-dimensional (2D) respectively) in contact with a rigid obstacle, presented in

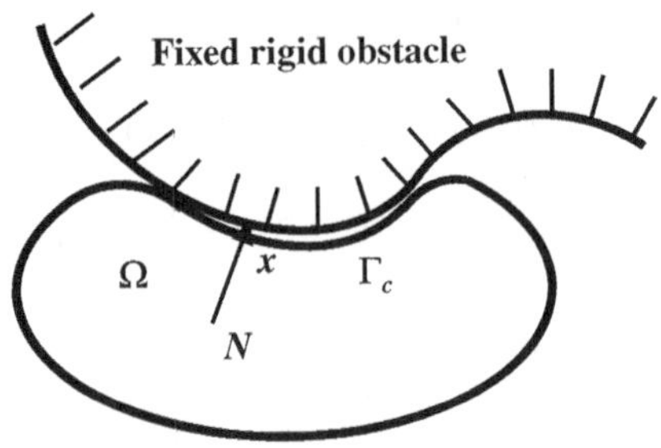

FIGURE 10.7 Contact to a fixed rigid obstacle—Signorini's problem.

Figure 10.7. In fact, only a portion of its boundary surface Γ_c contacts the obstacle. Γ_c could be a union of several simply or multiply connected surface regions (or curves respectively), including lines or single points as degenerated surfaces. Denote by $S(\boldsymbol{x}) = 0$ for the surface of obstacle and assume that S is a smooth function. Let $\boldsymbol{x}_c \in \Gamma_c$ be in contact with the obstacle. $\boldsymbol{x}_c$ satisfies the geometrical condition

$$S(\boldsymbol{x}_c) = 0. \tag{10.30}$$

Due to the existence of the obstacle, the motion of point $\boldsymbol{x}_c$ is restricted unilaterally, not to penetrate the surface $S(\boldsymbol{x}) = 0$.

This kinematic constraint condition is represented by the normal components of displacement and the compressive stress:

$$\begin{aligned} u_N &= u_j N_j \leq 0, \\ \sigma_N &= \sigma_{ij} N_i N_j \leq 0, \end{aligned} \tag{10.31}$$

where $\boldsymbol{N}$ is the unit outer normal vector of Γ_c at $\boldsymbol{x}_c$. σ_N is the normal component of the surface traction. When $\boldsymbol{x}_c$ stays on the obstacle surface, the compressive stress is exerted to the contact surface Γ_c. We have

$$\begin{aligned} u_N &= 0, \\ \sigma_N &\leq 0. \end{aligned} \tag{10.32a}$$

When $\boldsymbol{x}_c$ moves away from the obstacle surface, there is no more contact and the compressive contact stress drops to zero:

$$\begin{aligned} u_N &< 0, \\ \sigma_N &= 0. \end{aligned} \tag{10.32b}$$

From both of the above cases, we have the complementary condition, the Kuhn–Tucker condition,

$$u_N \sigma_N = 0. \tag{10.32c}$$

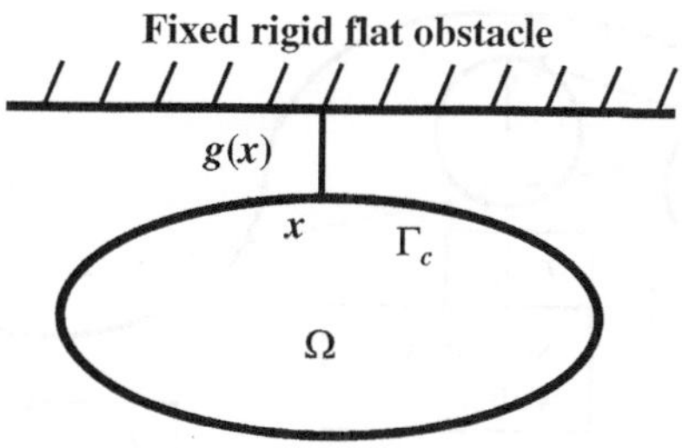

FIGURE 10.8 Contact model with gap function.

It means that the normal component of surface contact stress does not contribute to the work.

The contact condition (10.32) is only a constraint of no penetration for those points on Γ_c in contact with the obstacle. For points potentially moving to contact, we need additional control. On the other hand, (10.30) only checks the contact condition at current time. When combining these two aspects, a model with gap function, frequently seen in the literatures, is introduced; see Kikuchi and Oden (1988), Martins and Oden (1987), Eck et al. (2005), Wriggers (2002), and Zhong (1993). Assume that the distance from any point $\boldsymbol{x}_c \in \Gamma_c$ to the obstacle surface (measured in the outer normal direction of Γ_c) is a given function $g(\boldsymbol{x}) \geq 0$ at the beginning, illustrated in Figure 10.8. The nonpenetrability condition states that at any time the normal component of the displacement should not exceed $g(\boldsymbol{x})$. This condition can be defined as

$$\begin{aligned} u_N - g &\leq 0, & \text{if } u_N - g &< 0, & \sigma_N &= 0, \\ \sigma_N &\leq 0, & \text{if } u_N - g &= 0, & \sigma_N &\leq 0, \\ \sigma_N(u_N - g) &= 0. \end{aligned} \tag{10.33}$$

This is similar to (10.32). Note that (10.33) implies the contact condition (10.30). In the above discussion with (10.32), the satisfaction of (10.30) by the contact points is assumed. Now contact occurs only when u_N reaches g.

While the concept of gap function seems straightforward, it may be limited to the applications with small deformation, small gap, and simple geometry. For instance, when dealing with a flat rigid barrier as obstacle, condition (10.33) is simply perfect.

Note that $\boldsymbol{u}(t, X)$ is a measure of the displacement happened from time $= 0$ to t. In general, condition (10.31) or (10.33) about $\boldsymbol{u}(t, X)$ does not necessarily represent the impenetrability for all the time after contact occurs at time t. In cases of large deformation and large rotation with complex geometry, depicted in Figure 10.9, the displacement of large deformation can be irrelevant to the normal direction at the points in the surrounding area at current time. The normal of the material surface passing point $\boldsymbol{x}$ may not coincide with the normal of the obstacle surface where contact may happen. The motion of $\boldsymbol{x}$ along the normal of Γ_c may not be the most

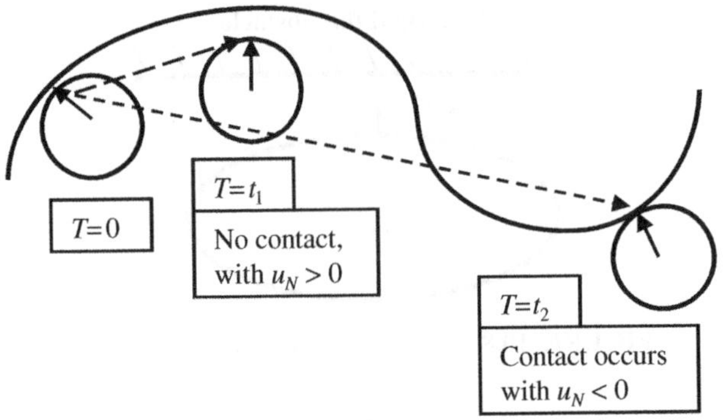

FIGURE 10.9 Possible contact during dynamic large deformation. (S.R. Wu, A variational principle for dynamic contact with large deformation. Elsevier 2009.)

possible motion leading to contact. Consequently, contact may not happen after a displacement with large positive normal component. On the other hand, a point $\boldsymbol{x}$ can go far away to contact the obstacle at an unpredictable location with large negative normal component of displacement. In this situation, the gap function $g(\boldsymbol{x})$ loses its relevance. The displacement is not really constrained by (10.31) or (10.33). The normal also changes direction due to large rotation accompanied with large deformation. The normal will lose its significance as a reference for measurement of large deformation by long time duration and the displacement condition becomes inappropriate. It will be even more challenging when the contact of two or more deformable bodies with large deformation is of concern.

In view of a general dynamic process, the contact may happen at some a priori unknown time and location during motion and deformation. We still consider point $\boldsymbol{x}_c \in \Gamma_c$ in contact with the obstacle surface, starting at time $t_1 \geq 0$. If $\boldsymbol{x}_c$ is going to stay on the obstacle surface (keeping in touch) for a while, from t_1 to t_2, the possible motion is always in tangential direction. Then for $t_1 \leq t \leq t_2$, we have

$$\begin{aligned} &\sigma_N \leq 0, \\ &v_N(\boldsymbol{x}_c, t) = \boldsymbol{v} \bullet \boldsymbol{N} = 0. \end{aligned} \tag{10.34a}$$

When $\boldsymbol{x}_c$ starts to move away from the obstacle surface at time t_2, we need a negative normal component of velocity and have

$$\begin{aligned} &\sigma_N = 0, \\ &v_N(\boldsymbol{x}_c, t_2) < 0. \end{aligned} \tag{10.34b}$$

The Kuhn–Tucker type of condition still has meaning and is written as

$$v_N \sigma_N = 0. \tag{10.34c}$$

Condition (10.34c) indicates that the rate of work done by the normal contact stress on the motion of contact point is zero.

The contact conditions of (10.34) are expressed in terms of velocity. In view of the contact situation and what will happen after contact, (10.34) appears to have the same form as (10.32) in terms of displacement. As a matter of fact, we can find their relation. For dynamics, we denote by $\boldsymbol{x} = \boldsymbol{r}(t, \boldsymbol{X})$ for the point originally located at $\boldsymbol{X}$ at $t = 0$, which at the current time is located at

$$\boldsymbol{x} = \boldsymbol{r}(t, \boldsymbol{X}) = \boldsymbol{X} + \boldsymbol{u}(t, \boldsymbol{X}), \tag{10.35}$$

with $\boldsymbol{u}$ being the displacement measured at time t. Considering the situation at current time, the motion after time t is viewed as additional displacement from t to $t + \Delta t$. If contact happens at time t, this increment of displacement satisfies contact condition (10.32), $(\boldsymbol{u}(t + \Delta t, \boldsymbol{X}) - \boldsymbol{u}(t, \boldsymbol{X})) \bullet \boldsymbol{N} \leq 0$. By taking limit $\Delta t \to 0$, we have the contact condition (10.34) for velocity. Note that the integration of the velocity condition does not yield good information about the displacement condition. Because the normal in large deformation is not a constant and the integration is not easy to carry out.

Usually, the velocity form of contact condition (10.34) is considered as the first order approximation of the displacement form (10.32). Taylor and Papadopoulos (1993) provided reasoning from mechanics point of view; see also Wriggers (2002) for more discussions. Meanwhile, the displacement form brings in additional difficulty to the mathematical assessment such as the existence of solutions to the dynamic contact problem. This might be just one of the reasons why contact problem has attracted so many mathematicians. For a long time, only few special problems in this category have been solved. Duvaut and Lions (1976) had detailed discussions on the mathematical theory about the existence and uniqueness of the solutions to this class of problems; see Eck et al. (2005) for more discussions and commentary on historical development.

Note that the contact condition (10.32) or (10.34) deals only with the constraint condition of no penetration. The condition of "in contact" such as $S(\boldsymbol{x}_c) = 0$ is assumed given, either by assumption or by additional condition but not examined yet. For large deformation problem, this is just the difficult point, because the contact is by no way to be given a priori or simply predictable. The additional condition is needed to find the material points "in contact" with the obstacle at any given time. The development of such numerical algorithm in the explicit finite element code is a great contribution to computational mechanics.

When velocity form is adopted, the gap function based on displacement form is no longer directly applicable. When penalty method is introduced, however, the concept of gap function finds its position in engineering applications.

10.2.2 Contact between Two Smooth Deformable Bodies

We now extend the concept developed in Section 10.2.1 to the case of contact between two deformable bodies. Denote Γ_x and Γ_y for the boundary surfaces of the two bodies.

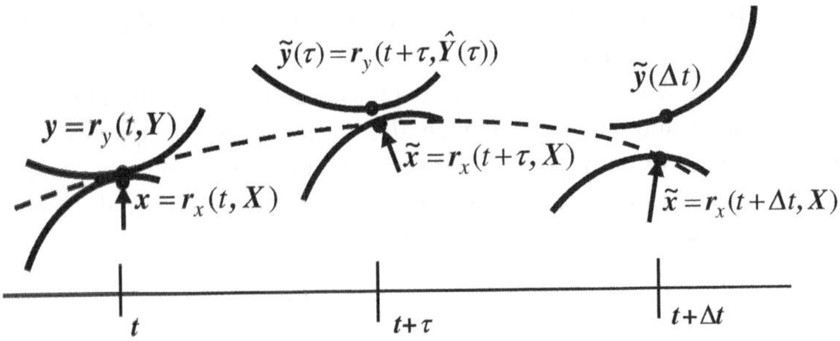

FIGURE 10.10 Trajectory of contact point.

The pair of contact points $\boldsymbol{x}$ and $\boldsymbol{y}$ are geometrically coincident. Let $\boldsymbol{N}_x$ and $\boldsymbol{N}_y$ be the unit outer normal vectors at $\boldsymbol{x}$ and $\boldsymbol{y}$, respectively. At this moment, $\boldsymbol{N}_y = -\boldsymbol{N}_x$, depicted in Figure 10.10. When the displacement and velocity in (10.32) to (10.34) are replaced by the relative displacement and relative velocity, the conditions for rigid obstacle may be updated to meet the nonpenetration requirement of the present situation. Written for the first body, we have the contact conditions similar to (10.32) to (10.34), respectively,

$$\begin{cases} u^x_{N_x} - u^y_{N_x} = (\boldsymbol{u}^x - \boldsymbol{u}^y) \bullet \boldsymbol{N}_x \le 0 \\ \sigma^x_{N_x} \le 0 \\ \sigma^x_{N_x}\left(u^x_{N_x} - u^y_{N_x}\right) = 0, \end{cases} \tag{10.36}$$

$$\begin{cases} u^x_{N_x} - u^y_{N_x} - g = (\boldsymbol{u}^x - \boldsymbol{u}^y) \bullet \boldsymbol{N}_x - g \le 0 \\ \sigma^x_{N_x} \le 0 \\ \sigma^x_{N_x}\left(u^x_{N_x} - u^y_{N_x} - g\right) = 0, \end{cases} \tag{10.37}$$

$$\begin{cases} v^x_{N_x} - v^y_{N_x} = (\boldsymbol{v}^x - \boldsymbol{v}^y) \bullet \boldsymbol{N}_x \le 0 \\ \sigma^x_{N_x} \le 0 \\ \sigma^x_{N_x}\left(v^x_{N_x} - v^y_{N_x}\right) = 0. \end{cases} \tag{10.38}$$

For large deformation- and large rotation-accompanied transient dynamics, we need to explore with slightly more details. Assume that at time t, $\boldsymbol{x} \in \Gamma_x(t)$ is in contact with $\Gamma_y(t)$ at $\mathbf{y}$. Their original positions at $t = 0$ are $\boldsymbol{X} \in \Gamma_x(0)$ and $\boldsymbol{Y} \in \Gamma_y(0)$, respectively. We have

$$\boldsymbol{x} = \boldsymbol{r}_x(t, \boldsymbol{X}) = \boldsymbol{y} = \boldsymbol{r}_y(t, \boldsymbol{Y}). \tag{10.39}$$

Consider the small motion from t to $t + \Delta t$. For any $\tau \in [0, \Delta t]$, $\boldsymbol{x}$ moves to $\tilde{\boldsymbol{x}} = \boldsymbol{r}_x(t + \tau, \boldsymbol{X})$. Regardless of whether it is still in contact with $\Gamma_y(t + \tau)$, we can find a point as the intersection of the normal at $\tilde{x}$ and the contact surface, $\tilde{y}(\tau) = N_x(t+\tau, X) \cap \Gamma_y(t+\tau)$. $\tilde{y}(\tau)$ is not necessarily the same point into which y is moving at the new location. No penetration requires

$$(\tilde{x} - \tilde{y}) \bullet N_x(t+\tau, X) \le 0. \tag{10.40}$$

We can identify $\hat{Y}(\tau)$ as the original location of $\tilde{y}(\tau)$ at time $t = 0$ such that $\tilde{y}(\tau) = r_y(t+\tau, \hat{Y}(\tau))$ with $\hat{Y}(0) = Y$ and $\tilde{y}(0) = y$. Its location at time t is denoted by, as a function of τ,

$$\psi(\tau) = r_y(t, \hat{Y}(\tau)), \quad 0 \le \tau \le \Delta t. \tag{10.41}$$

Note that $\psi(\tau)$ is not necessary the same point y. But we have

$$\psi(0) = r_y(t, Y) = y. \tag{10.42}$$

For continuum without rupture during the infinitesimal motion from t to $t + \Delta t$, $\hat{Y}(\tau)$ should be a smooth function. For small Δt, we also expect the points of all $\hat{Y}(\tau)$ to be close, but possibly all different. The same argument applies to $\psi(\tau)$. We have $\psi(\tau) \to y$ as $\tau \to +0$.

Using (10.39) and (10.42), we have

$$\begin{aligned} \tilde{x} - \tilde{y} &= \tilde{x} - x + x - y + y - \psi(\tau) + \psi(\tau) - \tilde{y} \\ &= r_x(t+\tau, X) - r_x(t, X) + \psi(0) - \psi(\tau) + r_y(t, \hat{Y}(\tau)) - r_y(t+\tau, \hat{Y}(\tau)) \\ &= (v(t, X) - \psi'(0) - v(t, \hat{Y}(\tau)))\tau. \end{aligned} \tag{10.43}$$

The no penetration condition (10.40) leads to

$$(\tilde{x} - \tilde{y}) \bullet N_x(t+\tau, X) = \tau(v(t, X) - v(t, \hat{Y}(\tau)) - \psi'(0)) \bullet N_x(t+\tau, X) \le 0.$$

Note that $\psi(\tau)$ represents the points on the contact surface $\Gamma_y(t+\tau)$. Hence, $\psi'(\tau)$ is in the tangent plane and orthogonal to the normal $N_y(t+\tau, \hat{Y}(\tau))$ at $\psi(\tau)$. For smooth contact surfaces, we have at time t or $\tau = 0$, $N_x(t, X) = -N_y(t, Y)$. Thus, as $\tau \to +0$, $\hat{Y}(\tau) \to Y$, we obtain

$$(v(t, X) - v(t, Y)) \bullet N_x(t, X) \le 0. \tag{10.44}$$

This is the velocity form of contact condition in (10.38).

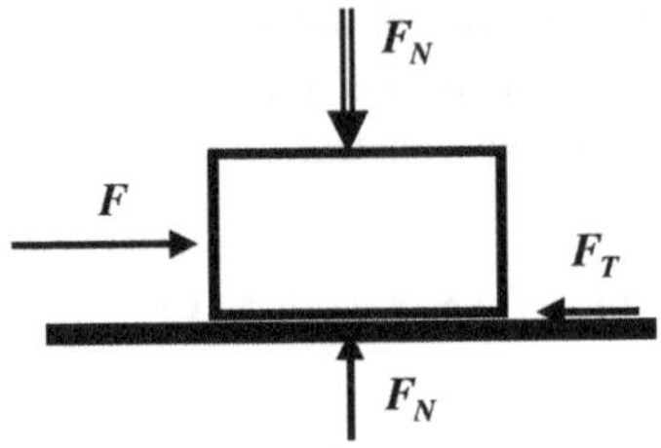

FIGURE 10.11 Friction between two material bodies in contact.

10.2.3 Coulomb's Law of Friction

Friction is a well-known phenomenon. When two material bodies are in contact, shown in Figure 10.11, the compressive contact force $\boldsymbol{F}_N$ in the normal direction is built up. Assume the lower body is fixed. If a lateral force $\boldsymbol{F}$ tries to drive the upper body to slide away, there is a resistance force $\boldsymbol{F}_T$ residing in the interface. This is the friction force $\boldsymbol{F}_T = -\boldsymbol{F}$, acting in the opposite direction of the driving force, and tangent to the contact surface. When the driving force increases, the friction force follows to increase until it reaches its maximal value $F_{T\max}$. If the driving force is smaller than the maximum $F_{T\max}$, the motion is not possible and the upper body is sticking on the lower body. The friction force just balances the driving force, passively, to maintain the equilibrium state. When the driving force exceeds the maximum $F_{T\max}$, relative sliding motion initiates. The maximum friction force is proportional to the normal compression force F_N. The proportionality, that is, the ratio of the maximal friction force over the normal contact force is called the static coefficient of friction:

$$\mu_s = F_{T\,\max}/F_N. \tag{10.45}$$

The next fundamental experimental finding is that the driving force F_{Td} needed to maintain the sliding motion is less than $F_{T\max}$, which is needed to initiate the motion. The ratio of the dynamic driving force over the normal contact force is called the dynamic coefficient of friction:

$$\mu_d = F_{Td}/F_N. \tag{10.46}$$

These coefficients μ_s and μ_d vary depending on the material properties under investigation. Further experiments found that for a given pair of materials in contact, while the static coefficient of friction is almost constant (within certain range), the dynamic coefficient of friction depends on the velocity as well as the normal contact force. The smooth transition from a static friction to a dynamic friction can be modeled by, including the dependence on the velocity, for example

$$\mu = \mu_d + (\mu_s - \mu_d)\mathrm{e}^{-c|\boldsymbol{v}|}, \tag{10.47}$$

where $\boldsymbol{v}$ is the velocity of the sliding motion.

To incorporate the friction in the mechanical system, a friction model was introduced by Coulomb (1781). Denote by F_N for the normal contact force, positive for compression, along the negative normal direction to the material contact surface. Duvaut and Lions (1976) introduced a mathematical model of Coulomb's law,

$$
\begin{aligned}
&|\boldsymbol{F}_T| \leq \mu F_N, \\
&|\boldsymbol{F}_T| < \mu F_N \Rightarrow \boldsymbol{v}_T = 0, \\
&|\boldsymbol{F}_T| = \mu F_N \Rightarrow \exists \lambda > 0, \quad \boldsymbol{v}_T = -\lambda \boldsymbol{F}_T.
\end{aligned}
\tag{10.48}
$$

The second relation of (10.48) is obvious, representing the "sticking" situation with static friction. The third relation has the simple explanation that the friction force is in the opposite direction of motion. Consider $-\boldsymbol{F}_T/|\boldsymbol{F}_T| = \boldsymbol{v}_T/|\boldsymbol{v}_T|$. Then $\boldsymbol{v}_T = -\boldsymbol{F}_T|\boldsymbol{v}_T|/|\boldsymbol{F}_T| = -\lambda \boldsymbol{F}_T$, with λ playing the role of $|\boldsymbol{v}_T|/|\boldsymbol{F}_T|$.

Friction involves microstructures at the material surface with a nature of random distribution, which is a subject beyond the scope of this book. We only consider a macroscopic model. In order to represent the contact condition (10.34), the traction (stress) and velocity at the contact surface need to be decomposed into normal components and tangential components. The normal direction $\boldsymbol{N}$ at the material contact surface Γ is used as reference. Define the surface traction (vector) σ_{Ni}, the normal traction σ_N, and the tangential traction (vector) σ_{Ti} by

$$
\begin{aligned}
&\sigma_{Ni} = \sigma_{ij} N_j, \\
&\sigma_N = \sigma_{ij} N_j N_i, \\
&\sigma_{Ti} = \sigma_{Ni} - \sigma_N N_i, \quad \sigma_T = \sqrt{\sigma_{Ti}\sigma_{Ti}}.
\end{aligned}
\tag{10.49}
$$

The components σ_N and σ_{Ti} then balance the contact forces:

$$
\begin{aligned}
&\sigma_N = -F_N, \\
&\sigma_{Ti} = F_{Ti}.
\end{aligned}
\tag{10.50}
$$

Note that the decomposition applies to general stress tensor. The decomposition of the displacement and velocity into the corresponding normal component and tangential components follows the same rule:

$$
\begin{aligned}
&u_N = u_i N_i, \qquad\qquad v_N = v_i N_i, \\
&u_{Ti} = u_i - u_N N_i, \quad v_{Ti} = v_i - v_N N_i, \quad v_T = \sqrt{v_{Ti} v_{Ti}}.
\end{aligned}
\tag{10.51}
$$

When extending Coulomb's law to the contact between two deformable bodies, we consider the velocity and friction force with the concept of relative motion. For simplicity without losing generality, we continue the discussion here for the case of a fixed rigid obstacle.

10.2.4 Conditions for "In Contact"

For large deformation discussed in Sections 10.2.1 and 10.2.2, the contact conditions (10.34) and (10.38) were the "no penetration" constraints, that the velocity and stress should satisfy. These conditions are for the motion and deformation to happen after contact occurred. The status of material body in contact with the obstacle or another material body or part of the same material body is not addressed by these conditions. We need additional effort to handle the "in contact" condition. Equation (10.30) seems simple and straightforward to play this role. It is the very basic requirement. Alternatively, we may take an even more fundamental requirement for point $\boldsymbol{x}$ to contact the obstacle surface Γ_o at any time t:

There exists $\boldsymbol{y} \in \Gamma_o$, such that $\boldsymbol{y} = \boldsymbol{x}$, or

$$\min_{\boldsymbol{y} \in \Gamma_o} ||\boldsymbol{y} - \boldsymbol{x}|| = 0. \tag{10.52}$$

No penetration requires for any point $\boldsymbol{y} \in \Gamma_o \cap \boldsymbol{N}_x$ (outer normal):

$$(\boldsymbol{x} - \boldsymbol{y}) \bullet \boldsymbol{N}_x \leq 0. \tag{10.53}$$

We cannot predict where and when the contact is to take place during the large deformation. We would have to check the contact condition at all the time periods before any better approach can be found.

10.2.5 Domain Contact

For truss, beam, membrane, plate, and shell, the structure is modeled by 1D or 2D configuration, reduced from the 3D geometry. For these mechanics models, like the load exerted on the body, the contact is considered to occur at part of the domain and the contact force is considered as a type of body force. The contact of thin shell structures is a good example. A 1D example was discussed in Sections 10.1.1 and 10.1.5. Elastohydrodynamic lubrication is another type of domain contact problems; see for example, Oden and Wu (1985) and Wu (1986). Contact at their boundary, the end points of 1D configuration and the edges of 2D configuration, are still possible and meaningful, cf. the example discussed in Section 10.1.4.

The contact conditions discussed so far can be extended to the domain contact. A major difference is the normal direction. For membrane, plate and shell, the normal is just like the normal to the boundary surface of a 3D body. Now the material body is a "surface." Because their geometry representation is a surface segment but not the surface of a closed volume in the previous discussions, the contact can happen at both sides of the surface. A closed surface, such as a sphere and a closed box, is the exception, for which only outer surface is possible for contact unless large deformation happens to bring the inner surface into contact with itself. For truss, rod, and beam, the normal at a specific point is not unique. It forms a plane instead. The contact can occur in any direction lying in this plane. The normal direction is to be determined in association with the obstacle. A special situation is when two 1D models are in contact, where the contact zone is a single point as part of the domain; see later discussion in Chapter 11.

On the other hand, the contact force plays a role of load and is viewed as a type of body force. Hence, its relation to stress on the surface is not as straightforward as the surface traction in the case of a regular 3D material body. Usually the contact force should be determined from solving the system.

10.3 VARIATIONAL PRINCIPLE FOR THE DYNAMIC CONTACT PROBLEM

Here we first discuss boundary contact as a constraint condition to the system. The domain contact will follow. We only discuss the contact to a rigid obstacle for simplicity. The contact between two deformable bodies will be a natural extension.

10.3.1 Variational Formulation for Frictionless Dynamic Contact Problem

Review the variational equations (2.9) or (2.10) for dynamic system (2.1) discussed earlier. Equation (2.10) is the result after applying Gauss–Green Theorem to (2.9). At that time, we did not include constraint condition. We now extend it to the dynamic frictionless contact.

We first consider a dynamic Signorini's problem, contact with rigid obstacle, using condition (10.34). Assume that part of the boundary, Γ_u and Γ_s, are subjected to prescribed displacement and traction respectively. Γ_u and Γ_s can never be in contact with the obstacle. In fact, the prescribed displacement or traction on the material surface Γ_u or Γ_s is usually realized mechanically by contacting the surface, so they cannot contact the obstacle again. The rest of the boundary, on which no displacement nor force is prescribed, can either be in contact with the obstacle or not, and the contact can happen at some time or at all the time. If a portion of such boundary is not in contact at time t, we consider that zero traction is applied to it. This kind of boundary may change in time. It is included in Γ_c as contact boundary. The whole boundary Γ is now decomposed into three disjoint parts:

$$\begin{aligned} &\bar{\Gamma} = \bar{\Gamma}_u \cup \bar{\Gamma}_s \cup \bar{\Gamma}_c \\ &\Gamma_u \cap \Gamma_s = \varphi, \quad \Gamma_u \cap \Gamma_c = \varphi, \quad \Gamma_s \cap \Gamma_c = \varphi, \end{aligned} \tag{10.54}$$

where $\bar{\Gamma}_u$ represents the closure of Γ_u, etc.

The dynamic frictionless contact is described below, with $\boldsymbol{N}$ as the normal of the contact boundary Γ_c.

Problem C1

$$\rho\ddot{u}_i - \sigma_{ij,j} = f_i(t,\boldsymbol{x})\ in\ \Omega, \quad i,j = 1,2,3, \tag{10.55a}$$

$$u_i(0,\boldsymbol{x}) = U_i^0(\boldsymbol{x}), \quad \dot{u}_i(0,\boldsymbol{x}) = U_i^1(\boldsymbol{x})\ in\ \Omega, \tag{10.55b}$$

$$u_i = U_i(t,\boldsymbol{x})\ on\ \Gamma_u, \tag{10.55c}$$

$$\sigma_{ij}n_j = g_i(t,\boldsymbol{x})\ on\ \Gamma_s, \tag{10.55d}$$

$$\begin{cases} \dot{u}_N \le 0, \quad \sigma_N \le 0, \quad \sigma_N\dot{u}_N = 0 \\ \sigma_{Ti} = 0 \end{cases} on\ \Gamma c, \tag{10.55e}$$

$$\sigma_{ij} = \sigma_{ij}(E, \nu, E_t, x_i, u_i, \dot{\varepsilon}_{ij}, \ldots), \tag{10.55f}$$

$$\dot{\varepsilon}_{ij} = (\dot{u}_{i,j} + \dot{u}_{j,i})/2. \tag{10.55g}$$

We need to update the variational principle now with contact conditions discussed so far. We can rewrite (2.10) with the contribution on Γ_c:

$$\int_\Omega (\rho\ddot{u}_i v_i + \sigma_{ij}v_{i,j})\mathrm{d}\Omega - \int_{\Gamma_c} \sigma_{ij}N_j v_i \mathrm{d}\Gamma = \int_\Omega f_i v_i \mathrm{d}\Omega + \int_{\Gamma_s} g_i v_i \mathrm{d}\Gamma. \tag{10.56}$$

When we applied Gauss–Green Theorem to (2.9), we used an assumption $v_i|_{\Gamma_u} = 0$ for the weight (test) function. Thus the solution satisfies boundary condition (10.55c) $u_i|_{\Gamma_u} = u_i(t)$, whereas the weight function satisfies a homogeneous boundary condition. They belong to different function sets, unless $u_i \equiv 0$. Alternatively, we can use Galerkin method with weight function $v_i - \dot{u}_i$ (or $v_i - u_i$ for the displacement form). Let $v_i|_{\Gamma_u} = \dot{u}_i$, which implies $(v_i - \dot{u}_i)|_{\Gamma_u} = 0$. Then we consider $\boldsymbol{v}$ and $\dot{\boldsymbol{u}}$ belonging to the same function set, that is, using the weight functions with the meaning of velocity. Thus (10.56) is valid with $\boldsymbol{v}$ replaced by $\boldsymbol{v} - \dot{\boldsymbol{u}}$:

$$\int_\Omega (\rho\ddot{u}_i(v_i - \dot{u}_i) + \sigma_{ij}(v_{i,j} - \dot{u}_{i,j}))\mathrm{d}\Omega - \int_{\Gamma_c} \sigma_{ij}N_j(v_i - \dot{u}_i)\mathrm{d}\Gamma$$
$$= \int_\Omega f_i(v_i - \dot{u}_i)\mathrm{d}\Omega + \int_{\Gamma_s} g_i(v_i - \dot{u}_i)\mathrm{d}\Gamma. \tag{10.57}$$

With the decomposition into normal and tangential components and the assumption of $\sigma_{Ti} = 0$, we have

$$\int_{\Gamma_c} \sigma_{ij} N_j (v_i - \dot{u}_i) \mathrm{d}\Gamma = \int_{\Gamma_c} (\sigma_N (v_N - \dot{u}_N) + \sigma_{Ti} (v_{Ti} - \dot{u}_{Ti})) \mathrm{d}\Gamma$$

$$= \int_{\Gamma_c} \sigma_N (v_N - \dot{u}_N) \mathrm{d}\Gamma. \tag{10.58}$$

For the frictionless Signorini's problem, the contact with rigid obstacle, (10.57) reduces to

$$\int_{\Omega} (\rho \ddot{u}_i (v_i - \dot{u}_i) + \sigma_{ij} (v_{i,j} - \dot{u}_{i,j})) \mathrm{d}\Omega - \int_{\Gamma_c} \sigma_N (v_N - \dot{u}_N) \mathrm{d}\Gamma$$

$$= \int_{\Omega} f_i (v_i - \dot{u}_i) \mathrm{d}\Omega + \int_{\Gamma_s} g_i (v_i - \dot{u}_i) \mathrm{d}\Gamma. \tag{10.59}$$

This approach is used by Lions and Magenes (1972) to treat the nonhomogeneous boundary value problems and Duvaut and Lions (1976) for the contact problems with nonhomogeneous boundary conditions.

Note that on the portion of boundary Γ_c without contact, $\sigma_N = 0$, but $\dot{u}_N \leq 0$ is not necessary. This portion of Γ_c has no contribution to (10.58). For simplicity, here we consider Γ_c as the portion of boundary, which is in contact with the obstacle.

The stress on Γ_c satisfies (10.34), $\sigma_N \leq 0$. The Kuhn–Tucker type of contact condition (10.34) gives, for the solution $\boldsymbol{u}$,

$$\sigma_N \dot{u}_N = 0.$$

For weight function $\boldsymbol{v}$, we require $v_N \leq 0$ by (10.34). It follows that

$$\sigma_N v_N \geq 0. \tag{10.60}$$

Note that the Kuhn–Tucker condition is satisfied only by the true solution, not by the arbitrary weight function. Hence, we obtain an inequality

$$\sigma_N (v_N - \dot{u}_N) \geq 0. \tag{10.61}$$

By using (10.61), we obtain from (10.59) an inequality as a reduced form

$$\int_{\Omega} (\rho \ddot{u}_i (v_i - \dot{u}_i) + \sigma_{ij} (v_{i,j} - \dot{u}_{i,j})) \mathrm{d}\Omega$$

$$\geq \int_{\Omega} f_i (v_i - \dot{u}_i) \mathrm{d}\Omega + \int_{\Gamma_s} g_i (v_i - \dot{u}_i) \mathrm{d}\Gamma. \tag{10.62}$$

Here, we have introduced a variational formula with inequality, named variational inequality (VI). The dynamic system (10.55) with frictionless contact to the obstacle (10.34) leads to the variational inequality (10.62), stated below,

Problem C1–VI *Denote by*

$$\begin{aligned} V_0 &= \{v_i \in H^1(\Omega),\ v_i|_{\Gamma_u} = U_i\}, \\ V_1 &= \{v_i \in H^1(\Omega),\ v_i|_{\Gamma_u} = \dot{U}_i\}, \end{aligned} \tag{10.63}$$

for the functions mapping $[0, T]$ *to functions on* Ω. *Define*

$$K = \{v_i \in V_1;\ v_i|_{\Gamma_c} \le 0\} \tag{10.64}$$

for the admissible functions satisfying the contact condition on velocity. Assume that $\sigma_{Ti}|_{\Gamma_c} = 0$. *For any time* $t \in [0, T]$ *find* $u_i \in V_0$, *which satisfies the initial condition (10.55b) and* $\dot{u}_i \in K$, *such that the variational inequality (10.62) is satisfied for any* $v_i \in K$.

The above discussion demonstrates that the solution of Problem C1 satisfies the variational principle (10.62), and therefore is a solution of Problem C1–VI. We show briefly the reverse side. Assume that (10.55f) and (10.55g) are satisfied. Assume that $u_i \in V_0$ is a solution of Problem C1–VI. Let $v_i = \dot{u}_i + w_i$, with $w_i \in C^\infty(\Omega)$ satisfying $w_i|_{\Gamma_u \cup \Gamma_c} = 0$. Thus $v_i \in K$. Then (10.62) gives

$$\int_\Omega (\rho \ddot{u}_i w_i + \sigma_{ij} w_{i,j}) \mathrm{d}\Omega \ge \int_\Omega f_i w_i \mathrm{d}\Omega + \int_{\Gamma_s} g_i w_i \mathrm{d}\Gamma. \tag{10.65}$$

Using $-w_i$ to replace w_i, (10.65) results in an equality

$$\begin{aligned} \int_\Omega (\rho \ddot{u}_i w_i + \sigma_{ij} w_{i,j}) \mathrm{d}\Omega &= \int_\Omega (\rho \ddot{u}_i - \sigma_{ij,j}) w_i \mathrm{d}\Omega + \int_{\Gamma_s} \sigma_{ij} N_j w_i \mathrm{d}\Gamma \\ &= \int_\Omega f_i w_i \mathrm{d}\Omega + \int_{\Gamma_s} g_i w_i \mathrm{d}\Gamma. \end{aligned} \tag{10.66}$$

Furthermore, by setting $w_i|_{\Gamma_s} = 0$, we obtain $\int_\Omega (\rho \ddot{u}_i - \sigma_{ij,j}) w_i \mathrm{d}\Omega = \int_\Omega f_i w_i \mathrm{d}\Omega$. What follows is $\rho \ddot{u}_i - \sigma_{ij,j} = f_i$, that is, (10.55a) is satisfied weakly, in the distribution sense. The remaining of (10.66) is then $\int_{\Gamma_s} \sigma_{ij} N_j w_i \mathrm{d}\Gamma = \int_{\Gamma_s} g_i w_i \mathrm{d}\Gamma$. This leads to $\sigma_{ij} N_j|_{\Gamma_s} = g_i$, that is, (10.55d) is satisfied in the distribution sense.

Turning back to (10.62) and using Gauss–Green theorem, with (10.55a) and (10.55d) satisfied, we have

$$\int_{\Gamma_c} \sigma_{ij} N_j (v_i - \dot{u}_i) \mathrm{d}\Gamma \ge 0. \tag{10.67}$$

The decomposition of components with the assumption $\sigma_{Ti}|_{\Gamma_c} = 0$ results in

$$\int_{\Gamma_c} \sigma_N (v_N - \dot{u}_N) \mathrm{d}\Gamma \geq 0. \tag{10.68}$$

We show by contradiction that $\sigma_N \dot{u}_N = 0$, almost everywhere on Γ_c. If this is not true, we can find $\sigma_N \dot{u}_N > 0$ (or <0 respectively) at some point $\boldsymbol{x} \in \Gamma_c$, and then a neighborhood δ_x with positive measure, where $\sigma_N \dot{u}_N > 0$ (or <0 respectively). Choose v_i such that $v_N|_{\Gamma_c \setminus \delta_x} = 0$. Let $v_N = 0$ or $v_N = 2\dot{u}_i$ in δ_x. Then we obtain $\int_{\Gamma_c} \sigma_N \dot{u}_N \mathrm{d}\Gamma = 0$, which is a contradiction. Therefore $\sigma_N \dot{u}_N = 0$, almost everywhere on Γ_c, that is, (10.55e) is satisfied.

Note that a solution of (10.55) (strong or weak) always satisfies (10.62). The solution of (10.62) is in the distribution sense, therefore satisfies (10.55) weakly. Whether (10.55) has a strong solution is not solidly proved yet.

The pioneer development of variational inequality traces back to the mid-1960s. VI is now a well-established mathematical tool, powerful in treating constrained systems. The fundamental theories and various applications of VI can be found in Stampacchia (1964), Brezis and Lions (1967), Lions and Stampacchia (1967), Lions (1969), Glowinski et al. (1981), Hlaváček et al. (1988) and the references cited there. Applications of variational inequality to the contact mechanics problems are extensively developed in Kikuchi and Oden (1988), Hlaváček et al. (1988), Han and Sofonea (2002), and Eck et al. (2005). Other applications of VI can be found in, for example, Johnson (1976), Li and Babuška (1996), and Han and Reddy (1999) for plasticity; Oden and Wu (1985) and Wu (1986) for elastohydrodynamic lubrication.

10.3.2 Variational Formulation for Frictional Dynamic Contact Problem

To consider friction in the dynamic system, we keep the tangential component σ_{Ti}. The condition for the normal component is still valid. We have the dynamic frictional contact problem described as

Problem C2

$$\rho \ddot{u}_i - \sigma_{ij,j} = f_i(t, x) \; in \; \Omega, \quad i, j = 1, 2, 3, \tag{10.69a}$$

$$u_i(0, \boldsymbol{x}) = U_i^0(\boldsymbol{x}), \quad \dot{u}_i(0, \boldsymbol{x}) = U_i^1(\boldsymbol{x}) \; in \; \Omega, \tag{10.69b}$$

$$u_i = U_i(t, \boldsymbol{x}) \; on \; \Gamma_u, \tag{10.69c}$$

$$\sigma_{ij} n_j = g_i(t, \boldsymbol{x}) \; on \; \Gamma_s, \tag{10.69d}$$

$$\begin{cases} \dot{u}_N \leq 0, \quad \sigma_N \leq 0, \quad \sigma_N \dot{u}_N = 0 \\ \sigma_T \leq \mu|\sigma_N|, \\ \sigma_T < \mu|\sigma_N| \Rightarrow \dot{u}_{Ti} = 0 \\ \sigma_T = \mu|\sigma_N| \Rightarrow \exists\lambda > 0, \quad \dot{u}_{Ti} = -\lambda\sigma_{Ti} \end{cases} \quad on\ \Gamma_c, \tag{10.69e}$$

$$\sigma_{ij} = \sigma_{ij}(E, \nu, E_t, x_i, u_i, \dot{\varepsilon}_{ij}, \ldots), \tag{10.69f}$$

$$\dot{\varepsilon}_{ij} = (\dot{u}_{i,j} + \dot{u}_{j,i})/2. \tag{10.69g}$$

With the first part of (10.58) and (10.61), (10.57) reduced to

$$\int_\Omega (\rho\ddot{u}_i(v_i - \dot{u}_i) + \sigma_{ij}(v_{i,j} - \dot{u}_{i,j}))\mathrm{d}\Omega - \int_{\Gamma_c} \sigma_{Ti}(v_{Ti} - \dot{u}_{Ti})\mathrm{d}\Gamma$$
$$\geq \int_\Omega f_i(v_i - \dot{u}_i)\mathrm{d}\Omega + \int_{\Gamma_s} g_i(v_i - \dot{u}_i)\mathrm{d}\Gamma. \tag{10.70}$$

Recall (10.48) or (10.69e) for Coulomb's friction model, $\boldsymbol{\sigma}_T$ is considered a function of the normal contact force σ_N. The friction force is in the opposite direction of the tangential velocity. Therefore,

$$\sigma_{Ti}\dot{u}_{Ti} = \begin{cases} 0 \text{ if } \sigma_T < \mu|\sigma_N| \\ -\lambda(\sigma_T)^2 = -\sigma_T|\dot{u}_T| \text{ if } \sigma_T = \mu|\sigma_N| \end{cases} \quad \text{on } \Gamma_c. \tag{10.71}$$

The negative sign indicates the dissipative nature of the work done by the friction. A functional to represent the virtual power of friction was introduced by Duvaut and Lions (1976),

$$j(\boldsymbol{u}, \boldsymbol{v}) = \int_{\Gamma_c} \mu|\sigma_N(\boldsymbol{u})|v_T\mathrm{d}\Gamma. \tag{10.72}$$

We can verify the following with the solution $\boldsymbol{u}$:

$$\mu|\sigma_N(\boldsymbol{u})|(v_T - \dot{u}_T) + \sigma_{Ti}(v_{Ti} - \dot{u}_{Ti}) \geq 0. \tag{10.73}$$

Denote by L for the left-hand side of (10.73). According to (10.48) and (10.71), if $\sigma_T < \mu|\sigma_N|$, $\dot{u}_{Ti} = 0$. Then $L = \mu|\sigma_N|v_T + \sigma_{Ti}v_{Ti}$. Meanwhile, $\mu|\sigma_N|v_T > \sigma_T v_T \geq \pm\,\sigma_{Ti}v_{Ti}$. Hence, $L > 0$ in this case. If $\sigma_T = \mu|\sigma_N|$, then there exists $\lambda > 0$, such that $\dot{u}_{Ti} = -\lambda\sigma_{Ti}$ and $\dot{u}_T = \lambda\sigma_T$. We have

$$L = \sigma_T(v_T - \lambda\sigma_T) + \sigma_{Ti}(v_{Ti} + \lambda\sigma_{Ti}) = \sigma_T v_T + \sigma_{Ti}v_{Ti} \geq 0.$$

Introducing (10.73) into (10.70), we obtain

$$\int_{\Omega} (\rho \ddot{u}_i(v_i - \dot{u}_i) + \sigma_{ij}(v_{i,j} - \dot{u}_{i,j}))\mathrm{d}\Omega + j(\boldsymbol{u}, \boldsymbol{v}) - j(\boldsymbol{u}, \dot{\boldsymbol{u}})$$
$$\geq \int_{\Omega} f_i(v_i - \dot{u}_i)\mathrm{d}\Omega + \int_{\Gamma_s} g_i(v_i - \dot{u}_i)\mathrm{d}\Gamma. \tag{10.74}$$

We state the variational inequality for the dynamic frictional problem below.

Problem C2–VI *For any time $t \in [0, T]$ find $u_i \in V_0$, which satisfies the initial condition (10.55b) and $\dot{u}_i \in K$, such that the variational inequality (10.74) is satisfied for any $v_i \in K$.*

The above discussion illustrated that the solution of Problem C2 is a solution of Problem C2–VI. We can also show the reverse side, that the solution of Problem C2–VI is a weak solution of Problem C2. We omit the details here, which can be found in Duvaut and Lions (1976).

We present the relation between the coefficient of friction and velocity for Coulomb's law in Figure 10.12a. When vibrating motion involving friction occurs, the friction jumps from one direction to the opposite direction. Such a discontinuity brings in additional difficulty to the contact problem. In fact, the friction functional

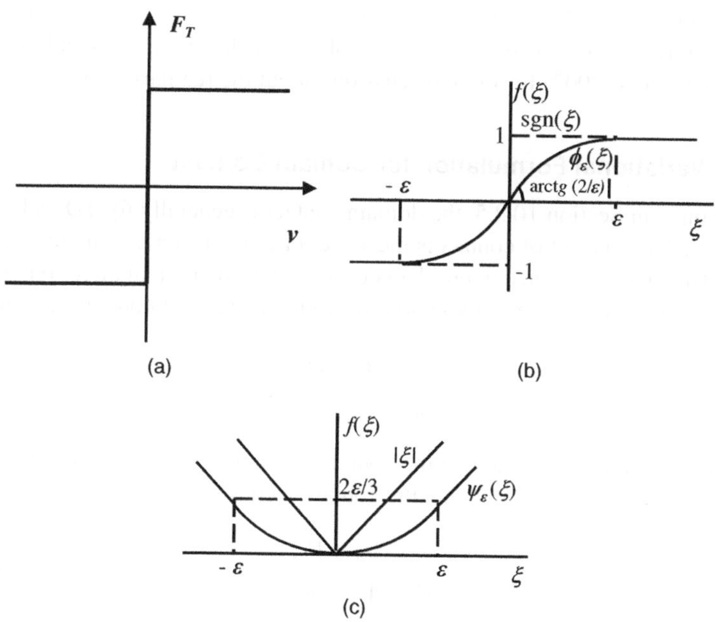

FIGURE 10.12 Friction model: (a) coulumb's law; (b) a regularization model: $\varphi_\varepsilon(v)$; (c) the function $\psi_\varepsilon(v)$, whose Gâteaux differential is $\varphi_\varepsilon(v)$. (Reprinted from Oden and Martins, 1985. Copyright (1985), with permission from Elsevier.)

defined in (10.72) is not differentiable and not monotone. To overcome this hardship, the regularization scheme is developed by Oden and coworkers, published in a series of papers; see for example, Pires (1982), Campos et al. (1982), Oden and Pires (1981, 1983a, 1983b), Demkowicz and Oden (1982), Oden and Martins (1985), and Martins and Oden (1983, 1987). For instance, let $F_T = \varphi(v)\mu F_N$,

$$\varphi(v) = \begin{cases} \dfrac{1}{\varepsilon}\left(2 - \left|\dfrac{\xi}{\varepsilon}\right|\right)\xi, & \text{if } |\xi(x)| \leq \varepsilon, \\ \dfrac{\xi}{|\xi|}, & \text{if } |\xi(x)| > \varepsilon. \end{cases} \tag{10.75}$$

The function φ is plotted in Figure 10.12b. This function is the Gâteaux differential of a convex functional below, shown in Figure 10.12c,

$$\psi(v) = \begin{cases} \varepsilon\left|\dfrac{\xi}{\varepsilon}\right|^2\left(1 - \dfrac{1}{3}\left|\dfrac{\xi}{\varepsilon}\right|\right), & \text{if } |\xi(x)| \leq \varepsilon, \\ \varepsilon\left(\left|\dfrac{\xi}{\varepsilon}\right| - \dfrac{1}{3}\right), & \text{if } |\xi(x)| > \varepsilon. \end{cases} \tag{10.76}$$

The regularization in friction provides convenience to the mathematical treatment of contact problem. Its legitimacy in application of elastodynamics is established by Martins and Oden (1983). The authors proved that when $\varepsilon \to 0$, the solution with regularized friction model converges to the solution with the original Coulomb's law; see also Wriggers (2002) for more discussions about the regularization.

10.3.3 Variational Formulation for Domain Contact

As mentioned in Section 10.2.5, the domain contact is generally for 1D and 2D configurations. The concept of contact is the same for domain contact and the boundary contact. The main difference in problem description is the normal direction. Now we view the domain consisting of two parts: contact zone Ω_1 and noncontact zone Ω_0,

$$\begin{aligned} \bar{\Omega} &= \bar{\Omega}_0 \cup \bar{\Omega}_1, \\ \Omega_0 \cap \Omega_1 &= \varphi. \end{aligned} \tag{10.77}$$

For simplicity, here we exclude the boundary contact associated with the domain contact problem. So we assume that the boundary consists only two parts:

$$\begin{aligned} \bar{\Gamma} &= \bar{\Gamma}_u \cup \bar{\Gamma}_s, \\ \Gamma_u \cap \Gamma_s &= \varphi. \end{aligned} \tag{10.78}$$

The body force, such as the gravity, applies to the whole domain. The contact zone can be viewed under the combined loads of the general body force and the contact force, whereas the governing equation of mechanics is the same for both zones. The contact force is also decomposed into normal and tangential components.

The frictionless and frictional domain contact problems similar to the boundary contact problems are stated below.

Problem C1′

$$\rho\ddot{u}_i - \sigma_{ij,j} = f_i(t,\boldsymbol{x}) + f_{ci}(t,\boldsymbol{x}) \; in \; \Omega, \quad i,j = 1,2,3, \tag{10.79a}$$

$$u_i(0,\boldsymbol{x}) = U_i^0(\boldsymbol{x}), \quad \dot{u}_i(0,\boldsymbol{x}) = U_i^1(\boldsymbol{x}) \; in \; \Omega, \tag{10.79b}$$

$$u_i = U_i(t,\boldsymbol{x}) \; on \; \Gamma_u, \tag{10.79c}$$

$$\sigma_{ij}n_j = g_i(t,\boldsymbol{x}) \; on \; \Gamma_s, \tag{10.79d}$$

$$\begin{cases} \dot{u}_N \le 0, \quad f_{cN} \le 0, \quad f_{cN}\dot{u}_N = 0; \\ f_{cTi} = 0 \end{cases} \; in \; \Omega_1, \quad f_{ci} \equiv 0 \; in \; \Omega_0, \tag{10.79e}$$

$$\sigma_{ij} = \sigma_{ij}(E, \nu, E_t, x_i, u_i, \dot{\varepsilon}_{ij}, \dots), \tag{10.79f}$$

$$\dot{\varepsilon}_{ij} = (\dot{u}_{i,j} + \dot{u}_{j,i})/2. \tag{10.79g}$$

Problem C2′

$$\rho\ddot{u}_i - \sigma_{ij,j} = f_i(t,\boldsymbol{x}) + f_{ci}(t,\boldsymbol{x}) \; in \; \Omega, \quad i,j = 1,2,3, \tag{10.80a}$$

$$u_i(0,\boldsymbol{x}) = U_i^0(\boldsymbol{x}), \quad \dot{u}_i(0,\boldsymbol{x}) = U_i^1(\boldsymbol{x}) \; in \; \Omega, \tag{10.80b}$$

$$u_i = U_i(t,\boldsymbol{x}) \; on \; \Gamma_u, \tag{10.80c}$$

$$\sigma_{ij}n_j = g_i(t,\boldsymbol{x}) \; on \; \Gamma_s, \tag{10.80d}$$

$$\begin{cases} \dot{u}_N \le 0, \quad f_{cN} \le 0, \quad f_{cN}\dot{u}_N = 0 \\ f_{cT} \le \mu|f_{cN}|, \\ f_{cT} < \mu|f_{cN}| \Rightarrow \dot{u}_{Ti} = 0 \\ f_{cT} = \mu|f_{cN}| \Rightarrow \exists\lambda > 0, \dot{u}_{Ti} = -\lambda\sigma_{Ti} \end{cases} \; in \; \Omega_1, \quad f_{ci} \equiv 0 \; in \; \Omega_0, \tag{10.80e}$$

$$\sigma_{ij} = \sigma_{ij}(E, \nu, E_t, x_i, u_i, \dot{\varepsilon}_{ij}, \dots), \tag{10.80f}$$

$$\dot{\varepsilon}_{ij} = (\dot{u}_{i,j} + \dot{u}_{j,i})/2. \tag{10.80g}$$

The corresponding variational principles can be derived in a similar way.

Problem C1′–VI *Define*

$$K_1 = \{v_i \in V_1, \quad v_N \le 0, \; in \; \Omega_1\} \tag{10.81}$$

for the admissible functions satisfying the contact condition on velocity. Assume that $f_{cTi}|_{\Omega_1} = 0$. *For any time* $t \in [0,T]$ *find* $u_i \in V_0$, *which satisfies the initial condition (10.79b) and* $\dot{u}_i \in K_1$*, such that the following variational inequality is satisfied for any* $v_i \in K_1$*:*

$$\int_\Omega (\rho \ddot{u}_i(v_i - \dot{u}_i) + \sigma_{ij}(v_{i,j} - \dot{u}_{i,j}))\mathrm{d}\Omega \\ \geq \int_\Omega f_i(v_i - \dot{u}_i)\mathrm{d}\Omega + \int_{\Gamma_s} g_i(v_i - \dot{u}_i)\mathrm{d}\Gamma. \tag{10.82}$$

Problem C2′–VI *For any time* $t \in [0, T]$ *find* $u_i \in V_0$*, which satisfies the initial condition (10.80b) and* $\dot{u}_i \in K_1$*, such that the following variational inequality is satisfied for any* $v_i \in K_1$*:*

$$\int_\Omega (\rho \ddot{u}_i(v_i - \dot{u}_i) + \sigma_{ij}(v_{i,j} - \dot{u}_{i,j}))\mathrm{d}\Omega + j_1(\boldsymbol{u}, \boldsymbol{v}) - j_1(\boldsymbol{u}, \dot{\boldsymbol{u}}) \\ \geq \int_\Omega f_i(v_i - \dot{u}_i)\mathrm{d}\Omega + \int_{\Gamma_s} g_i(v_i - \dot{u}_i)\mathrm{d}\Gamma, \tag{10.83}$$

where the friction functional is defined as

$$j_1(\boldsymbol{u}, \boldsymbol{v}) = \int_{\Omega_1} \mu |f_{cN}(\boldsymbol{u})| v_T \mathrm{d}\Omega. \tag{10.84}$$

The equivalence of the solutions of the constrained partial differential equation systems and the corresponding variational inequalities can be shown in a similar way, which is not repeated here.

10.4 PENALTY METHOD AND THE REGULARIZATION OF VARIATIONAL INEQUALITY

The penalty method has been applied to variational inequalities such as those, which have arisen in contact problems. The penalty method is suitable for the explicit finite element method because of its iterative nature.

10.4.1 Concept of Penalty Method

To cast the idea of penalty method for contact problem, we recall the example discussed in Section 10.1.1. For the unconstrained problem (10.1), it is understood that the principle of minimum potential energy is equivalent to the PDE. We state as

Problem M_0 *Find the minimizer of the potential energy functional*

$$F(v) = \int_0^{2L} (Tv'^2/2 + fv)\mathrm{d}x, \quad v(0) = v(2L) = 0. \tag{10.85}$$

For the constrained problem (10.3), a domain contact problem, the principle of minimum potential energy is still applicable but with constraint condition. Note that the contact force does not contribute work in this case. Oden and Kikuchi (1980) stated the equivalence of the PDE system for Signorini problem and the energy principle,

Problem M_1 *Find $u \in K_1 = \{v \in V | v \geq -d\}$, such that*

$$F(u) \leq F(v) = \int_0^{2L} (Tv'^2/2 + fv)\mathrm{d}x, \quad \forall v \in K_1. \tag{10.86}$$

Here, the constraint is represented by the admissible subset K_1. For simplicity, we define the functional space $V = H_0^1(0, 2L)$. The contact force makes no contribution to the energy functional, therefore, does not appear in (10.86). These authors formulated a variational inequality and showed its equivalence to Problem M_I.

Problem V_1 *Find $u \in K_1 = \{v \in V | v \geq -d\}$, such that*

$$\int_0^{2L} (-Tu'' + f)(v - u)\mathrm{d}x \geq 0, \quad \forall v \in K_1. \tag{10.87}$$

In fact, the variational inequality has the same solution of the PDE (10.3) and the constrained Problem M_1.

It is a challenging task to solve the variational inequality directly. Although several iterative schemes have been reported, we do not believe to have seen them implemented in commercial software as a practical algorithm. The penalty method has been proven as a powerful tool to regularize the variational inequality and then use traditional approach to solve a variational principle on the whole space. Here, we introduce a penalty term for (10.87) and formulate the penalty method.

Problem V_p *Find $u_\varepsilon \in V$, such that, $\forall v \in V$*

$$F_\varepsilon(u_\varepsilon) \leq F_\varepsilon(v) = \int_0^{2L} (Tv'^2/2 + fv + (v + d)_-^2/2\varepsilon)\mathrm{d}x. \tag{10.88}$$

Comparing to (10.86), there is an additional term in the functional. Here, we denote by $(\bullet)_-$ for the negative part of a function

$$g(x)_- = \min(0, g(x)) = (g(x) - |g(x)|)/2 (\leq 0). \tag{10.89}$$

If the ideal solution u_ε is found belonging to K_1, then $(u_\varepsilon + d)_- = 0$ and it is in fact the solution of (10.86). Since we are looking for a minimizer in the whole space V, the solution does not have to satisfy the constraint condition. If the minimizer $u_\varepsilon \notin K_1$, there is an additional term $(u_\varepsilon + d)_-^2/2\varepsilon > 0$. It looks like a penalty to the minimal energy.

The importance of the penalty method is the relation between u_ε and u as $\varepsilon \to 0$. The penalty energy is proportional to ε^{-1}. It requires $(u_\varepsilon + d)_- \to 0$ to avoid a high energy penalty. Subsequently, u_ε approaches to the solution of the variational inequality so that the variational principle makes sense. This is however by no means a trivial task. For certain class of contact problems, the convergence of the penalty method has been established; see Oden and Kikuchi (1980) and Kikuchi and Oden (1988) for elastic contact. The similar theory for certain type of dynamic contact problem was developed in Lions (1969). The method has been extended to other types of contact problem; see Wu (1986) for the elastohydrodynamic lubrication.

We do not intend to describe the mathematical details here, but will instead continue the discussion of the above example to illustrate the concept of penalty method as used in practical applications.

Use the usual minimization procedure $\delta F(v) = 0$ for (10.88), we arrive at

$$\int_0^{2L} (Tu'_\varepsilon v' + fv + (u_\varepsilon + d)_- v/\varepsilon)\mathrm{d}x = 0, \quad \forall v \in V$$

$$-Tu''_\varepsilon + f + (u_\varepsilon + d)_-/\varepsilon = 0. \tag{10.90}$$

By symmetry we consider only half of the string $[0, L]$. Assume there is only one contact point $x_{1\varepsilon} \in [0, L]$, $u_\varepsilon(x_{1\varepsilon}) = -d$; $u_\varepsilon > -d$ for $x \in [0, x_{1\varepsilon})$ and $u_\varepsilon < -d$ for $x \in (x_{1\varepsilon}, L]$. We can rewrite (10.90)

$$\begin{aligned} &-Tu''_\varepsilon + f = 0, \quad \text{in } \Omega_0 = [0, x_{1\varepsilon}], \\ &-Tu''_\varepsilon + f + (u_\varepsilon + d)/\varepsilon = 0, \quad \text{in } \Omega_1 = [x_{1\varepsilon}, L], \\ &u_\varepsilon(0) = 0, \quad u'_\varepsilon(L) = 0. \end{aligned} \tag{10.91}$$

The loading condition is illustrated in Figure 10.13. Note that the "additional" term in the second equation plays the role of the penalty contact force with $1/\varepsilon$ as the stiffness and $(u_\varepsilon + d)$ as the elastic deformation. This resembles the Winkler Foundation model.

We can find a solution, cf. Wu (2001),

$$\begin{aligned} u_\varepsilon &= fx^2/2T + C_{1\varepsilon}x, \quad \text{in } \Omega_0, \\ u_\varepsilon &= C_{2\varepsilon}\mathrm{ch}\lambda(L - x) - (d + \varepsilon f), \quad \text{in } \Omega_1, \end{aligned} \tag{10.92}$$

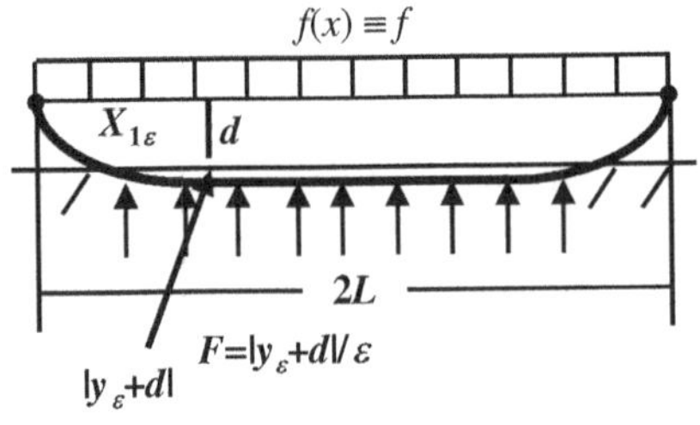

FIGURE 10.13 Penalty method for the string under uniform loading with a rigid obstacle.

where $\lambda = 1/\sqrt{\varepsilon T}$. The C^1 continuity at the contact point $x_{1\varepsilon}$ leads to

$$\begin{aligned} fx_{1\varepsilon}^2/2T + C_{1\varepsilon}x_{1\varepsilon} &= -d = C_{2\varepsilon}\text{ch}\lambda(L - x_{1\varepsilon}) - (d + \varepsilon f), \\ fx_{1\varepsilon}/T + C_{1\varepsilon} &= -C_{2\varepsilon}\lambda\text{sh}\lambda(L - x_{1\varepsilon}) = -\eta. \end{aligned} \tag{10.93}$$

It follows that

$$C_{2\varepsilon} = \varepsilon f/\text{ch}\lambda(L - x_{1\varepsilon}),$$

$$\eta = \varepsilon f \lambda \text{th}\lambda(L - x_{1\varepsilon}).$$

We can express the solution of (10.93) by

$$\begin{aligned} x_{1\varepsilon} &= -\frac{T\eta}{f} + \sqrt{\frac{T^2\eta^2}{f^2} + \frac{2Td}{f}}, \\ C_{1\varepsilon} &= -\sqrt{\eta^2 + \frac{2fd}{T}}. \end{aligned} \tag{10.94}$$

For small $\varepsilon, \lambda = 0(\varepsilon^{-1/2}), \varepsilon\lambda = 0(\varepsilon^{1/2})$, and $\eta = o(\varepsilon^{1/2})$. We simply obtain asymptotically

$$\begin{aligned} x_{1\varepsilon} &= \sqrt{2Td/f} + O(\eta), \\ C_{1\varepsilon} &= -\sqrt{2fd/T} + O(\eta^2). \end{aligned} \tag{10.95}$$

Their limits as $\varepsilon \to 0$ turn out to be the same as (10.6), the solution of the Signorini problem (10.3). In addition, we find the error bounds

$$\begin{aligned} x_{1\varepsilon} - x_1 &= O(\eta) = O(\varepsilon^{1/2}), \\ C_{1\varepsilon} - C_1 &= O(\varepsilon), \\ |u_\varepsilon - u| &= O(\varepsilon). \end{aligned} \tag{10.96}$$

We can also verify that the contact point $x_{1\varepsilon}$ is unique. When checking the second derivative at the contact point, we find

$$u_\varepsilon''(x_{1\varepsilon} - 0) = f/T = u_\varepsilon''(x_{1\varepsilon} + 0). \tag{10.97}$$

But $u_\varepsilon'''(x_{1\varepsilon} - 0) \neq u_\varepsilon'''(x_{1\varepsilon} + 0)$. The C^2 continuity, higher than that of the original problem, is another feature of the penalty method.

10.4.2 Penalty Method for Nonlinear Dynamic Contact Problem

Consider the contact with a rigid obstacle. Recall the variational inequality (10.82) and (10.83) for the dynamic contact problem, without and with friction, respectively. As the continuation of the discussion in Section 10.4.1, we propose a penalty term in the sense of rate form for the contact point $\boldsymbol{x} \in \Gamma_c$,

$$\frac{1}{\varepsilon_v}(\dot{\boldsymbol{u}})_{N+}v_N. \tag{10.98}$$

Here for any function f, we denote its positive part by

$$f_+ = (f + |f|)/2. \tag{10.99}$$

Similar to the case discussed in Section 10.4.1, $(\dot{u})_{N+} > 0$ means penetration and results in a penalty force. $1/\varepsilon_v$ has the meaning of the viscosity.

As discussed in Section 10.2.4, we need the "in-contact" condition (10.52) to determine the status of contact. The condition regarding position vector can be also treated by a penalty method. According to the nature of the penalty method, the information regarding where penetration occurs is needed so that the penalty force is applied at the right place. The contact condition (10.34) of velocity form gives only the restriction on current velocity. There is no information about current location of contact or the location where penetration has already occurred. Penetration is allowed by the penalty method and is exactly the key to the penalty method. A geometrical penalty term for penetration is recently proposed by Wu (2009). For any point $\boldsymbol{x} \in \Gamma_c$, where contact occurs, find the impact location $\boldsymbol{y} = \boldsymbol{y}(\boldsymbol{x})$ as the point on the obstacle surface, lying on the normal $\boldsymbol{N}$ at $\boldsymbol{x} \in \Gamma_c$. The penalty term is defined by

$$\frac{1}{\varepsilon_x}(\boldsymbol{x} - \boldsymbol{y})_{N+}v_N. \tag{10.100}$$

Here $1/\varepsilon_x$ represents stiffness. In fact, this penalty term corresponds to condition (10.30). If we consider $\boldsymbol{y}$ as the impacted point,

$$\boldsymbol{y}{:}S(\boldsymbol{y}) = 0, \quad \boldsymbol{y} \in \boldsymbol{N}(\boldsymbol{x}), \tag{10.101}$$

then $(\boldsymbol{x} - \boldsymbol{y}) \bullet \boldsymbol{N} = (\boldsymbol{x} - \boldsymbol{y})_N > 0$ represents penetration. The penetration results in a penalty contact force by (10.100). Considering $\boldsymbol{x} = \boldsymbol{r}(t, \boldsymbol{X}) = \boldsymbol{X} + \boldsymbol{u}(t, \boldsymbol{X})$, (10.100) is in fact related to the total displacement, which is however not formulated as the main variable here. This is a major difference to the small deformation problem.

In this way, the portions of Γ_c with or without contact to the obstacle are identified. As a matter of fact, the situation mentioned in Section 10.3.1 is thus resolved without difficulty.

As discussed in Section 10.2, we should not simply claim that $(x - y)_{N+} > 0$ means penetration, although the reverse side is generally true: that is, $(x - y)_{N+} < 0$ means no penetration at $\boldsymbol{y}$ by $\boldsymbol{x}$. Hence, the application of this penalty term for large deformation needs more careful examination. It is important to make sure that the candidate contact point $\boldsymbol{x}$ is in the small neighborhood of $\boldsymbol{y}$ of the contacted surface. The overall distance should be small, before considering the normal distance; see more discussions in Chapter 11 for the development of numerical procedures.

In summary, the variational principle for the nonlinear frictional dynamic problem with the proposed penalty method is as following:

Problem C2–VP *For any time $t \in [0, T]$ find $u_{\varepsilon i} \in V_0$, which satisfies the initial condition (10.80b) and $\ddot{u}_{\varepsilon i} \in V_1$, such that the following variational equation is satisfied for any $v_{\varepsilon i} \in V = \{v_{\varepsilon i} \in H^1(\Omega),\ v_{\varepsilon i}|_{\Gamma_u} = 0\}$,*

$$\int_{\Omega} ((\rho(\ddot{\boldsymbol{u}}_{\varepsilon})_i(\boldsymbol{v}_{\varepsilon})_i + (\boldsymbol{\sigma}_{\varepsilon})_{ij}(\boldsymbol{v}_{\varepsilon})_{i,j}))\mathrm{d}\Omega - \int_{\Gamma_c} (\boldsymbol{\sigma}_{\varepsilon})_{Ti}(\boldsymbol{v}_{\varepsilon})_{Ti}\mathrm{d}\Gamma$$
$$+ \int_{\Gamma_c} \left(\frac{1}{\varepsilon_v}(\dot{\boldsymbol{u}}_{\varepsilon})_{N+}(\boldsymbol{v}_{\varepsilon})_N + \frac{1}{\varepsilon_x}(\boldsymbol{x} - \boldsymbol{y}(\boldsymbol{x}))_{N+}(\boldsymbol{v}_{\varepsilon})_N\right)\mathrm{d}\Gamma$$
$$= \int_{\Omega} f_i(\boldsymbol{v}_{\varepsilon})_i\mathrm{d}\Omega + \int_{\Gamma_s} g_i(\boldsymbol{v}_{\varepsilon})_i\mathrm{d}\Gamma. \tag{10.102}$$

Here, $\boldsymbol{\sigma}_{\varepsilon}$ is determined by using (10.80f).

In view of the variational inequality of Problem C2–VI, if the penetration occurs, the penalty terms (10.98) and (10.100) are active. Their contributions to the left-hand side of (10.102) may be positive and consequently the variational inequality (10.74) will not be valid by $u_{\varepsilon i} \in V_0$. This gives the meaning to the penalty method for the nonlinear dynamic contact problems.

It is worth noting that the penalty term in (10.88) for minimizing the potential energy of the constrained string is a quadratic form, which results in a linear term in the penalty equation. The penalty terms of (10.98) and (10.100) are similar to that of (10.88). In fact, the penalty term is not restricted by using the quadratic form. As long as the penalty functional is bounded, the penalty method is valid for a general type of variational inequalities. An unbounded penalty functional, however, can cause instability due to the time step size requirement by explicit scheme. For theories about penalty method, we are referred to Oden and Kikuchi (1980) and Oden (1985).

Note that $\boldsymbol{x} = \boldsymbol{X} + \boldsymbol{u}(t, \boldsymbol{X})$. $\boldsymbol{x}$ can be considered as function of displacement, as also with the case of $\boldsymbol{y}$. Since the normal changes in time, the displacement condition (10.32) for normal component of displacement cannot be integrated from velocity condition (10.34) directly. On the other hand, in view of the first order approximation using Taylor expansion, we have the function and its first derivative, which resemble the two penalty terms in (10.102). Verifying the location or using the gap function is often seen in the literature, for example, Kikuchi and Oden (1988), Jones and Papadopoulos (2001), and Kloosterman et al. (2001). This is the basic approach implemented in a few of software; see later discussions in Chapter 11. Inclusion of both conditions: location and velocity, in the variational equations was discussed in Heinstein et al. (2000) and Belytschko et al. (2000), among others. The variational principle including both penalty terms was demonstrated in the recent study by Wu (2009).

For the domain contact with friction, Problem C2'–VI, the corresponding variational principle with the proposed penalty method is as follows,

Problem C2′–VP *For any time $t \in [0, T]$ find $u_{\varepsilon i} \in V_0$, which satisfies the initial condition (10.80b) and $\dot{u}_{\varepsilon i} \in V_1$, such that the following variational equation is satisfied for any $v_{\varepsilon i} \in V = \{v_{\varepsilon i} \in H^1(\Omega),\ v_{\varepsilon i}|_{\Gamma_u} = 0\}$.*

$$\begin{aligned}
&\int_\Omega ((\rho(\ddot{\boldsymbol{u}}_\varepsilon)_i(\boldsymbol{v}_\varepsilon)_i + (\boldsymbol{\sigma}_\varepsilon)_{ij}(\boldsymbol{v}_\varepsilon)_{i,j}))\mathrm{d}\Omega - \int_\Omega (\boldsymbol{\sigma}_\varepsilon)_{Ti}(\boldsymbol{v}_\varepsilon)_{Ti}\mathrm{d}\Omega \\
&\quad + \int_\Omega \left(\frac{1}{\varepsilon_v}(\dot{\boldsymbol{u}}_\varepsilon)_{N+}(\boldsymbol{v}_\varepsilon)_N + \frac{1}{\varepsilon_x}(\boldsymbol{x} - \boldsymbol{y}(\boldsymbol{x}))_{N+}(\boldsymbol{v}_\varepsilon)_N\right)\mathrm{d}\Omega \\
&\quad = \int_\Omega f_i(\boldsymbol{v}_\varepsilon)_i\mathrm{d}\Omega + \int_{\Gamma_s} g_i(\boldsymbol{v}_\varepsilon)_i\mathrm{d}\Gamma.
\end{aligned} \tag{10.103}$$

Remark 10.5 In general, variational formulation contains only the variables of interest, such as stress and velocity, which are treated as unknowns. Here in (10.102) and (10.103), besides the primary variables, we have included the locations of points. The locations involve the entire histories of deformation, which do not explicitly appear in the system of differential equations, for example, (10.69) and (10.80). The implementation in software is therefore not trivial.

10.4.3 Explicit Finite Element Procedure with Penalty Method for Dynamic Contact

Comparing to the system discussed in Chapter 2, we have included contact force and frictional force in (10.102) and (10.103) by penalty method. The variational equations for the penalized systems have just a few more terms than the classical formulation (2.9) without contact. The requirements on the variables and the test functions are basically the same as those for (2.9).

Conceptually the location where contact occurs is found by checking the violation to no-penetration condition (10.53). We propose to accomplish it point-wisely without solving any system equation. In the discretized system, this can be done by checking all the nodes. Once the contact location is identified, the penalty terms are applied and the friction force is calculated. This procedure yields the contact force as resistant force to the motion. The action is also taken point-wisely. The related finite element equations are updated from (2.19) and (2.21):

$$\begin{aligned}
\partial_{t^2}(\boldsymbol{u}_\varepsilon^h)_n &= \boldsymbol{M}^{-1}\boldsymbol{F}_n, \\
(\boldsymbol{F}_n)_i^M &= -\int_\Omega (\sigma_{ij}\Phi_{M,j})\mathrm{d}\Omega + \int_\Omega f_i\Phi_M\mathrm{d}\Omega + \int_{\Gamma_s} g_i\Phi_M\mathrm{d}\Gamma + F_{Mi}^{\text{contact}}, \\
F_{Mi}^{\text{contact}} &= -\left(\frac{1}{\varepsilon_x}(\boldsymbol{x}_M - \boldsymbol{y}_M)_{N+} + \frac{1}{\varepsilon_v}(\dot{u}(\boldsymbol{x}_M) - \dot{u}(\boldsymbol{y}_M))_{N+}\right)(\boldsymbol{N}_{x_M})_i \\
&\quad - |\boldsymbol{F}(\boldsymbol{x}_M)_t|(\boldsymbol{t}_{x_M})_i,
\end{aligned} \tag{10.104}$$

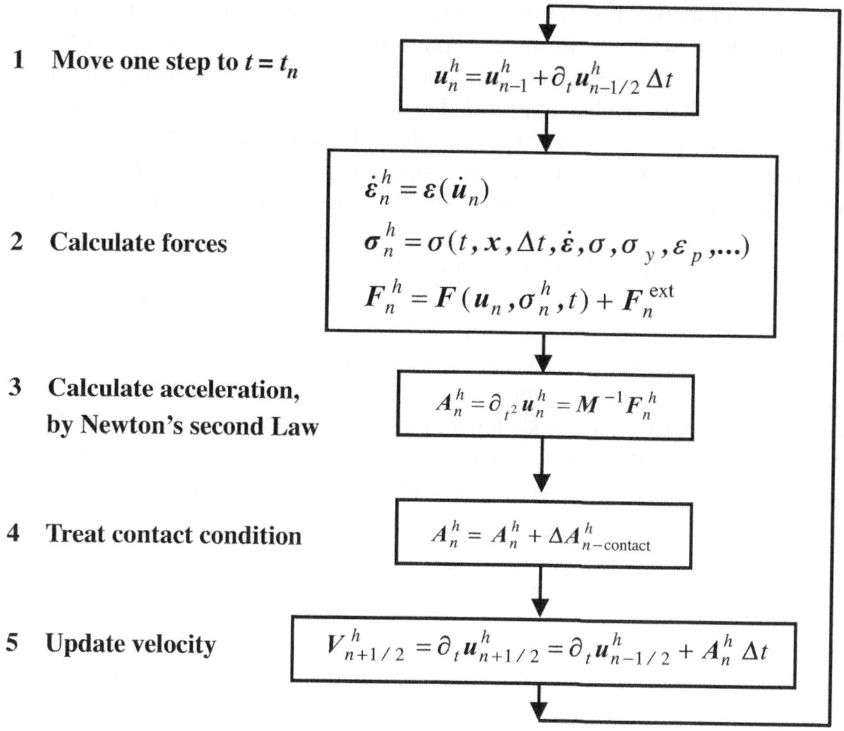

FIGURE 10.14 Explicit finite element procedure—with contact (extended from Figure 7.10).

where F_{Mi} represents the impact force, applied at the node where the penetration is found. For node without contact and penetration, there is no such additional nodal contact force. $\boldsymbol{y}_M$ represents the location impacted by $\boldsymbol{x}_M$. $\boldsymbol{N}_{xM}$ and $\boldsymbol{t}_{xM}$ are the normal and tangential directions at $\boldsymbol{x}_M$ respectively. The explicit finite element procedure with the contribution from the contact calculation is proposed in Figure 10.14. This is updated from Figure 7.10.

The numerical process is in the incremental form. Because of the relatively small time step size used in the explicit scheme, any possible penetration by the penalty method in each time step is expected to be small. With each time step, calculation is made to track the penetrated nodes. If the contact force calculated in the last time step is not strong enough to push the penetrated nodes out of contact, the contact force will be continuously exerted on them. This process acts quite differently from the traditional constraint method, which does not allow penetration at all and also bears great difficulty in the computation, particularly for cases of large dynamic deformation. Note that the Lagrange multiplier method, cf. Zhong (1993), Belytschko et al. (2000), and Wriggers (2002), may maintain contact without penetration. Usually this can be done by constraining the impact node with a localized impact zone. In explicit calculation, however, even the equation for Lagrange multiplier is satisfied instantly, the equation of motion is still possibly not exactly satisfied due to central difference approach. The system is still under certain disturbance and the no-penetration is not a final state. Penalty method does not aim no-penetration instantly, and therefore fits well with the explicit approach. For linear elasticity with small deformation, Kikuchi and Oden (1988) proved the equivalence between the penalty method and the perturbed Lagrangian method. The later approach converges to Lagrange multiplier method. Hence, the penalty method and the Lagrange multiplier method are closely related.

When considering shell structure with its mid-surface used for defining the shell element, most engineers will take into account the thickness of the structure and treat the shell as a 3D volume. This is particularly the case when progressive collapse of a thin-walled component occurs and there are multilayer contacts, the thickness of the shell structure is therefore important to overall impact behavior. The idea then is that contact happens when a node touches the outer (or inner) surface instead of the mid-surface, that is,. the surface of the original geometry. When both contact objects are shell structures, the half thickness from both sides is taken into account. The penalty term (10.100) can be modified

$$v_N((\boldsymbol{x} - \boldsymbol{y})_N - g)_+/\varepsilon_x, \; g = (t_1 + t_2)/2. \tag{10.105}$$

In this case, the distance from outer surface to mid-surface, that is, half the thickness plays the role of a gap function discussed in Section 10.2.1. By the nature of penalty method, however, there is no restriction that the node should not penetrate the mid-surface.

CHAPTER 11

NUMERICAL PROCEDURES FOR THREE-DIMENSIONAL SURFACE CONTACT

In early development of finite element method for contact problems, (10.32) or (10.34) was used as constraint conditions. In many applications, the iterative schemes were built in the finite element code to search for unknown contact points and to calculate unknown contact forces. Wilson and Parsons (1970) proposed a differential displacement method for elastic axisymmetric problem involving contact. For a pair of contact nodes, the difference of their displacement was assumed (comparing to non contact situation). Then the stiffness matrix was condensed for half of the paired contact points. The contact force was recovered after solving the reduced system. Conry and Seireg (1971) developed an iterative procedure based on energy method to solve the contact problem for small elastic deformation. The pair of contact points was identified. The contact condition similar to (10.32) was defined as constraint and was treated iteratively. Chan and Tuba (1971) studied the plane problem of linear elasticity involving contact. The nodal contact at any unknown location of the element side was taken into account. Coulomb's law of friction was also introduced. The no-penetration constraint was solved iteratively by using an overrelaxation procedure. The contact points, except those given initially, were found sequentially through an iterative process. Hughes et al. (1976) investigated static and dynamic contact problems in several applications. The constraints, like (10.32) for static and (10.34) for dynamic cases, were solved iteratively. The contact points were assumed paired in case if two deformable material bodies were in contact. Hughes et al. (1977) further discussed the situation when contact happened at any unknown location in

Introduction to the Explicit Finite Element Method for Nonlinear Transient Dynamics, First Edition.
Shen R. Wu and Lei Gu.

the element segment. The contact location was considered as an additional unknown and solved along with the system.

The software implementation for large deformation of nonlinear analysis with contact has been a challenging task. Hallquist et al. (1985) presented a general sliding interface algorithm for two-dimensional (2D) static and dynamic contact in finite element method. The algorithm was applicable for large deformation problems. The method could handle the situations such as pressing into contact with friction or bouncing back (pulling out of contact). The contact was considered a constraint to be satisfied with the mechanics equations and was solved iteratively. A penalty method for surface contact was also proposed. It was followed by a further development from Benson and Hallquist (1990), named single surface contact algorithm, which is applicable to three-dimensional (3D) shell structures.

There have been many numerical methods and algorithms developed with continuous improvement to tackle the contact and impact problems. The survey articles published by researchers all over the world are valuable resources. For instance, see Zhong and Mackerie (1994), Wriggers (1995), and Bourago (2002). We briefly introduce some of these developments in this chapter. It is believed that the contact algorithm has been greatly improved in speed and robustness. In early stage of applications of explicit software, the contact algorithm could take more than 70% of the total computing time. Nowadays, it takes less than 30% of total CPU time. These technical achievements are important components of the explicit finite element software. Their contribution is critical to the success of many applications to become more practical and affordable. These important developments have greatly supported the explicit finite element method and promoted many important applications. As a matter of fact, many problems once perceived to be difficult due to contact issues have become solvable.

11.1 A CONTACT ALGORITHM WITH SLAVE NODE SEARCHING MASTER SEGMENT

Here, we illustrate the fundamentals of the surface contact algorithm mainly referring to Benson and Hallquist (1990). Basically the contact algorithm has two parts. The first step is to identify the contact location, therefore, a geometrical procedure. The second step is to calculate the corresponding contact force, as a mechanical procedure.

Certain preparations need to be in place now. Take into consideration that the contact is an event of an impactor hitting a target. In finite element analysis, due to meshed geometry, the surface contact is viewed as a node hitting a surface segment. This is a paired relation, but not necessarily unique. We adopt the terminologies commonly used in practical applications. The impacting node is called a slave node, whereas the target surface segment is called a master segment. The nodes of the master segment (e.g., 3 or 4 nodes) are called master nodes. The surface segments associated with the slave node are also called slave segments. We collect all the nodes of the impactor as the set of slave nodes and all the segments and the nodes of the target surface as the sets of master segments and master nodes, respectively. The impactor and the target can be clearly declared in some cases, but can also be just the same piece of the material structure. The latter situation is usually called self-contact, whereas the former one is called slave–master contact. When the grouping of slave nodes and master nodes/segments is prepared, these two cases conceptually and essentially make no difference.

In view of a node impacting a segment, the normal at the slave node does not enter our equation. In stead, unlike the situation discussed in Chapter 10, the normal of the master segment will be used for reference. This normal is on the opposite direction of the impactor surface.

11.1.1 Global Search

To find the location where contact occurs, the first important step is to find the candidate impacting node and the paired target surface segment. It is then followed by verifying the contact condition. A natural and rather effective way is to find the nearest master node for any given slave node first, and then to find the right contact segment from the surrounding master segments.

Denote by $\boldsymbol{S}$ for a node in the set of slave nodes. Its nearest master node, denoted by $\boldsymbol{M}$, can be simply found by comparing the distance from the slave node to the master nodes. Pick one master segment, which has $\boldsymbol{M}$ as one of its nodes. Let the neighboring nodes be $\boldsymbol{M}_1$ and $\boldsymbol{M}_2$, depicted in Figure 11.1a. For simplicity, assume that the four-node segment has no warping. Let

$$N = \boldsymbol{r}_{MM_1} \times \boldsymbol{r}_{MM_2}, \quad N = N/|N|. \tag{11.1}$$

Assume that $\boldsymbol{P}$ is the projection of $\boldsymbol{S}$ in the master segment. In the plane containing the master segment, rotate the vector $\boldsymbol{r}_{\mathrm{MP}}$ 90° counterclockwise about $\boldsymbol{M}$ to $\boldsymbol{P}'$, depicted in Figure 11.1b. It is easy to find the angles formed by these line segments satisfying:

$$\alpha' = \alpha + 90^\circ \in [90^\circ, 270^\circ], \quad \beta' = 90^\circ - \beta \in [-90^\circ, 90^\circ].$$

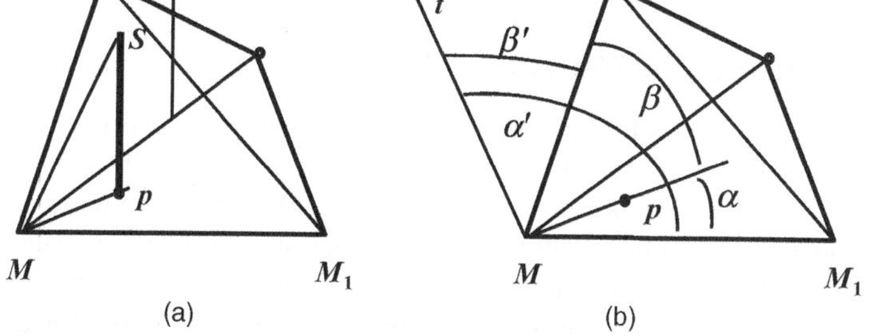

FIGURE 11.1 Projection of a slave node in a master segment: (a) projection of a slave node; (b) a first check for the projection to lie inside the segment.

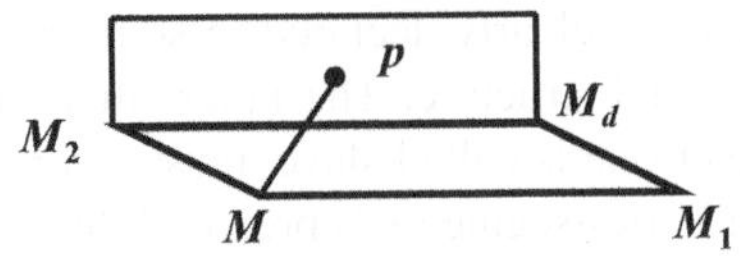

FIGURE 11.2 Misleading with a skewed quadrilateral.

Hence, $\cos\alpha' \leq 0$ and $\cos\beta' \geq 0$. We then derive a criterion for the first check:

$$\begin{aligned} \boldsymbol{r}_{MP'} &= \boldsymbol{N} \times \boldsymbol{r}_{MP}, \\ \boldsymbol{r}_{MP'} \bullet \boldsymbol{r}_{MM_1} &\leq 0, \\ \boldsymbol{r}_{MP'} \bullet \boldsymbol{r}_{MM_2} &\geq 0. \end{aligned} \tag{11.2}$$

After the determination of the normal vector, this process needs only one cross product and two scalar products of vectors. In fact, (11.2) promises that $\boldsymbol{P}$ is inside the corner area formed by lines $\boldsymbol{MM}_1$ and $\boldsymbol{MM}_2$, otherwise it will fall outside the segment. Since $\boldsymbol{M}$ is the nearest master node of $\boldsymbol{S}$, it is expected to be nearest to $\boldsymbol{P}$. Hence, we expect that when (11.2) is satisfied, the projection $\boldsymbol{P}$ will most possibly fall inside the segment. Checking all the master segments surrounding $\boldsymbol{M}$, we can find the candidate that satisfies (11.2). This method applies to three-node triangular segment too.

Note that this method may fail to work for severely skewed element with large aspect ratio, as shown in Figure 11.2. To resolve this issue, it is necessary to spend more CPU time and do more checking. We can use other master nodes to repeat the checking process. A good choice is the master node $\boldsymbol{M_d}$ on the opposite side of the diagonal, shown in Figure 11.2. For triangular segment, any of the other two master nodes can serve the purpose. The intersection of the two corner areas covers the whole segment. If the projection $\boldsymbol{P}$ is also inside the $\boldsymbol{M_d}$–corner area, we can conclude that $\boldsymbol{P}$ is inside the segment. Of course if $\boldsymbol{P}$ is outside the $\boldsymbol{M_d}$–corner area, $\boldsymbol{P}$ is outside the segment and this master segment drops out of the candidate list. Then, we can use the connected master segments to continue searching process.

11.1.2 Bucket Sorting Method

A challenging problem is that we are facing heavy workload involved in searching for the nearest master node corresponding to each slave node, by comparing the distance between the slave nodes and master nodes. Obviously the master segments far away from the slave node will not be contacted and should be temporarily excluded from the search list. The domain where the master nodes as well as the slave nodes reside is then divided into many subsets, named buckets. We hope, for each slave node, the searching is only executed in few buckets with as less master nodes as possible.

A simple approach is to evenly divide the domain of interest into several subsets. For instance, let

$$x_1 = \min\{x^J \mid J = 1, 2, \ldots, \text{number of nodes}\},$$
$$x_2 = \max\{x^J \mid J = 1, 2, \ldots, \text{number of nodes}\}.$$

Then, set $\Delta x = (x_2 - x_1)/K$ and obtain K subdivisions in x-direction. We can similarly work out y- and z-directions. This step results in many buckets in the 3D domain. The master nodes and slave nodes are binned to these buckets according to their coordinates in a natural way. The searching process in the bucket has much less workload than what is needed for searching the whole domain. The method is called bucket sorting.

Because of the complexity of the structures in practice, some of the buckets are end up without any node while others have too many. In addition, if the mesh is not uniform in the domain of interest, then generally the uniform division cannot achieve an even distribution. Certain approach along with the concept of buckets is needed to narrow down the search list.

Here, we brief the efficient method of bucket sorting proposed by Benson and Hallquist (1990). We examine a one-dimensional (1D) case first. We take into consideration a set of line segments and let L be the length of the longest segment. Divide the domain into B buckets with equal width

$$B = (X_2 - X_1)/W + 1, \quad W = L/3. \tag{11.3}$$

With such a division, we expect all the segments to lie within three neighboring buckets (impossible to touch the fourth bucket). In fact, if both nodes of a segment lie outside the three neighboring buckets, then these two nodes have to be on one side of these buckets and the segment has no chance to intersect with these buckets at all. For each of the nodes, we can find a bucket containing it. Meanwhile, we can count the number of nodes falling in each of the buckets. The next step is to find the overlapping pairs.

For a given slave node S in bucket J, depicted in Figure 11.3, we examine all master nodes in the neighboring three buckets $J - 1$, J, and $J + 1$, excluding

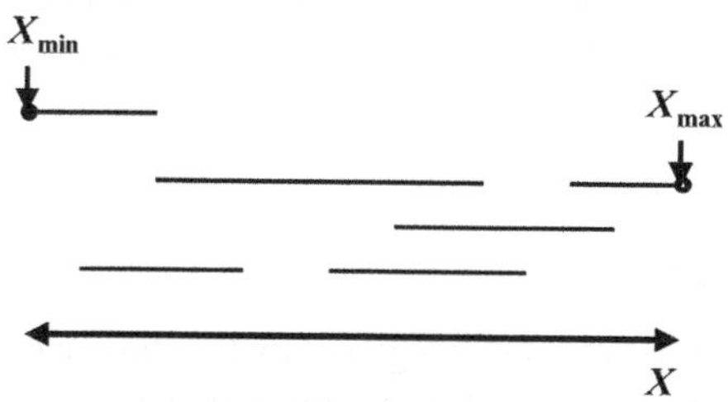

FIGURE 11.3 Concept of three buckets. (Reprinted from Benson and Hallquist, 1990. Copyright (1990), with permission from Elsevier.)

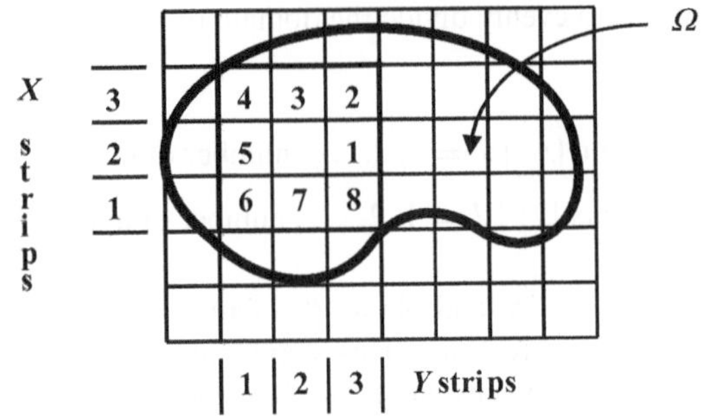

FIGURE 11.4 Process in 2D domain. (Reprinted from Benson and Hallquist, 1990. Copyright (1990), with permission from Elsevier.)

$\boldsymbol{S}$ self and its neighboring nodes, which form segment(s) with $\boldsymbol{S}$. Simply, a node does not impact the segment containing itself. If contact is going to happen, the master segment should have overlapped with the three buckets of the slave node $\boldsymbol{S}$. For any master node $\boldsymbol{M}$ in these three buckets, also find its neighboring node $\bar{\boldsymbol{M}}$, which forms a segment S_M with $\boldsymbol{M}$. If $X(\boldsymbol{S})$ lies inside the interval between $X(\boldsymbol{M})$ and $X(\bar{\boldsymbol{M}})$, node $\boldsymbol{S}$ and segment S_M are considered an overlapping pair. Only the overlapping pairs are considered for contact candidate and treated in the next phase of searching.

When extending the method to 2D or 3D cases, Benson and Hallquist (1990) developed an approach checking the overlapping pairs sequentially in $x/y/z$ directions. After dividing the domain into buckets in $x/y/z$ directions accordingly, the checking at bucket (I, J, K) goes with x-direction for y/z slab I and finds the overlapping pairs in the neighboring three slabs with $I-1, I$, and $I+1$. Then, it processes in y-direction for every Z-slice with the neighboring three slices $J-1, J$, and $J+1$; and finally the z-direction for the three neighboring buckets with $K-1$, K, and $K+1$. A 2D case is illustrated in Figure 11.4.

Obviously for each slave node, the comparisons to be performed are much less than what is needed for searching the whole domain of interest. In addition, if the mesh is relatively uniform, for a specific bucket, the workload of this step has little difference for each node. Although the workload is still not so even with respect to all the buckets, it needs only a simple comparison of coordinates and does not take substantial portion of CPU time. Hence, the computation should be easy for vectorization and parallelization.

11.1.3 Local Search

After identifying the candidate master segment for the slave node, we move to the next step: searching the contact point in this master segment. We consider the impacting in normal direction as normal contact and the impacting in tangential direction as sliding. The contact point hence turns out to be the projection of the slave node.

At this point, we process the quadrilateral segment for the general case, that is, including warping. The position vector of a generic point in the segment can be expressed by using the isoparametric transform with shape functions

$$\boldsymbol{r}_P = \sum \boldsymbol{r}_M \Phi_M(\xi, \eta). \tag{11.4}$$

The normal is perpendicular to the segment, that is, perpendicular to the two tangent vectors $\partial \boldsymbol{r}/\partial \xi_\alpha$. When close enough, the slave node lies on the normal at the contact point. Hence,

$$\boldsymbol{r}_{PS} \bullet \partial \boldsymbol{r}/\partial \xi|_P = \boldsymbol{r}_{PS} \bullet \partial \boldsymbol{r}/\partial \eta|_P = 0, \quad \boldsymbol{r}_{PS} = \boldsymbol{r}_S - \boldsymbol{r}_P. \tag{11.5}$$

Note that the normal varies over the segment but not a constant vector unless the segment is flat. The normal $\boldsymbol{N}$ used in Section 11.1.1 is a representative normal vector of the segment, which is only used for a quick global search. With bilinear interpolation and the shape functions defined in (3.43), we have

$$\begin{aligned}
&\left(\boldsymbol{r}_S - \sum \boldsymbol{r}_N \left(1 + \Lambda_N^1 \xi + \Lambda_N^2 \eta + \Gamma_N \xi \eta\right)\right) \bullet \sum \boldsymbol{r}_M \left(\Lambda_M^1 + \Gamma_M \eta\right) = 0, \\
&\left(\boldsymbol{r}_S - \sum \boldsymbol{r}_N \left(1 + \Lambda_N^1 \xi + \Lambda_N^2 \eta + \Gamma_N \xi \eta\right)\right) \bullet \sum \boldsymbol{r}_M \left(\Lambda_M^2 + \Gamma_M \xi\right) = 0.
\end{aligned} \tag{11.6}$$

This is a nonlinear algebraic system and can be solved iteratively.

The solution (ξ, η) of (11.6) represents the location of $\boldsymbol{P}$, the projection of $\boldsymbol{S}$, shown in Figure 11.5. By interpolation, point $\boldsymbol{P}$ falls inside the segment if and only if

$$|\xi| \leq 1, \quad |\eta| \leq 1. \tag{11.7}$$

This is a second checkpoint. In general, if the check using the second master node is performed as described in Section 11.1.1, $\boldsymbol{P}$ is more possibly falling inside the segment.

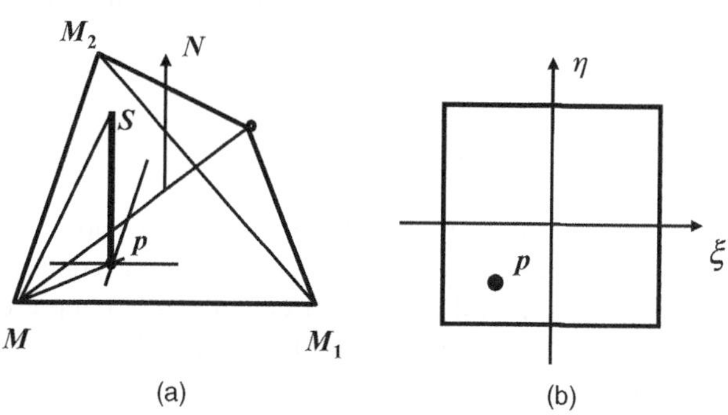

FIGURE 11.5 A second check for the projection to lie inside the segment: (a) the physical domain; (b) the reference domain.

For a flat segment (no warping), the local search is simplified. For instance, we can use the normality conditions $\boldsymbol{r}_{PS} \bullet \boldsymbol{r}_{MM_1} = \boldsymbol{r}_{PS} \bullet \boldsymbol{r}_{MM_2} = 0$ to replace (11.5) and use the following equations to replace (11.6):

$$\begin{aligned} \left(\boldsymbol{r}_S - \sum \boldsymbol{r}_N \left(1 + \Lambda_N^1 \xi + \Lambda_N^2 \eta + \Gamma_N \xi \eta\right)\right) \bullet \boldsymbol{r}_{MM_1} &= 0, \\ \left(\boldsymbol{r}_S - \sum \boldsymbol{r}_N \left(1 + \Lambda_N^1 \xi + \Lambda_N^2 \eta + \Gamma_N \xi \eta\right)\right) \bullet \boldsymbol{r}_{MM_2} &= 0. \end{aligned} \tag{11.6a}$$

This quadratic system can be solved directly.

As an approximate approach, we can split the quadrilateral into two or four triangles and check each of the triangular segments, using $\boldsymbol{r}_{PS} \bullet \boldsymbol{r}_{MM_1} = \boldsymbol{r}_{PS} \bullet \boldsymbol{r}_{MM_2} = 0$,

$$\begin{aligned} (\boldsymbol{r}_S - (\boldsymbol{r}_1(1 - \xi - \eta) + \boldsymbol{r}_2 \xi + \boldsymbol{r}_3 \eta)) \bullet \boldsymbol{r}_{MM_1} &= 0, \\ (\boldsymbol{r}_S - (\boldsymbol{r}_1(1 - \xi - \eta) + \boldsymbol{r}_2 \xi + \boldsymbol{r}_3 \eta)) \bullet \boldsymbol{r}_{MM_2} &= 0. \end{aligned} \tag{11.8}$$

This is a linear system and can be solved directly. The conditions for $\boldsymbol{P}$ falling inside the triangular segment are

$$0 \le \xi \le 1, \quad 0 \le \eta \le 1, \quad 0 \le 1 - \xi - \eta \le 1. \tag{11.9}$$

As discussed in Section 3.4.3, we can use (3.43) to measure the severity of warping. If the warping is small, the approximate approach can be legitimate. It is up to the user to balance the saving in CPU time and gaining more accuracy in the searching process. More or less, this depends on the experience of applications.

11.1.4 Penalty Contact Force

We need to further check and reconfirm that the contact or penetration does occur. For simplicity, we use $\boldsymbol{N}$ for the normal vector at impact point. Note that the normal vector of the contact boundary was used in the discussions in Chapter 10. Here, we are using the normal of the target or the obstacle surface. This normal is in the opposite direction. When applying the penalty method, we are interested in the distance from the slave node to the master segment, in view of (10.105) with a gap function g and (10.100) (as $g = 0$),

$$\begin{aligned} D &= (\boldsymbol{S} - \boldsymbol{P}) \bullet \boldsymbol{N}, \\ p &= -(D - g)_-. \end{aligned} \tag{11.10}$$

Here, we denote $(f)_- = (f - |f|)/2 \le 0$ for the negative part of a function. This is the third checkpoint. If $p = 0$, the slave node is just coming to contact the surface of the shell structure. Penetration of the surface occurs if $p > 0$. p represents the depth of penetration, that is, the severity of penetration. The penalty contact force applied to the slave node then becomes

$$F_S = \frac{p}{\varepsilon_x} N = \frac{-(D - g)_-}{\varepsilon_x} N. \tag{11.11}$$

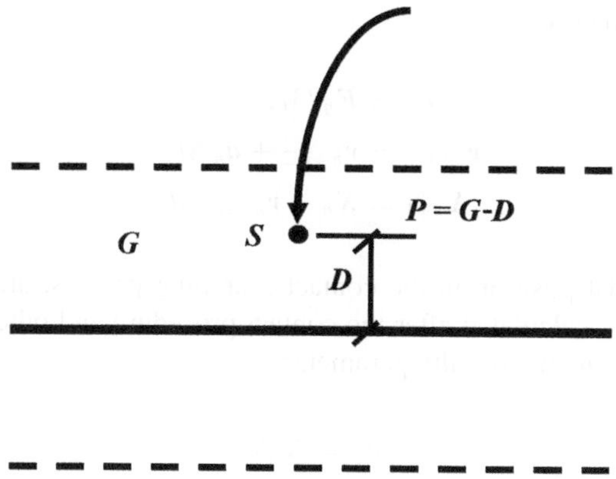

FIGURE 11.6 Penetration of the slave node.

It works like a spring element, with stiffness $K = 1/\varepsilon_x$, illustrated in Figure 11.6. In this artificial spring, p is the compression and $\boldsymbol{F}_S$ is the resistance force;

$$\begin{aligned} K &= 1/\varepsilon_x, \\ \boldsymbol{F}_S &= KpN. \end{aligned} \tag{11.11a}$$

A pure tension, however, does not generate resistance force. This is the nature of the unilateral contact, as well as the penalty method.

Note that the gap g plays the role of gap function in the compliance method, where the relation of deformation and contact force is assumed based on certain laboratory test or experience. For the large deformation of a specific structural component made of specific material, however, to obtain compliance itself is a significant challenge. In fact, the contact force and the material deformation should be balanced according to the material's physical and geometrical characteristics. On the other hand, the penalty method only controls the contact condition, that is, if there is no penetration then there is no force. The penalty contact force, like the external load, will result in additional deformation of the structure by its mechanical property. We now leave the compliance relation alone to let the process of penalty method meet the physical requirement. The method works for general materials and structures, no matter if the structural components are built in bulk shape or thin-walled shape.

We now extend the concept to include the second penalty term in (10.98) regarding velocity. At the contact point, if the normal velocity's direction is toward the obstacle, penetration will either happen or become even deeper. This activates a penalty force $F_{S(v)} = -(v_N)\_N/\varepsilon_v$. A simple approach is to combine the two penalty forces with the contribution of position and velocity. That means, if we use the updated nodal position from calculation after the current time step instead of the nodal position in the last time step, the effect of the velocity then is naturally realized in the formulation.

For instance, we can use

$$\begin{aligned} \boldsymbol{a}_n &= \boldsymbol{F}_n/M, \\ \boldsymbol{v}_{n+1/2} &= \boldsymbol{v}_{n-1/2} + \boldsymbol{a}_n\Delta t, \\ \boldsymbol{X}_{n+1} &= \boldsymbol{X}_n + \boldsymbol{v}_{n+1/2}\Delta t, \end{aligned} \tag{11.12}$$

to replace the nodal position in the contact searching process, although their real positions should be calculated after the contact procedure and other procedures. In this case, we are using the penalty parameter

$$1/\varepsilon_v = \Delta t/\varepsilon_x. \tag{11.13}$$

We can also use the velocity of last time step. By the nature of explicit scheme, the difference of one time step may not cause significant difference in the result. Of course, this is a point to pay attention to. When doubtful situation arises, appropriate changes in the software implementation should follow.

Contact is an interaction, therefore by Newton's third law, it is revealed to us that an impact force is applied to the master segment:

$$\boldsymbol{F}_M = -\boldsymbol{F}_S = -Kp\boldsymbol{N}. \tag{11.14}$$

The impact force is distributed to the nodes of the master segment. A simple method is using the shape functions:

$$\boldsymbol{F}_J = -Kp\boldsymbol{N}\Phi_J(\xi, \eta). \tag{11.15}$$

The next step is to calculate the friction force in the tangential direction. The tangential velocity of the slave node and the contact point in the master segment are obtained by decomposition:

$$\begin{aligned} \boldsymbol{v}_{St} &= \boldsymbol{v}_S - (\boldsymbol{v}_S \bullet \boldsymbol{N})\boldsymbol{N}, \\ \boldsymbol{v}_{Mt} &= \boldsymbol{v}_M - (\boldsymbol{v}_M \bullet \boldsymbol{N})\boldsymbol{N}. \end{aligned} \tag{11.16}$$

Here, $\boldsymbol{v}_M$ can be obtained by using interpolation. We use the Coulomb's law of friction (10.48) with the relative tangential velocity to calculate the friction force. The friction action is to reduce the relative velocity. We first assume the friction force to be in the direction of relative tangential velocity:

$$\begin{aligned} \boldsymbol{v}_t &= \boldsymbol{v}_{St} - \boldsymbol{v}_{Mt}, \quad \boldsymbol{t} = \boldsymbol{v}_t/|\boldsymbol{v}_t|, \\ \boldsymbol{F}_{Mt} &= \mu F_N\boldsymbol{t}, \\ \boldsymbol{F}_{St} &= -\boldsymbol{F}_{Mt} = -\mu F_N\boldsymbol{t}. \end{aligned} \tag{11.17}$$

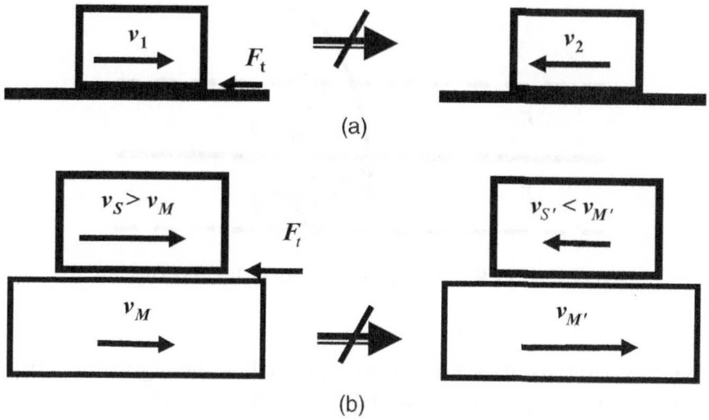

FIGURE 11.7 Tangential motion with friction: (a) the contact with an obstacle; (b) the contact with another moving body.

This is true for sliding with reduced velocity. Physically speaking, the friction force is of a passive nature, that is, it can only reduce the tangential velocity but not reverse the direction of motion, which is illustrated in Figure 11.7. In our calculation, the friction force is restricted not to reverse the relative motion in one time step (or several time steps). At least this fact is required by stability. We can assume that the contribution of the friction force to the slave node is limited by $F_{st} = m^s|\boldsymbol{v}_t|/\Delta t$. This corresponds to the limit of sticking situation of the slave node. For a general sticking situation, we have $\boldsymbol{v}_t = 0$ and $|\boldsymbol{F}_T| < \mu F_N$. Then,

$$\boldsymbol{F}_{St} = -\min(\mu F_N, m^s|\boldsymbol{v}_t|/\Delta t)\boldsymbol{t}. \tag{11.18}$$

This explains the implementation of Coulomb's law of friction (10.48).

This is a simplified approach. The situation becomes more complicated if the impact of the master segment with multiple slave nodes and the interaction of several neighboring master segments are to be considered.

As discussed earlier, for the shell element as a simplified representation of thin-walled structure, we consider the thickness in constructing a volume with the shell element as the mid-surface. Penetrating the physical surface of the thin-walled structure is then considered as penetrating the shell element. The thickness plays the role of the gap. Also, the penetration could occur at both sides of the shell element. The earlier discussion in this section can be easily extended to include two-sided contact. A special situation is that when a slave node penetrates deeper than the gap, that is, crossing the geometrical mid-surface by an amount of $\delta < g$, and arrives at the other half side of the thickness. As shown in Figure 11.8, the slave node can be viewed as penetrating $g + \delta$ from one side, or $g - \delta$ from the other side. It is therefore important to keep the history and trace the penetrated slave node once it is identified as an impacting node. On the other hand, we generally do not consider two-sided contact for the solid element surface.

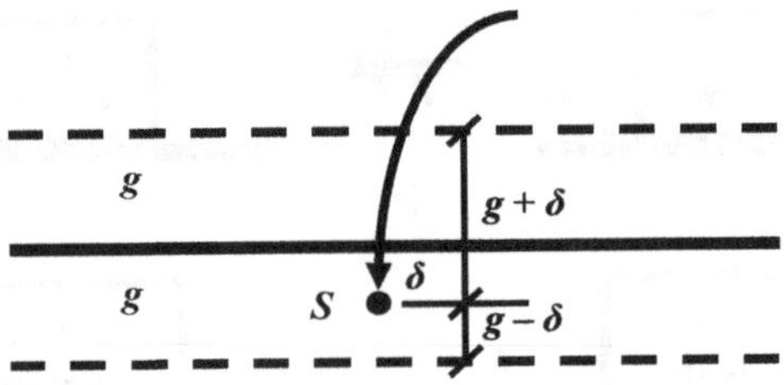

FIGURE 11.8 Penetration through the mid-surface.

From what has been discussed so far, we believe that the algorithm applies to quite general situations of surface contact, including shell and brick elements; possibilities of one-sided or two-sided contact for shell structures; "slave–master" contact with two surfaces or "self" contact with a single surface; etc. Besides the fundamental search algorithm, we only need to take care of certain data structure in the software implementation.

11.2 A CONTACT ALGORITHM WITH MASTER SEGMENT SEARCHING SLAVE NODE

When several layers of shell elements are in contact with each other, it is possible that the nearest master node does not lie in the contact segment but a nearby segment. A typical case is demonstrated in Figure 11.9, where searching for the nearest master node but misses the right candidate, therefore leads to a wrong segment. To alleviate this difficulty, Heinstein et al. (2000) proposed another global search algorithm, which used the master segment to search the candidate slave nodes. The algorithm will be briefly described here.

11.2.1 Global Search with Bucket Sorting Based on Segment's Capture Box

The bucket sorting is designed differently. The physical domain is uniformly divided into cells, with the cell size determined by the minimal size of the master segment (instead of the maximum used in Benson and Hallquist (1990)).

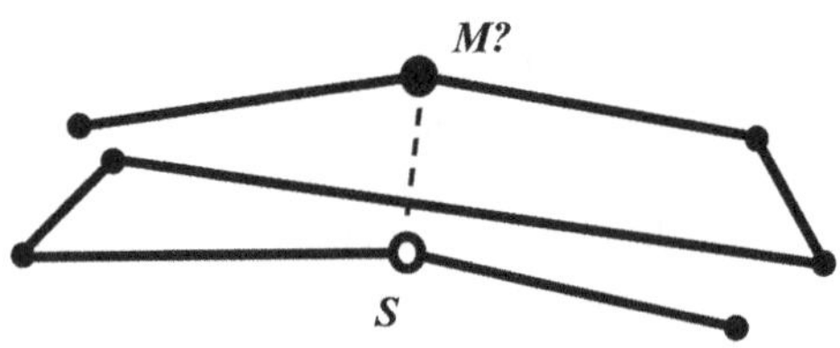

FIGURE 11.9 Mistaken master node.

For each of the master segments, a bounding box is defined in the global *x*/*y*/*z* directions. The bounding box should contain all the nodes of the master segment, at current location and their possible new locations. The new location is predicted by one step of motion with the nodal velocity of the master segment under investigation. The bounding box provides the boundary for the master segment in one time step motion. Furthermore, from the possible motion in one time step of the slave nodes, a capture box is defined by expanding the bounding box with the one step traveling distance (with the maximum velocity in positive and negative *x*/*y*/*z* directions). Only the slave nodes currently lying in the capture box could possibly impact the master segment.

We check the buckets after these preparations. Find the buckets that contain the corner points of the capture box first. The range of buckets associated with the capture box is then determined. All the slave nodes located inside these associated buckets are considered possible to contact the master segment.

11.2.2 Local Search with the Projection of Slave Point

It comes to the same point as local search of Section 11.1.3. We check the slave nodes in the capture box of the master segment to see if any one has projection inside the segment. Then, use the normal distance to the master segment to confirm if the impacting to the master segment will occur in one time step. Because we have one single master segment for one capture box, many of the computations for the set of slave nodes in the capture box have the same data structure. This is different from the method of searching the nearest master node for the slave node, where the different master segments are processed and many computations have no common database. The local search of this algorithm is faster.

11.3 METHOD OF CONTACT TERRITORY AND DEFENSE NODE

Zhong (1993) developed another approach for searching the slave nodes by master segments. We discuss the fundamentals below. A surface is perceived as the union of segments, edges, and nodes and is geometrically represented by the finite elements. The concept of contact territory is then proposed for the surface. It is a volume associated with normal distance. The contact territory is similarly defined for the edges (using normal vectors) and the nodes (using directional vectors). In the positive direction, the distance D_c plays the role of the gap for contact to take action. In the negative direction, the distance D_p represents the allowed penetration depth. The territory of the contact surface is formed by collecting the contact territories of the segments, edges, and nodes of the contact surface. A contact territory of the 2D elements is presented in Figure 11.10. It is a closed volume. Contact is considered to occur when any slave node enters this volume. The concept applies to any contact subject; see Zhong and Nilsson (1996) for more discussions. Here, we discuss the surface segments only.

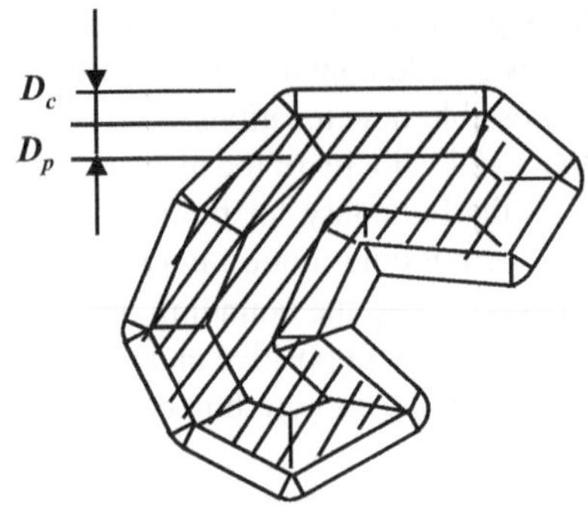

FIGURE 11.10 The contact territory. (Reprinted from Zhong and Nilsson, 1996. Copyright (1996), with permission from Elsevier.)

11.3.1 Global Search with Bucket Sorting Based on Segment's Territory

For the master segment, find the range of all of its nodes $x_j^{\min}$ and $x_j^{\max}$. A box named hierarchy territory is defined, which contains all of these nodes. An expanded territory is introduced by expanding the box in all six directions by an amount no less than the maximum distance D_c, which is used in defining the contact territory. This box is larger than the hierarchy territory and should contain the contact territory of the master segment. Any slave node outside this box has no chance to contact the segment in one time step.

The global search starts with a bucket sorting. Similar to the methods discussed in Sections 11.1 and 11.2, the domain of interest is divided into buckets in three dimensions. It takes only simple calculations to sort the nodes (both master and slave) into the buckets. The next step is for a specific master contact surface (object) to find the buckets, which intersect with its expanded territory, by comparing the coordinates of the corner points. Then determine the range of the buckets' indices numbered in three dimensions. All the buckets within the range are considered as the test cells. All the slave nodes in these test cells are then checked for contact candidate of this contact surface.

11.3.2 Local Search in the Territory

For any test pair of a slave node and a master segment, contact is not going to happen if the slave node is too far away in the normal direction. Next step of checking, we use normal directions at the master nodes of the master segment, depicted in Figure 11.11. If several master segments share the master node, an average normal vector is calculated with consideration of warping. Now, we define for edge ***ab***

$$\begin{aligned} \boldsymbol{V}_g &= \boldsymbol{r}_{ab} \times (\boldsymbol{N}_a + \boldsymbol{N}_b), \\ \boldsymbol{V}_g &= \boldsymbol{V}_g / |\boldsymbol{V}_g|. \end{aligned} \tag{11.19}$$

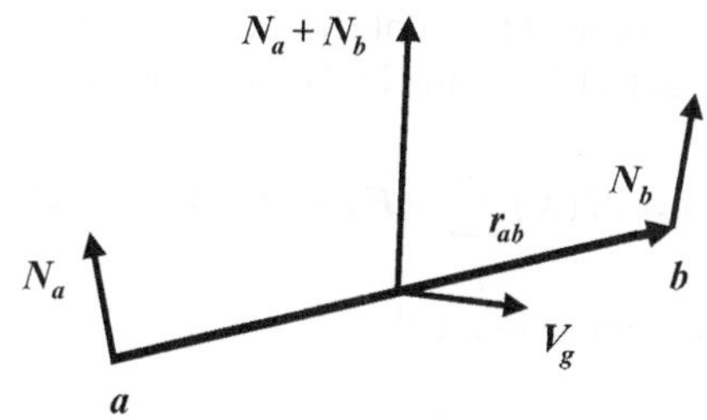

FIGURE 11.11 Nodal normals. (Reprinted from Zhong and Nilsson, 1996. Copyright (1996), with permission from Elsevier.)

For any slave node $\boldsymbol{S}$, check all the edges of the master segment to see if

$$\boldsymbol{r}_{aS} \bullet \boldsymbol{V}_g \le 0. \tag{11.20}$$

If (11.20) holds for all the edges, the projection of the slave node is inside the master segment, and the test pair is identified as potential contact pair. Finally, if the node has a normal distance d within the contact distance, $-D_p \le d \le D_c$, it is identified as a contact pair.

Similar to the method discussed in Section 11.2.2, the preparation for the master segment is used for all the associated slave nodes defined in the test pairs. Hence, many of the computations are saved.

11.3.3 Defense Node and Contact Force

A constraint method was presented in Zhong (1993) with the concept of defense node, briefly described below. A pair of interaction forces is developed at the contact location. To apply the Newton's second law, a pseudo node $\boldsymbol{N_D}$ at the impact point $\boldsymbol{X}$ in the master segment is introduced, called defense node. An effective mass M and the effective force $\boldsymbol{F}$ from the element associated with the contact surface are assigned to the defense node. Then, we have

$$\boldsymbol{F}(\boldsymbol{X}) + \boldsymbol{f}(\boldsymbol{X}) = M(\boldsymbol{X})\boldsymbol{A}(\boldsymbol{X}), \tag{11.21}$$

where $\boldsymbol{f}$ is the contact force and $\boldsymbol{A}$ is the acceleration at $\boldsymbol{X}$. By the usual interpolation,

$$\boldsymbol{A}(\boldsymbol{X}) = \sum\nolimits_J \boldsymbol{A}_J \Phi_J(\boldsymbol{X}). \tag{11.22}$$

On the other hand, for the nodes of the master segment, we have

$$\boldsymbol{F}_J + \boldsymbol{f}_J(\boldsymbol{X}) = M_J \boldsymbol{A}_J. \tag{11.23}$$

Here, f_J are the contribution of the contact force f and are distributed to the master nodes. Plugging (11.22 and 11.23) into (11.21), we obtain

$$F(X) + f(X) = M(X) \sum_J (F_J + f_J(X))\Phi_J(X)/M_J. \tag{11.24}$$

The following relations are suggested:

$$\begin{aligned} F(X) &= M(X) \sum_J F_J \Phi_J(X)/M_J, \\ f(X) &= M(X) \sum_J f_J(X)\Phi_J(X)/M_J. \end{aligned} \tag{11.25}$$

We need a mechanism to distribute the contact force to the nodes like the method of interpolation:

$$\begin{aligned} f_J(X) &= f(X)\Psi_J(X), \\ \sum_J f_J(X) &= f(X). \end{aligned} \tag{11.26}$$

In addition, we need the unity property

$$\begin{aligned} \Psi_J(X_I) &= \delta_{IJ}, \\ \sum_J \Psi_J(X) &\equiv 1. \end{aligned} \tag{11.27}$$

Equation (11.25) leads to

$$M(X) \sum_J \Psi_J(X)\Phi_J(X)/M_J = 1. \tag{11.28}$$

It is clear that the element shape functions Φ_J do not satisfy the requirement of (11.28) for Ψ_J. A solution in simple form is suggested:

$$\begin{aligned} \Psi_J(X) &= M_J \Phi_J(X)/M(X)\bar{\Phi}(X), \\ \bar{\Phi}(X) &= \sum_J (\Phi_J(X))^2. \end{aligned} \tag{11.29}$$

Then, (11.27) leads to

$$M(X) = \sum_J M_J \Phi_J(X)/\bar{\Phi}(X). \tag{11.30}$$

Obviously, $\bar{\Phi}(X_K) = 1$ and $M(X_K) = M_K$.

The constraint condition requires no penetration at contact. The increase of normal distance between the slave node and the defense node reduces penetration. Let the future penetration be zero,

$$P_{n+1} = P_n - ((u_S)_{n+1} - (u_S)_n - (u(X)_{n+1} - u(X)_n)) = 0, \tag{11.31}$$

where all the variables are in the normal direction. The subscript n and $n+1$ indicate time points before and after the computation of contact. With the central difference method in time domain, assuming constant time step, we have

$$A_n = (V_{n+1/2} - V_{n-1/2})/\Delta t, \quad V_{n+1/2} = (U_{n+1} - U_n)/\Delta t. \tag{11.32}$$

By Newton's second law, we have the following for the slave node and the defense node:

$$\begin{aligned} F_S + f_S &= M_S(A_S)_n, \\ F + f_D &= M(X)(A(X))_n. \end{aligned} \tag{11.33}$$

Note that the impact force f_D is in the negative normal direction and the reaction force f_S is in the positive normal direction. Using (11.31), (11.32), and the interaction of contact force, we obtain

$$f_S = -f_D = \frac{M_S M(X)}{M_S + M(X)} \left(\frac{P_n}{\Delta t^2} - \frac{F_S}{M_S} - \frac{(V_S)_{n-1/2}}{\Delta t} + \frac{F}{M(X)} + \frac{(V_D)_{n-1/2}}{\Delta t} \right). \tag{11.34}$$

For the three-node triangular element, the shape functions are coincident with the area coordinates. For the four-node quadrilateral elements, a set of simplified shape functions can also be defined by splitting the quadrilateral into four triangles and using the areas of the triangles.

11.4 PINBALL CONTACT ALGORITHM

Before successfully achieving good performance in vectorization and parallelization of contact algorithms, a pinball contact was developed by Belytschko and Neal (1991). Conceptually, a solid element is represented by a ball, which has the center at the element center and has the same volume of the element. When two pinballs are in contact, the elements are considered in contact, and of course with their surfaces in contact. For two balls, contact occurs when the distance of their centers is smaller than the summation of their radii. This natural criterion is easy to check. Hence, the pinball contact algorithm is expected to run faster than the node-surface algorithm so far discussed. Belytschko and Yeh (1992, 1993) extended the method to shell elements, discussed below.

11.4.1 The Pinball Hierarchy

For a pair of pinballs centered at C_1 and C_2 with radii R_1 and R_2, respectively, the central distance can be calculated from their locations. The no penetration condition is expressed as

$$\begin{aligned} D &= |\boldsymbol{r}_{C_1} - \boldsymbol{r}_{C_2}|, \\ p &= R_1 + R_2 - D < 0. \end{aligned} \tag{11.35}$$

For a shell element, however, the equivalent volume does not represent the element. Belytschko and Yeh (1992, 1993) presented a splitting pinball algorithm. For four-node quadrilateral, let the center of the ball be coincident with the element center. Let the radius be the maximum distance from the center to its nodes:

$$R = \max\{|r_C - r_{N_J}|, \quad J = 1, \ldots, 4\}. \tag{11.36}$$

When the contact of two pinballs is found, the balls are split by one-to-four as the second level. The rule is to seat the centers of the second level balls at isoparametric coordinates ($\pm 1/2$, $\pm 1/2$). Also, imagine that the element is uniformly split into four in the reference frame. Such a division does not result in equal areas in physical domain due to skew geometry. Hence, the radii of the second level balls for the four-node quadrilateral element are adjusted.

The one-to-four splitting for the three-node triangular element results in four equal triangles. The centers of the balls are set at the centroids of the triangles. The radius of the triangular element is defined by the rule of equal area:

$$R = \sqrt{A/\pi}. \tag{11.37}$$

If any pair of the second level pinballs is in contact, they are further split into four balls of level 3. The splitting is recommended to end when the diameter reaches the magnitude of one to two times of the shell thickness. From the programming point of view, the pinball architecture can be defined at the beginning with ball centers' positions recalculated every time when contact happens. But it is not necessary to recalculate the radii frequently.

11.4.2 Penalty Contact Force

The penetration is calculated by (11.35) for $p > 0$. The contact force by penalty method was proposed in Belytschko and Yeh (1992, 1993):

$$\begin{aligned} f &= \min(f_1, f_2), \\ f_1 &= \begin{cases} \dfrac{\rho_1 \rho_2 R_1^3 R_2^3}{\rho_1 R_1^3 + \rho_2 R_2^3} \dfrac{\dot{p}}{\Delta t}, & \text{if} \quad \dot{p} > 0; \\ 0, & \text{if} \quad \dot{p} = 0; \end{cases} \\ f_2 &= \frac{\mu_1 \mu_2}{\mu_1 + \mu_2} \sqrt{\frac{R_1 R_2}{R_1 + R_2}} p^{\frac{3}{2}}, \end{aligned} \tag{11.38}$$

where for $\alpha = 1, 2$, ρ_α, μ_α, and R_α were the densities, shear moduli, and radii of the two pinballs.

The distribution of contact force to the nodes can use element interpolation as discussed earlier for node-surface contact. Here, we need to record the levels and the isoparametric coordinates of the centers of the pinballs in contact.

As the splitting scheme has the same data structure of mesh refinement, the h-adaptive method can be applied with pinball contact algorithm very well.

11.5 EDGE (LINE SEGMENT) CONTACT

Under certain circumstances, due to the linear interpolation of the finite element discretization, the surfaces are modeled by piecewise polygons. The surface-to-surface contact may turn out to be the edge-to-edge contact as depicted in Figure 11.12a. The algorithm searching for node-to-surface contact fails to find this kind of contact. In certain cases, when the structures with open surface segments move to contact each other, edge-to-edge contact is naturally a kind of contact with physical meaning, as illustrated in Figure 11.12b. The edge contact can also include truss and beam elements. This is considered as a general situation of contact of line segments.

11.5.1 Search for Line Contact

We remind ourselves that the truss or beam element is defined as a two-node line element. The line segment indeed represents a physical bar with a cross section of certain shape. For simplicity, we only consider cylinder as representative of the bars. The contact of two line segments deals with an intersection of two lines at a common point. In the finite element method, this kind of geometrical intersection without a reasonable tolerance is almost impossible. Similar to the case of surface contact, we introduce a gap as the summation of two radii of the bars and define two finite

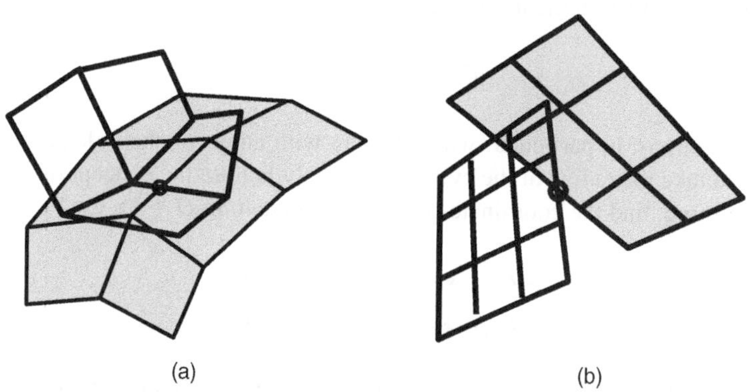

FIGURE 11.12 Edge contact of shell elements: (a) edges in the discrete shell model; (b) edges of open shell structure.

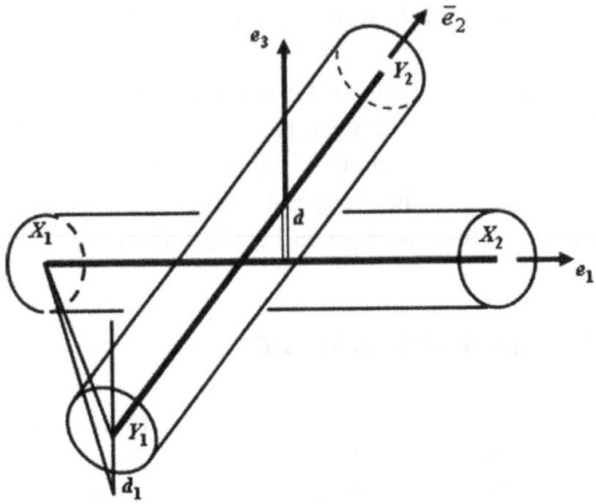

FIGURE 11.13 Contact of two cylinders, representation of line elements.

cylinders for the bar elements. When two of such cylinders have intersection, the edge-to-edge contact is considered to occur.

Let us examine two finite cylinders depicted in Figure 11.13, and check the possibility of edge contact. Denote their end points by $\boldsymbol{X}_1, \boldsymbol{X}_2, \boldsymbol{Y}_1$, and $\boldsymbol{Y}_2$; their radii by r_x and r_y, respectively. Define a unit vector $\boldsymbol{e}_1$ along $\boldsymbol{X}_1$ and $\boldsymbol{X}_2$, and let L_x be the length of element X. Define a unit vector $\bar{\boldsymbol{e}}_2$ along $\boldsymbol{Y}_1$ and $\boldsymbol{Y}_2$, and $\boldsymbol{e}_3 = \boldsymbol{e}_1 \times \bar{\boldsymbol{e}}_2$, normalized, then $\boldsymbol{e}_2 = \boldsymbol{e}_3 \times \boldsymbol{e}_1$. Construct a 2D coordinate system $O - \xi\eta$ with the origin at $\boldsymbol{X}_1$, ξ-axis in direction $\boldsymbol{e}_1$ and η-axis in direction $\boldsymbol{e}_2$. Now, we can find

$$\begin{aligned} d_1 &= \boldsymbol{r}_{X_1Y_1} \bullet \boldsymbol{e}_3, \\ d_2 &= \boldsymbol{r}_{X_1Y_2} \bullet \boldsymbol{e}_3. \end{aligned} \tag{11.39}$$

We should see $d_1 = d_2$, within the computer's precision. This is the distance between the two line segments. There is no contact if

$$|d_1| > g = r_x + r_y. \tag{11.40}$$

In applications, in particular, when dealing with edges of shell elements, we do not have to take the radius of the cylinder seriously but use the concept of gap.

Furthermore, find the coordinates in the local system in $O - \xi\eta$ for $\boldsymbol{Y}_1$ and $\boldsymbol{Y}_2$

$$\begin{aligned} \xi_1 &= \boldsymbol{r}_{X_1Y_1} \bullet \boldsymbol{e}_1, \quad \eta_1 = \boldsymbol{r}_{X_1Y_1} \bullet \boldsymbol{e}_2, \\ \xi_2 &= \boldsymbol{r}_{X_1Y_2} \bullet \boldsymbol{e}_1, \quad \eta_2 = \boldsymbol{r}_{X_1Y_2} \bullet \boldsymbol{e}_2. \end{aligned} \tag{11.41}$$

The projections of the two line elements in $O - \xi\eta$ plane are presented in Figure 11.14. It is understood if $\eta_1\eta_2 > 0$, then $\boldsymbol{Y}_1$ and $\boldsymbol{Y}_2$ are on the same side of $\boldsymbol{X}_1\boldsymbol{X}_2$, and there is no intersection.

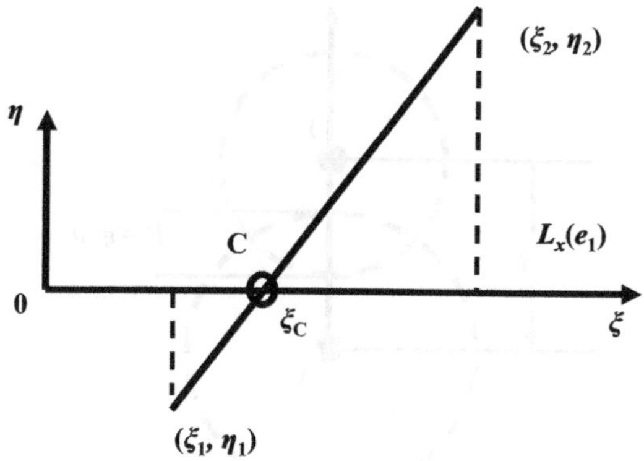

FIGURE 11.14 Projection of two line segments.

For the case with $\eta_1\eta_2 \leq 0$, we can find intersection point $\boldsymbol{C}$ $(\xi_C, 0)$ by direct calculations:

$$\frac{\xi_C - \xi_1}{-\eta_1} = \frac{\xi_2 - \xi_C}{\eta_2}, \quad \xi_C = \frac{\xi_1\eta_2 - \xi_2\eta_1}{\eta_2 - \eta_1}. \tag{11.42}$$

If $\xi_C < 0$ or $\xi_C > L_x$, then there is no contact.

11.5.2 Penalty Contact Force of Edge-to-Edge Contact

As discussed in Section 11.5.1, contact occurs when the distance between the two lines is smaller than the gap, as shown in Figure 11.15. The penalty method produces the contact force exerted on these two elements:

$$\begin{aligned} p &= (g - |d|)_+ \text{sgn}(d), \\ \boldsymbol{f}_y &= Kp\boldsymbol{e}_3, \\ \boldsymbol{f}_x &= -Kp\boldsymbol{e}_3. \end{aligned} \tag{11.43}$$

Here, K is the penalty parameter.

Departing from the meaning of surface contact, where a normal direction by default is defined by the right-hand rule, the line segment has no specific definition of its normal direction. The impact direction is determined by the intersecting line segments during the calculation. As illustrated in Figure 11.13, viewing in $\boldsymbol{e}_3$ direction, if $\boldsymbol{X}_1\boldsymbol{X}_2$ is "below" $\boldsymbol{Y}_1\boldsymbol{Y}_2$, then the projection d_1 of $\boldsymbol{r}_{X_1Y_1}$ on $\boldsymbol{e}_3$ is positive, therefore the impact force exerted on $\boldsymbol{Y}_1\boldsymbol{Y}_2$ is "upward." The reaction force applied on $\boldsymbol{X}_1\boldsymbol{X}_2$ is "downward." If $\boldsymbol{X}_1\boldsymbol{X}_2$ is above $\boldsymbol{Y}_1\boldsymbol{Y}_2$, then the projection d_1 is negative and the directions of the contact forces are reversed. $\boldsymbol{e}_3$ plays the role of the normal direction. The impact

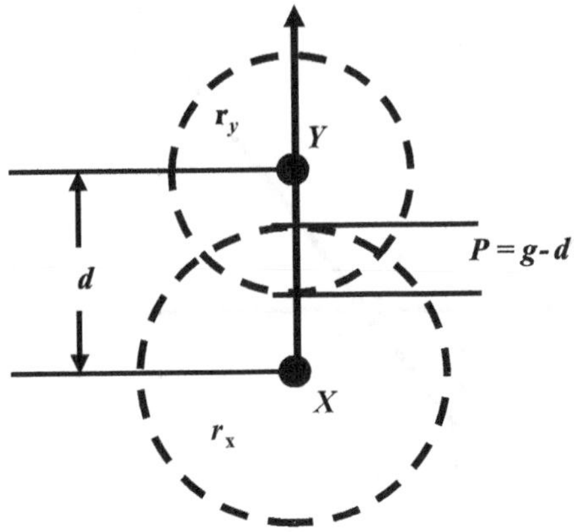

FIGURE 11.15 Penetration in line contact.

forces in the e_3 direction are then distributed to the nodes, by using interpolation for instance,

$$
\begin{aligned}
f_{y_1} &= f_y \frac{\eta_2}{\eta_2 - \eta_1}, \quad f_{x_1} = f_x \frac{L_x - \xi_C}{L_x}, \\
f_{y_2} &= f_y \frac{-\eta_1}{\eta_2 - \eta_1}, \quad f_{x_2} = f_x \frac{\xi_C}{L_x}.
\end{aligned} \tag{11.44}
$$

In order to process the friction force, we need the "tangent" components of the velocities (decomposed with respect to e_3 as the "normal" direction) of these two elements at the contact point. The procedure similar to that used for the surface contact can be properly adopted.

11.6 EVALUATION OF CONTACT ALGORITHM WITH PENALTY METHOD

Example 11.1 A rod impacting the fixed flat rigid obstacle by penalty contact method The problem was described in Section 10.1.4 with analytical solution. Here we perform the finite element analysis using contact algorithm with penalty method, discussed in Section 10.4. We use a set of uniform meshes of truss elements with element size ranging from 2 to 200 mm and the penalty coefficient $\varepsilon = 0.01$, or the contact stiffness (terminology of the software) $K = 100$. A mesh with 4,000 elements is used for generating a reference solution, which is supposed to be close to the exact solution of the penalty method. It is also expected to be close to the solution of the original Signorini's problem. To avoid loss of accuracy, we use 10% of the critical time step size calculated from elements to perform the computation. The cross-section area of 100 mm^2 is defined for the truss elements.

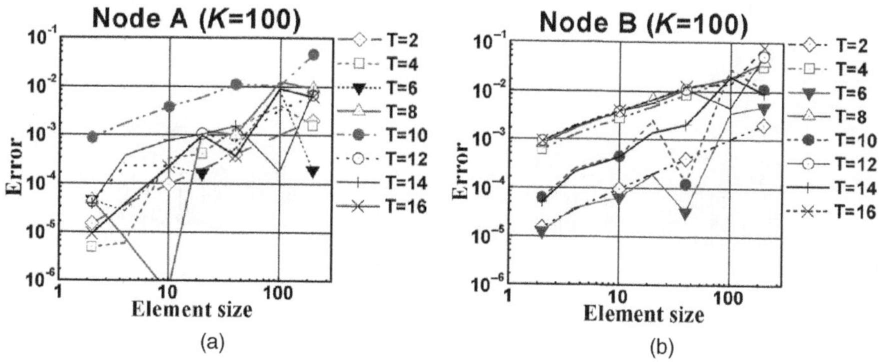

FIGURE 11.16 Error analysis—convergence of penalty method with respect to refined meshes, with $K = 100$.

The errors of the displacement of the end points at a few key time points, with respect to the reference solution, are plotted in Figure 11.16. The trend of convergence with respect to the reduced element size is clearly presented. It is observed that the error curves in the log–log scale asymptotically have a slope equal to 3/4 or larger.

Using a uniform mesh with 200 elements in the next step, we set the contact stiffness $K = 0.01, 0.1, \ldots$, to examine the robustness of the penalty method. The result of rigid obstacle ($K = \infty$, see more discussion later in Example 11.2) is used for reference. The error of the displacement of the end points at the key time is depicted in Figure 11.17. With $K \to \infty$ or the penalty coefficient $\varepsilon = 1/K \to 0$, the result tends to converge. Furthermore, in log–log scale, the error curves asymptotically have a slope equal to one or higher. Note again that the case with $K = \infty$ does not provide an exact solution to the original problem, but a finite element approximation.

Example 11.2 Impact of a component in a progressive collapse deformation mode with severe contact We continue the analysis of Example 7.3 with much larger plastic deformation and severe material self-contact. The commercial software LS-DYNA V971 with contact algorithm that uses penalty method is employed to

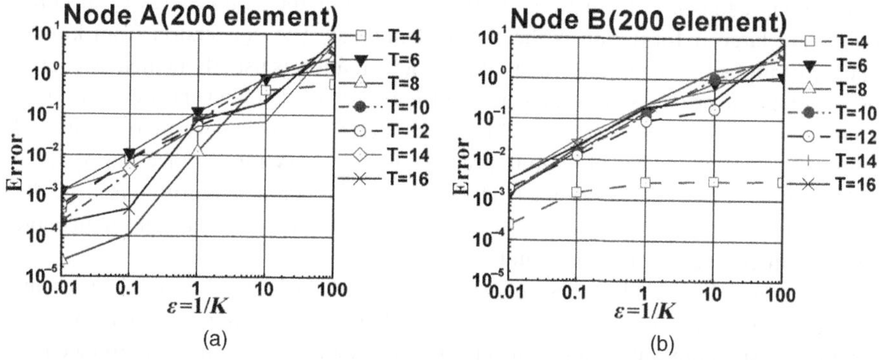

FIGURE 11.17 Error analysis—robustness of penalty method, with 200-elements mesh.

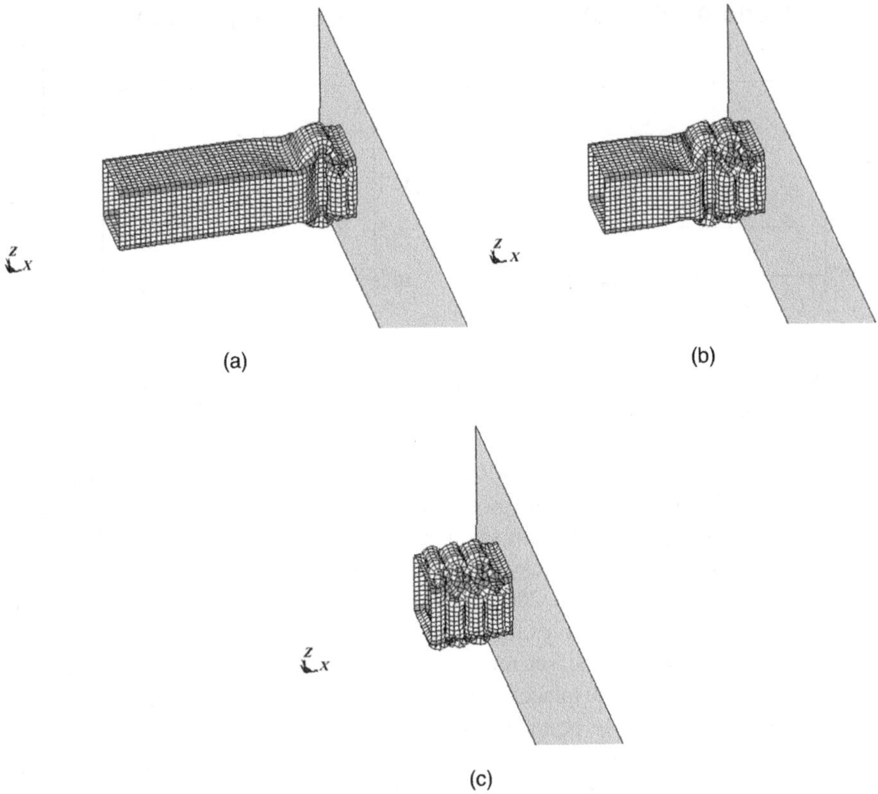

FIGURE 11.18 Large deformation of the component computed by using B-T element with 5 mm mesh: (a) at 5 ms; (b) at 10 ms; (c) at 15 ms.

perform the analysis. The deformation at 15 ms calculated by using B-T element with the 5 mm mesh is depicted in Figure 11.18. This is a progressive collapse mode of large plastic deformation. For impact problem, the strain energy contributed by plastic large deformation is the focus of interest. The conversion from kinetic energy of impact to the strain energy due to deformation is usually named as energy absorption. The progressive collapsing deformation is for excellent energy absorption. Due to large deformation, material contact inevitably occurs in several locations and at certain time, which are difficult to predict. In fact, every fold formed during large deformation induces severe contact. The robustness of contact algorithm is critically important to the reliable analysis. In this example, there is basically no noticeable material interpenetration to be found.

As shown in Figure 11.18, the deformation computed with the 5 mm mesh presents severe warping. For comparison, the result calculated by using B-T element with 2.5 mm mesh is presented in Figure 11.19a. The corresponding results produced by

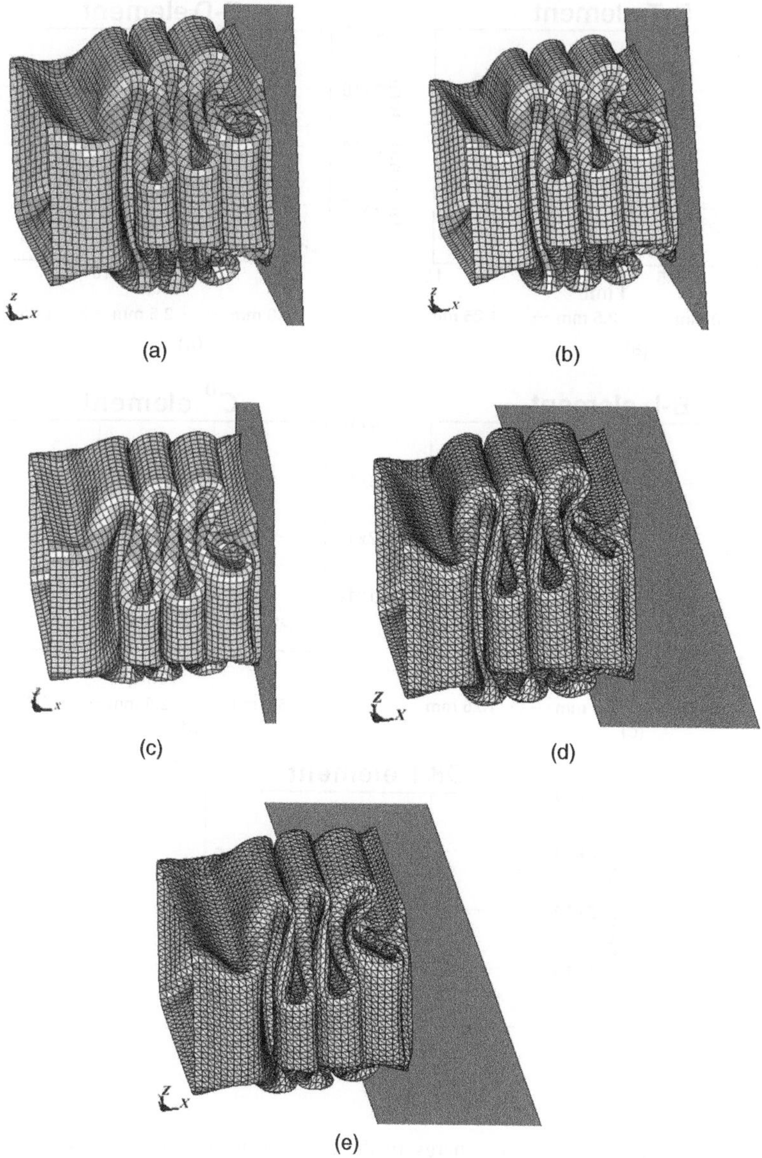

FIGURE 11.19 Large deformation at 15 ms computed by using various elements with 2.5 mm mesh: (a) result of B-T element; (b) result of B-D element; (c) result of B-L element; (d) result of C^0 element; (e) result of DKT element.

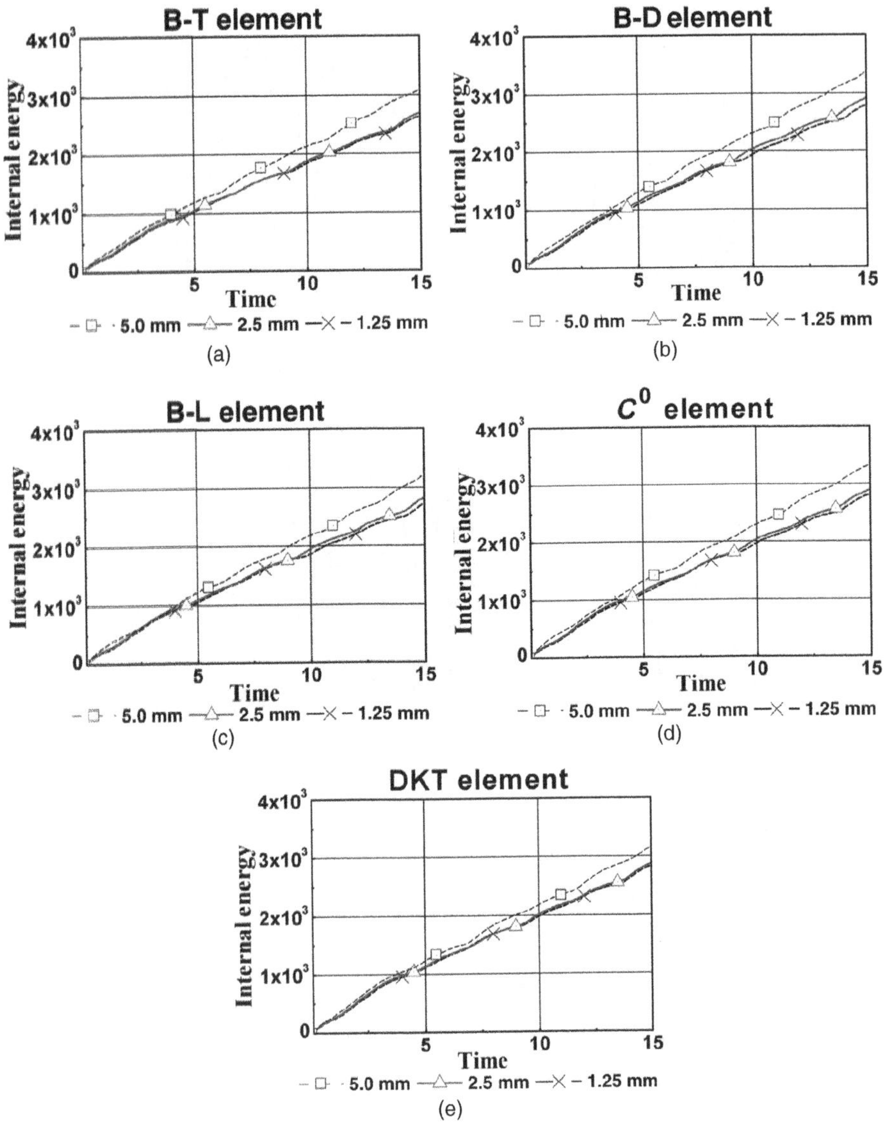

FIGURE 11.20 Computed internal energy of the component by using various shell elements: (a) B-T element; (b) B-D element; (c) B-L element; (d) C^0 triangular element; (e) DKT element.

the quadrilateral B-D element and B-L element are also depicted in Figure 11.19. The deformation of these cases is close to each other. As discussed in Example 7.3, warping mainly happens in the folding line area formed as the buckling mechanism. In progressive collapse deformation, warping also happens progressively. The area of warping reduces with mesh refinement. A further refined mesh with 1.25 mm

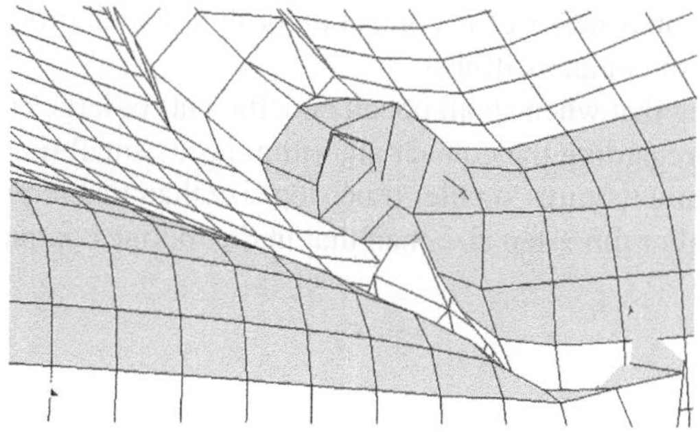

FIGURE 11.21 Deformation with a contact coefficient $C = 0.1$.

is also used for investigation and the study is extended to triangular C^0 and DKT elements. There is no significant penetration observed for all of the five types of shell elements calculated with the three meshes. The material penetration is controlled by the contact algorithm. The internal energy calculated from deformation after 15 ms impact in these cases is presented in Figure 11.20. The results of the five types of elements all have tendency to converge, and with some difference from each other. For more discussion about convergence behavior for the large deformation impact problems, we refer to Wu (2001) and Wu et al. (2004).

To investigate the robustness of contact algorithm, we continue the discussion related to the effect of penalty coefficient/contact stiffness. We apply a scalar coefficient $C = 0.1$, 1, 10, and 100 to the default contact stiffness set by the software. The comparison is performed with B-T element and the 2.5 mm mesh. The results show that the penetration occurs with small scalar value $C = 0.1$ at the first few milliseconds, depicted in Figure 11.21. However, it does not keep growing. With increasing of C value, the strength to resist interpenetration increases and the penetration reduces. The energy absorption calculated from these cases is presented in Figure 11.22. With $C = 0.1$, the calculated energy absorption is a bit lower due to penetration. The results derived from $C = 1$, 10, and 100 are very close. The differences are within 0.3%.

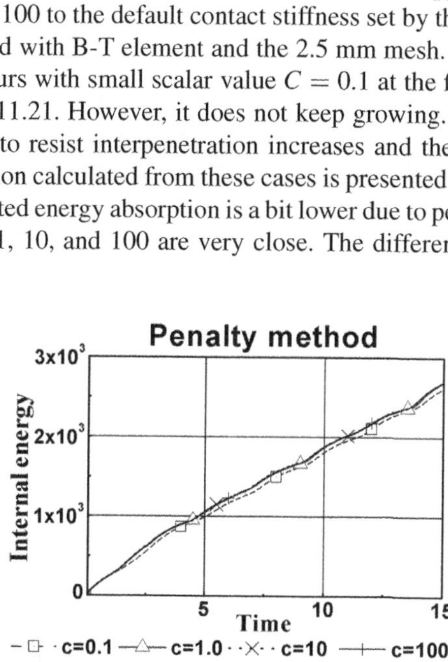

FIGURE 11.22 Internal energy calculated with various contact coefficients.

This indicates the robustness of the contact algorithm, which produces close results for a certain range of contact stiffness.

It is worth noting that when small penalty coefficient (or large contact stiffness) is used, the stability regarding the contact algorithm needs small time step size. When the contact procedure requires smaller time step size than that required by elements, we should use smaller time step size than that in the routine computation.

CHAPTER 12

KINEMATIC CONSTRAINT CONDITIONS

For transient dynamic analyses, several kinds of constraints can be defined for certain applications. The constraint conditions are used not only for simulating the real constraints but also for modeling mechanical behavior of some complex structures.

In the following, we will discuss the fundamentals of rigid wall and rigid body. For simplicity, we assume that the constraint conditions are prescribed in the global coordinate system. We adopt the concept of slave nodes introduced in Chapter 11. The readers are referred to Hallquist (2006) and ALTAIR (2009) for more discussions.

Boundary conditions for prescribed displacement and load are trivial, and will not be discussed here. The prescribed load can be discretized and applied to the nodes. It is treated for external forces after element calculation for internal forces.

12.1 RIGID WALL

Let us revisit the Signorini's unilateral contact problem. We start with an infinitely large flat rigid obstacle called a rigid wall (or barrier). We assume that the contact to the rigid wall is of the nature of plastic impact. The slave node impacting the rigid wall will not bounce off by itself until other force pulls it out of contact. Here, we discuss the constraint method, whereas the penalty contact method is discussed in Chapters 10 and 11. We then extend the study to moving obstacle, the finite-sized barrier, and the barrier with curved surfaces. The constraint of rigid wall can be used to model the rigid impactor in the impact tests.

Introduction to the Explicit Finite Element Method for Nonlinear Transient Dynamics, First Edition.
Shen R. Wu and Lei Gu.

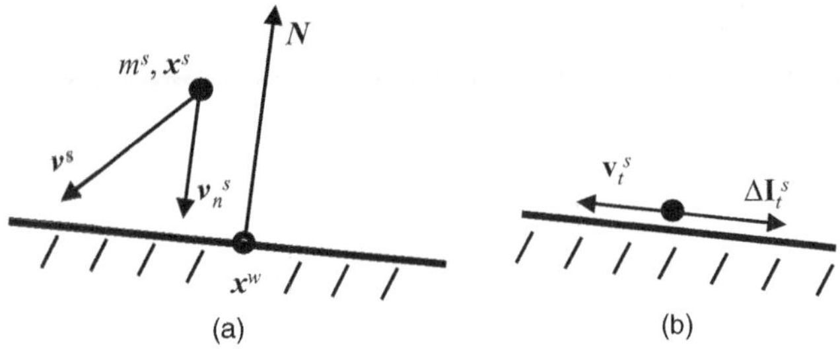

FIGURE 12.1 Rigid wall as a unilateral constraint: (a) Normal action; (b) Tangential action.

The unilateral contact condition of no penetration is the fundamental requirement of the rigid wall:

$$(\boldsymbol{x}^s - \boldsymbol{x}^w) \bullet \boldsymbol{N} > 0. \tag{12.1}$$

Here, the superscript w stands for the rigid wall and s stands for slave node. This is the same requirement as the one by surface contact.

12.1.1 A Stationary Flat Rigid Wall

We consider a fixed rigid wall. The interaction between the rigid wall and the slave nodes is individual with each of the nodes, which tend to penetrate. There is no interaction among the slave nodes. The computation is simply to search the potential penetration and to generate a resistance force preventing the penetration.

We assume that the slave nodes have not penetrated the obstacle at current time. Similar to the contact algorithm, we can use one-step motion as prediction. As illustrated in Figure 12.1, the numerical procedure can be designed, for each slave node:

$$\bar{\boldsymbol{x}}^s = \boldsymbol{x}^s + \boldsymbol{v}^s \Delta t. \tag{12.2}$$

We pick a point $\boldsymbol{x}^w$ on the wall as the reference position of the wall. If

$$\bar{d} = (\bar{\boldsymbol{x}}^s - \boldsymbol{x}^w) \bullet \boldsymbol{N} \leq 0, \tag{12.3}$$

then a resistance force or an impulse $\Delta \boldsymbol{I}_n^s$ should be generated to prevent the penetration. For instance, we can simply use $\Delta \boldsymbol{I}_n^s = m^s \mid \boldsymbol{v}^s \bullet \boldsymbol{N} \mid \boldsymbol{N}$ to stop it without moving closer. Here, m^s is the mass of the slave node. We also assume that the impact is plastic impact so that the impacting node will not rebound by itself. By the conservation of momentum, the impulse delivered to the barrier due to interaction is

$$\Delta \boldsymbol{I}_n^w = -\sum \Delta \boldsymbol{I}_n^s. \tag{12.4}$$

This will be accounted for the impact force to the barrier, $\boldsymbol{f}_n^w = \partial \boldsymbol{I}_n^w / \partial t$.

When friction is involved, we need to consider the tangential motion of the impacting slave nodes. The tangential velocity is determined by decomposition:

$$\begin{aligned} v_n^s &= \boldsymbol{v}^s \bullet \boldsymbol{N}, \\ \boldsymbol{v}_t^s &= \boldsymbol{v}^s - v_n^s \boldsymbol{N}. \end{aligned} \tag{12.5}$$

Define the tangential direction with

$$\boldsymbol{t}_s = \boldsymbol{v}_t^s / \left| \boldsymbol{v}_t^s \right|. \tag{12.6}$$

By Coulomb's law of friction, we apply a tangential impulse $\Delta \boldsymbol{I}_t^s = -\mu \Delta I_n^s \boldsymbol{t}_s$.

The friction force is of passive nature. For the sake of stability, the tangential force exerted on the slave node should be limited. It can at most stop the tangential motion (a sticking state) but not force the slave node to reverse its sliding motion in one (or several) time step. Therefore, we have a restriction $|\Delta \boldsymbol{I}_t^s| \leq m^s |\boldsymbol{v}_t^s|$ and

$$\begin{aligned} \Delta \boldsymbol{I}_t^s &= -\min\left(\mu \Delta I_n^s, m^s \left| \boldsymbol{v}_t^s \right|\right) \boldsymbol{t}_s, \\ \Delta \boldsymbol{I}_t^w &= -\sum \Delta \boldsymbol{I}_t^s. \end{aligned} \tag{12.7}$$

This result gives an explanation to (10.48), the mathematical description of Coulomb's law of friction by Duvaut and Lions (1976).

The algorithm can be extended to the case of a fixed finite-sized flat rigid barrier, such as a rectangle and a circle. The additional requirement is to exclude the slave nodes, which land outside the barrier. This can be accomplished by checking the projection of the slave nodes in the region that the rigid wall occupies.

12.1.2 A Moving Flat Rigid Wall

Because of the motion of the rigid wall, the impact by every slave node affects the wall's motion. Hence, there is indirect interaction among the impacting slave nodes. An infinitely large rigid wall is considered now. In a class of applications, it is used to simulate the contact surface of an impactor. Due to the heavy weight and large size of the impactor and the short time duration of the impact, its rotation is negligible. We then assume that the rigid wall can have translation only. But the motion is not necessary to be in its normal direction. The assumption of plastic impact means that the impacting slave node will follow the motion of the rigid wall in its normal direction until other force pulls it out of contact.

We can device an algorithm similar to what is discussed in Section 12.1.1. Here, we consider the location of slave node with respect to the current position of the rigid wall $\boldsymbol{x}^w$ and the relative motion:

$$\begin{aligned} \bar{\boldsymbol{x}}^s &= \boldsymbol{x}^s + \boldsymbol{v}^s \Delta t, \\ \bar{\boldsymbol{x}}^w &= \boldsymbol{x}^w + \boldsymbol{v}^w \Delta t. \end{aligned} \tag{12.8}$$

If $\bar{d} = (\bar{\boldsymbol{x}}^s - \bar{\boldsymbol{x}}^w) \bullet \boldsymbol{N} \leq 0$, an impulse $\Delta \boldsymbol{I}_n^s$ should be applied to the slave node $\boldsymbol{x}^s$. All the identified impacting slave nodes will move along with the rigid wall with a common normal velocity $\tilde{v}_n = \tilde{v}_n^w = \tilde{v}_n^s$. We have the normal component of the current velocities:

$$\begin{aligned} v_n^w &= \boldsymbol{v}^w \bullet \boldsymbol{N}, \\ v_n^s &= \boldsymbol{v}^s \bullet \boldsymbol{N}. \end{aligned} \tag{12.9}$$

The impulses are expressed as following with the mass of the wall:

$$\begin{aligned} \Delta \boldsymbol{I}_n^s &= m^s \left(\tilde{v}_n - v_n^s\right)\boldsymbol{N}, \\ \Delta \boldsymbol{I}_n^w &= m^w \left(\tilde{v}_n - v_n^w\right)\boldsymbol{N}. \end{aligned} \tag{12.10}$$

By conservation of momentum $\Delta \boldsymbol{I}_n^w + \sum \Delta \boldsymbol{I}_n^s = 0$, we obtain a solution

$$\tilde{v}_n^w = \left(m^w v_n^w + \sum m^s v_n^s\right) \Big/ \left(m^w + \sum m^s\right). \tag{12.11}$$

The normal impact forces are

$$\begin{aligned} \boldsymbol{f}_n^s &= \partial \boldsymbol{I}_n^s / \partial t, \\ \boldsymbol{f}_n^w &= \partial \boldsymbol{I}_n^w / \partial t. \end{aligned} \tag{12.12}$$

The tangential motion needs an update if the friction is significant. We adopt the Coulomb's law of friction described in Duvaut and Lions (1976) as discussed in Chapters 10 and 11. Assume that the friction is along the direction of relative tangential motion, shown in Figure 12.2. The tangential velocities are

$$\begin{aligned} \boldsymbol{v}_t^w &= \boldsymbol{v}^w - v_n^w \boldsymbol{N}, \\ \boldsymbol{v}_t^s &= \boldsymbol{v}^s - v_n^s \boldsymbol{N}. \end{aligned} \tag{12.13}$$

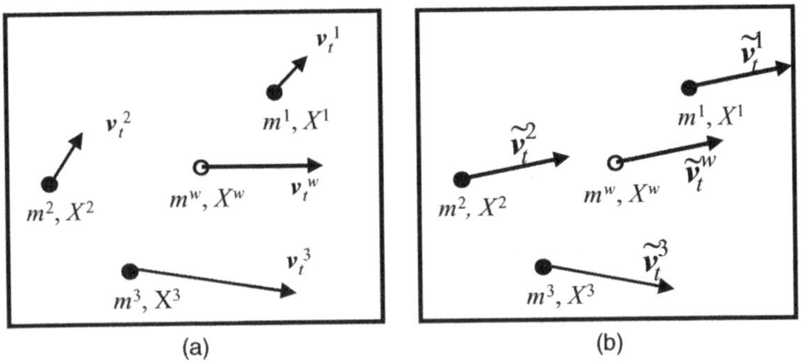

FIGURE 12.2 Tangential actions of a moving rigid wall: (a) before calculation of rigid wall; (b) after calculation of rigid wall: s^1 and s^2 are sticking with the wall, s^3 is sliding.

Let

$$t_s = \left(v_t^s - v_t^w\right) / \left|v_t^s - v_t^w\right|. \tag{12.14}$$

For the identified impacting slave nodes, we apply a tangential impulse with limited action as discussed in Section 12.1.1:

$$\Delta I_t^s = -C_n^s \mu \left|\Delta I_n^s\right| t^s + C_t^s m^s \left(\tilde{v}_t^w - v_t^s\right). \tag{12.15}$$

If $\mu \Delta I_n^s \leq m^s |v_t^s - v_t^w|$, then the relative tangential motion will "slow down" and the slave node will be sliding on the rigid wall. We set $C_n^s = 1$ and $C_t^s = 0$ for this case. On the other hand, if $\mu \Delta I_n^s > m^s |v_t^s - v_t^w|$, then the relative tangential motion will stop and the slave node will be sticking on the rigid wall. We set $C_n^s = 0$ and $C_t^s = 1$. In this case, the slave node will have the same tangential velocity of the rigid wall, $\tilde{v}_t^w$. Note that the sticking condition is verified by using the current state, but the action with $\tilde{v}_t^w$ is assumed for the post state.

Taking into account of the impulse exerted on the rigid wall, $\Delta I_t^w = m^w(\tilde{v}_t^w - v_t^w)$; and using the conservation of momentum, $\Delta I_t^w + \sum \Delta I_t^s = 0$, we obtain a solution

$$\tilde{v}_t^w = \left(m^w v_t^w + \sum \left(C_n^s \mu \left|\Delta I_n^s\right| t^s + C_t^s m^s v_t^s\right)\right) / \left(m^w + \sum C_t^s m^s\right). \tag{12.16}$$

As demonstrated in Figure 12.2b, after this procedure, some impacting slave nodes will be sticking with the rigid wall and moving with $\tilde{v}_t^w$. Others will be sliding with different tangential velocity.

When extended to a finite-sized moving flat rigid wall, we need the condition for the slave node to impact the inside of the surface similar to the fixed rigid wall discussed in Section 12.1.1.

12.1.3 Rigid Wall with a Curved Surface

For a rigid wall defined by a curved surface, such as sphere, ellipsoid, and cylinder, we can define the constraint conditions similar to (12.1). For example, we check the penetration condition for ellipsoid depicted in Figure 12.3,

$$\left(\frac{x^s - x^w}{a}\right)^2 + \left(\frac{y^s - y^w}{b}\right)^2 + \left(\frac{z^s - z^w}{c}\right)^2 < 1, \tag{12.17}$$

where x^w represents the position of the center of the ellipsoid. a, b, and c are the semi radii. For a cylinder with radius r, axial direction L, and a reference point x^w on the

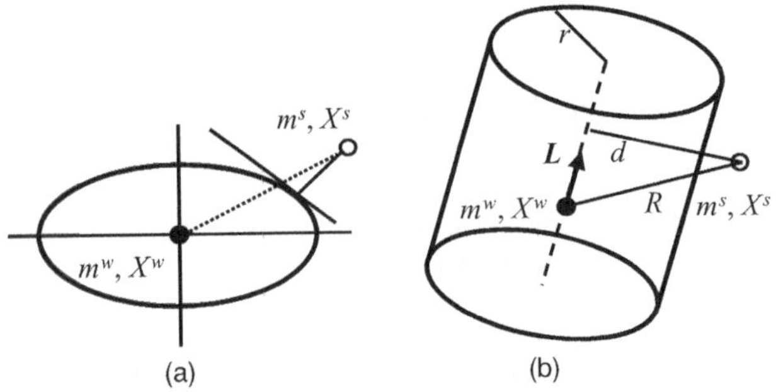

FIGURE 12.3 Rigid wall with a curved surface: (a) ellipsoid; (b) cylinder.

axis, we check the penetration condition

$$\begin{aligned} \boldsymbol{R} &= \boldsymbol{x}^s - \boldsymbol{x}^w, \\ \boldsymbol{d} &= \boldsymbol{R} - (\boldsymbol{R} \bullet \boldsymbol{L})\boldsymbol{L}, \\ |\boldsymbol{d}| &< r. \end{aligned} \tag{12.18}$$

Similar to the computation in the normal direction for the flat wall, using the same notations as described before, we have

$$\Delta \boldsymbol{I}_n^s = m^s \left(\tilde{v}_n^s - v_n^s\right) \boldsymbol{N}^s. \tag{12.19}$$

The decomposition of the velocities into components in normal and tangential directions is similar to the case discussed before. In this case, however, the normal direction at impact varies from point to point and is no longer a constant, as depicted in Figure 12.4. The normal is calculated from the contact point $\boldsymbol{\xi}_s$ on the obstacle surface $S(\boldsymbol{x}) = 0$. $\boldsymbol{\xi}_s$ is supposed to be closest to the slave node $\boldsymbol{x}_s$. We have $S(\boldsymbol{\xi}_s) = 0$ and $(\boldsymbol{x}_s - \boldsymbol{\xi}_s)//\nabla S(\boldsymbol{\xi}_s)$. $\boldsymbol{\xi}_s$ can be solved iteratively for simple geometry. For a cylinder or sphere, the radial direction equals the normal direction, and then $\boldsymbol{\xi}_s$ can be obtained directly by a scaling. Thus, the calculation for the wall with curved surface needs extra work. We keep the assumptions that the contact is a plastic impact and the

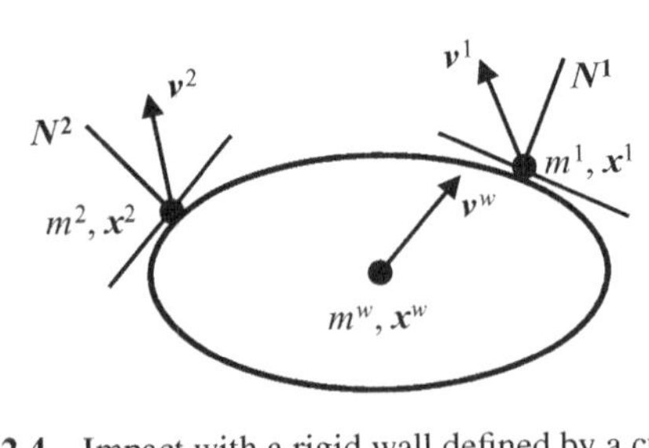

FIGURE 12.4 Impact with a rigid wall defined by a curved surface.

motion of the wall is a translation only. Hence, the normal direction at an impact point is the same before and after impact.

We have the same action as (12.15) in the tangential direction. We now use the vector form instead of a single normal component used for the flat wall. The total impulse acting on the rigid wall is

$$\Delta \boldsymbol{I}^w = m^w(\tilde{\boldsymbol{v}}^w - \boldsymbol{v}^w) = -\sum \left(\Delta \boldsymbol{I}_n^s + \Delta \boldsymbol{I}_t^s\right)$$
$$= -\sum \left(m^s \left(\tilde{v}_{n(s)}^w - v_{n(s)}^s\right) \boldsymbol{N}^s - C_n^s m^s \mu \mid \tilde{v}_{n(s)}^w - v_{n(s)}^s \mid \boldsymbol{t}_s + C_t^s m^s \left(\tilde{\boldsymbol{v}}_{t(s)}^w - \boldsymbol{v}_{t(s)}^s\right)\right).$$

Here, $\tilde{v}_{n(s)}^w = \tilde{\boldsymbol{v}}^w \bullet \boldsymbol{N}$ and $\tilde{\boldsymbol{v}}_{t(s)}^w = \tilde{\boldsymbol{v}}^w - \tilde{v}_{n(s)}^w \boldsymbol{N}$. According to the unilateral contact condition, the contact force applied to the slave node is in the positive normal direction of the wall. Hence, $f_n^s \Delta t = \Delta I_n^s = m^s(\tilde{v}_{n(s)}^w - v_{n(s)}^s) = m^s(\tilde{\boldsymbol{v}}^w - \boldsymbol{v}^s) \bullet \boldsymbol{N}^s > 0$.

It is worth noting that when friction is considered, the new velocity $\tilde{\boldsymbol{v}}^w$ of the rigid wall is affected by both the normal and tangential actions of all the impact slave nodes. $\tilde{\boldsymbol{v}}^w$ cannot be determined by the normal actions alone. Consequently, the scalars C_n^s and C_t^s for frictional actions are undetermined too, depending on the solution $\tilde{\boldsymbol{v}}^w$. We rewrite the above equation,

$$m^w \tilde{\boldsymbol{v}}^w + \sum m^s \left(\tilde{v}_{n(s)}^w \boldsymbol{N}^s - C_n^s \mu \tilde{v}_{n(s)}^w \boldsymbol{t}_s + C_t^s \tilde{\boldsymbol{v}}_{t(s)}^w\right)$$
$$= m^w \boldsymbol{v}^w + \sum m^s \left(v_{n(s)}^s \boldsymbol{N}^s - C_n^s \mu v_{n(s)}^s \boldsymbol{t}_s + C_t^s \boldsymbol{v}_{t(s)}^s\right). \qquad (12.20)$$

An iterative process may be needed.

For a frictionless situation, dropping the terms related to the tangential actions, we have

$$m^w \tilde{\boldsymbol{v}}^w + \sum m^s \tilde{v}_{n(s)}^w \boldsymbol{N}^s = m^w \boldsymbol{v}^w + \sum m^s v_{n(s)}^s \boldsymbol{N}^s. \qquad (12.21)$$

In components, this is a system of three linear equations, for example, the x-equation is

$$m^w \tilde{v}_x^w + \sum m^s \left(\tilde{v}_x^w N_x^s + \tilde{v}_y^w N_y^s + \tilde{v}_z^w N_z^s\right) N_x^s$$
$$= m^w v_x^w + \sum m^s \left(v_x^s N_x^s + v_y^s N_y^s + v_z^s N_z^s\right) N_x^s. \qquad (12.22)$$

The solution is straightforward. If the normal is a constant, it recovers the case of a flat wall, cf. (12.11).

In fact, the solution of (12.21) can be used as the initial guess for the frictional case (12.20). Using it to determine the scalars C_n^s and C_t^s, we can find an updated solution $\tilde{\boldsymbol{v}}^w$; see Hallquist (2006) and ALTAIR (2009) for discussions about other approaches such as penalty method and Lagrange multiplier method.

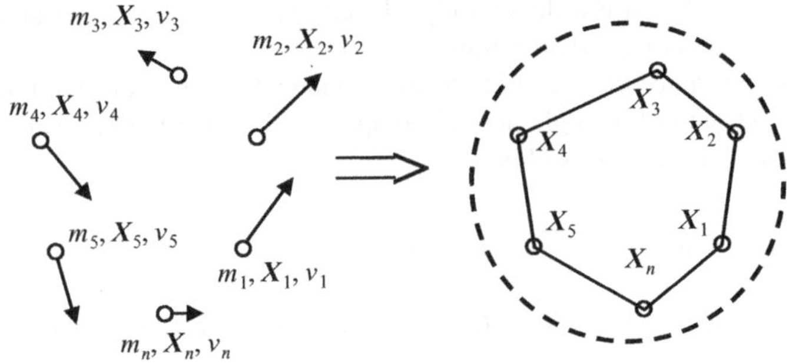

FIGURE 12.5 A set of nodes moving like a rigid body.

12.2 RIGID BODY

In classical mechanics, the motion of a rigid body is composed of a translation and a rotation, as illustrated in Figure 12.5. In practical applications, certain part of the structure can be modeled as a rigid body when deformation at its location is negligible. A set of nodes (or elements) can be defined as a rigid body. The constraint condition requires these nodes to be rigidly connected particles moving as a rigid body.

Following the course of classical mechanics, we can define the center of gravity (CG) of the group with the total mass:

$$\begin{aligned} M_B &= \sum m_s, \\ \boldsymbol{x}_c &= \sum m_s \boldsymbol{x}_s / M_B. \end{aligned} \tag{12.23}$$

Here, we use the subscript s for the set of (slave) nodes of the rigid body. The moment of inertia with respect to CG is then defined as

$$\boldsymbol{I}_B = \sum (\boldsymbol{I}_s + m_s(\boldsymbol{r}_s \bullet \boldsymbol{r}_s \boldsymbol{I} - \boldsymbol{r}_s \otimes \boldsymbol{r}_s)), \tag{12.24}$$

where $\boldsymbol{I}_s$ represents the nodal moment of inertia of the slave nodes introduced from R-M plate theory. We denote for the position vector relative to CG by

$$\boldsymbol{r}_s = \boldsymbol{x}_s - \boldsymbol{x}_c. \tag{12.25}$$

The initial value of the total momentum and angular momentum of this set of nodes are

$$\begin{aligned} \boldsymbol{p}_0 &= \sum m_s \boldsymbol{v}_0^s, \\ \boldsymbol{H}_0 &= \sum \left(\boldsymbol{I}_s \boldsymbol{\omega}_0^s + m_s \boldsymbol{r}_s \times \boldsymbol{v}_0^s\right). \end{aligned} \tag{12.26}$$

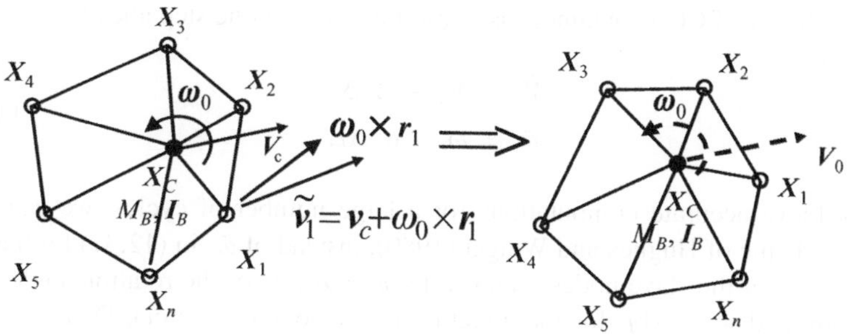

FIGURE 12.6 Motion of rigid body with the group of slave nodes.

From conservation of momentum, we find the velocity and angular velocity for the CG

$$\begin{aligned} \boldsymbol{v}_0 &= \boldsymbol{p}_0/M_B, \\ \boldsymbol{\omega}_0 &= \boldsymbol{I}_B^{-1}\boldsymbol{H}_0. \end{aligned} \tag{12.27}$$

We consider the CG, $\boldsymbol{v}_0$, and $\boldsymbol{\omega}_0$ as the property of the rigid body. In practical applications, the input data of initial velocity and angular velocity of the node set may not satisfy the condition of moving as a group rigidly connected. The new velocities are assigned to these slave nodes here so that they can move as being rigidly connected, shown in Figure 12.6,

$$\begin{aligned} \boldsymbol{v}_s &= \boldsymbol{v}_0 + \boldsymbol{\omega}_0 \times \boldsymbol{r}_s, \\ \boldsymbol{\omega}_s &= \boldsymbol{\omega}_0. \end{aligned} \tag{12.28}$$

After the adjustment, usually the initial kinetic energy reduces from what is calculated from the initial conditions inputted at these slave nodes. The important point is the conservation of momentum. This is just like a plastic impact freezing the node set as a rigid body.

In the computation of each time step, the treatment of rigid body is an exercise of classical mechanics. The resultants of forces and moments are calculated as a summation of all the internal and external forces exerted on these slave nodes:

$$\begin{aligned} \boldsymbol{F} &= \sum \boldsymbol{F}_s, \\ \boldsymbol{M} &= \sum (\boldsymbol{M}_s + \boldsymbol{r}_s \times \boldsymbol{F}_s). \end{aligned} \tag{12.29}$$

Then, from Newton's second law

$$\begin{aligned} \boldsymbol{F} &= M_B \boldsymbol{A}_c \\ \boldsymbol{M} &= \mathrm{d}\boldsymbol{H}/\mathrm{d}t = \boldsymbol{I}_B \boldsymbol{\alpha}_c + \boldsymbol{\omega}_c \times \boldsymbol{I}_B \boldsymbol{\omega}_c \end{aligned} \Rightarrow \begin{aligned} \boldsymbol{A}_c &= \boldsymbol{F}/M_B \\ \boldsymbol{\alpha}_c &= \boldsymbol{I}_B^{-1}(\boldsymbol{M} - \boldsymbol{\omega}_c \times \boldsymbol{I}_B \boldsymbol{\omega}_c). \end{aligned} \tag{12.30}$$

The velocity of CG is obtained, as a result of moving one step ahead,

$$\begin{aligned}\tilde{\boldsymbol{V}}_C &= \boldsymbol{V}_C + \boldsymbol{A}_C\Delta t,\\ \tilde{\boldsymbol{\omega}}_C &= \boldsymbol{\omega}_C + \boldsymbol{\alpha}_C\Delta t.\end{aligned} \tag{12.31}$$

To achieve accurate computations for a large number of cycles, we can use a rotation scheme of Hughes and Winget (1980), instead of $\tilde{\boldsymbol{\omega}}_C$ in (12.31) for the rigid body motion of the slave nodes (denoted by $\boldsymbol{\eta} = \boldsymbol{\omega}_C\Delta t$ for the rotation angle of the rigid body and $\boldsymbol{n} = \boldsymbol{\eta}/|\boldsymbol{\eta}|$ for the direction of the rotation vector). Define a vector $\boldsymbol{\lambda} = 2\tan(|\boldsymbol{\eta}|/2)\boldsymbol{n}$ and an antisymmetric matrix

$$\Omega = \begin{bmatrix} 0 & -\lambda_3 & \lambda_2 \\ \lambda_3 & 0 & -\lambda_1 \\ -\lambda_2 & \lambda_1 & 0 \end{bmatrix}. \tag{12.32}$$

Then, the rotation matrix is formed as

$$\boldsymbol{R} = (\boldsymbol{I} - \Omega/2)^{-1}(\boldsymbol{I} + \Omega/2) \approx \boldsymbol{I} + (\Omega + \Omega^2/2)/(1 + \lambda^2/4). \tag{12.33}$$

We have the relative rotation of the slave nodes about the CG:

$$\tilde{\boldsymbol{r}}_s = \boldsymbol{R}\boldsymbol{r}_s. \tag{12.34}$$

This yields a motion $\tilde{\boldsymbol{x}}_s = \tilde{\boldsymbol{X}}_C + \tilde{\boldsymbol{r}}_s$ with $\tilde{\boldsymbol{X}}_C = \boldsymbol{X}_C + \boldsymbol{V}_C\Delta t$.

Note that for the nodes of the rigid body, no stiffness appears in the equation except for the rigid connection. Hence, there is no stability requirement on the time step size.

12.3 EXPLICIT FINITE ELEMENT PROCEDURE WITH CONSTRAINT CONDITIONS

After the calculation of constraint conditions such as rigid wall and rigid body discussed in this chapter, we need to modify the calculated nodal velocity and acceleration. In this way, after the computation for constraint conditions is done, additional change in velocity or acceleration will break the constraint condition. Such a situation may occur if a slave node belongs to a second constraint condition, which we then need to modify the velocity and acceleration again. In this case, the common slave node is expected to satisfy both conditions, which however are treated individually, even if two constraint conditions are of the same kind.

In fact, explicit integration is executed in the step-by-step cyclic process; there is no system equation solver. The equations of the constraint conditions are not "solved" with the equation of motion, and are not "solved" all together. Therefore, generally any node should not be defined in two constraint conditions. Examples are depicted

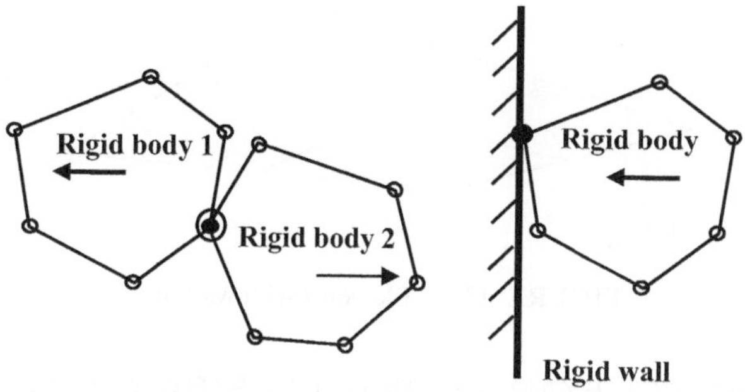

FIGURE 12.7 Conflict constraint conditions.

in Figure 12.7 for illustration. It is important to pay attention to the error message generated by software regarding the conflict conditions, cf. Hallquist (2006).

This concern also suggests that the constraint condition is better calculated in the last step of the cyclic procedure. The program flow chart with constraint conditions is illustrated in Figure 12.8.

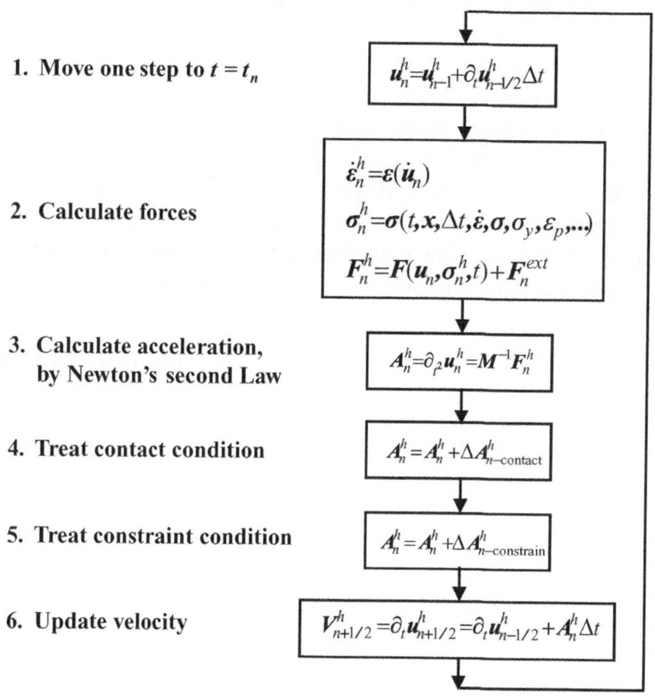

FIGURE 12.8 Explicit finite element procedure—with constraint condition (extended from Fig. 10.14).

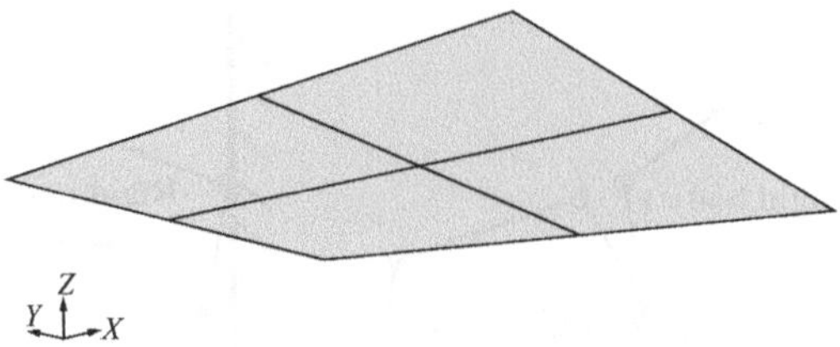

FIGURE 12.9 Elements with warping.

12.4 APPLICATION EXAMPLES WITH CONSTRAINT CONDITIONS

Example 12.1 Rotation of rigid body As mentioned in Section 3.5.3, the hourglass control method of B-T element is not orthogonal to rigid body rotation. This may result in unexpected force for a warped element under rigid rotation. We consider a patch of four warped shell elements, depicted in Figure 12.9. The element size is 10 mm $\times$ 10 mm, aligned with xy-plane. Four corner nodes are shifted by ± 1 mm in z-direction to form warping. The center node of the patch is fixed. An initial velocity v_z is assigned to one side $y = -10$ mm/ms, to start the rotational motion. Mild steel material is assumed with Young's modulus $E = 210$ kN/mm^2, mass density $\rho = 7.8 \times 10^{-6}$ kg/mm^3, and Poisson ratio $\nu = 0.3$.

The computation uses LS-DYNA V971 with B-T elements. The result shows a rotation about z-axis along with the rotation about y-axis. The top view at $T = 93$ ms is depicted in Figure 12.10a, showing a nearly 45° rotation. The x-displacement of a corner node is depicted in Figure 12.10b, demonstrating the history of this type of motion. On the other hand, the computation using B-D element and B-L element results in the proper rotation about y-axis without meaningful z-rotation.

Now, we define a rigid body as discussed in Section 12.2. The CG of the rigid body, at the center node of the patch, is fixed like the above case. The computation is excised up to 1,000 ms. The y-displacement of a corner node is presented in Figure 12.11, with the time duration from 900 to 1,000 ms. Its x-displacement only

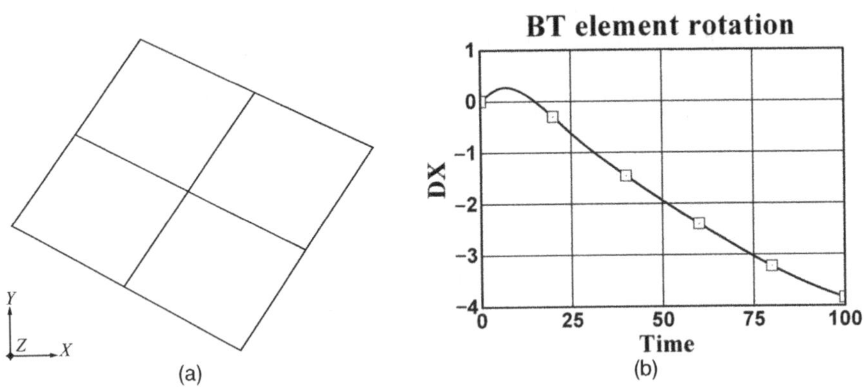

FIGURE 12.10 Elements with warping: (a) position of the B-T elements at 93 ms; (b) time history of x-displacement of an end node.

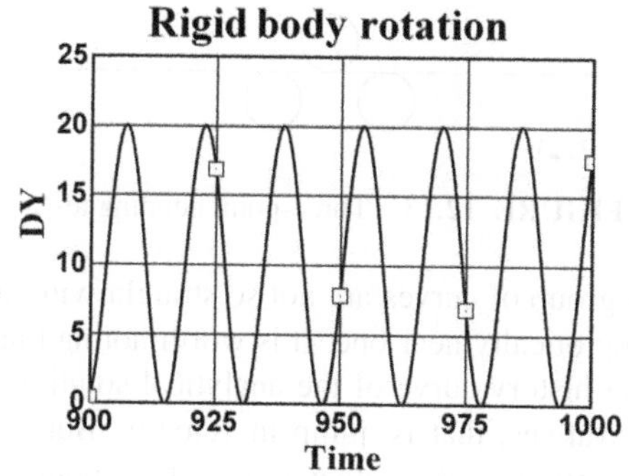

FIGURE 12.11 Time history of y-displacement of a corner node in rigid rotation.

grows up to an order of 10^{-7}. The period of rotation is about 15.770 ms at the first revolution and 15.775 ms at the last revolution of the 1,000 ms time duration. The rigid rotation about y-axis is stable.

Example 12.2 A rod impacting the fixed flat rigid obstacle—rigid wall method The problem was discussed in Example 11.1 with the penalty method. Here, we use rigid wall described in Section 12.2 to model the rigid obstacle as a constraint condition. Similar to Example 11.1, a set of uniform meshes is employed for investigation. To avoid the possible loss of accuracy, smaller time step as 10% of the critical value determined by elements is used. The finite element solutions are compared to the analytical solution of the original Signorini's problem.

The errors of the displacement of the end points at few key time points are presented in Figure 12.12 in log–log scale. With decreasing element size, the error of the nodal displacements decreases. The convergence tendency is clearly presented. The error curves of time equal to 2 represent a free fall before impacting the obstacle, calculated by the central difference method in time domain. We can verify that this error is proportional to the time step, which in turn is set to be in proportion to the element size. Therefore, they have a slope equal to one. Furthermore, a group of the error curves, at certain time points, are quite straight and have slope asymptotically

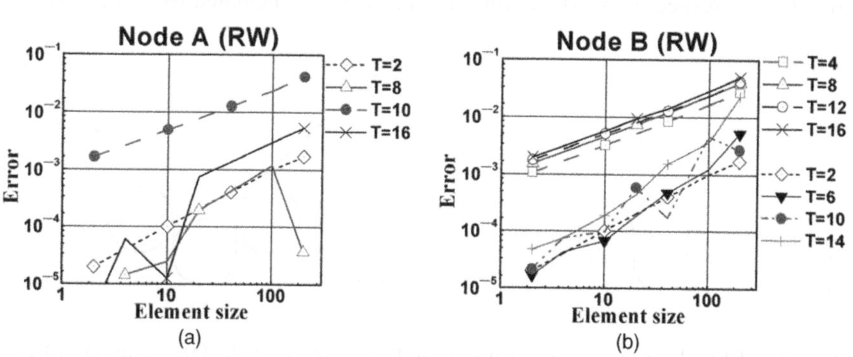

FIGURE 12.12 Error analysis—convergence of rigid wall method with respect to the refined meshes.

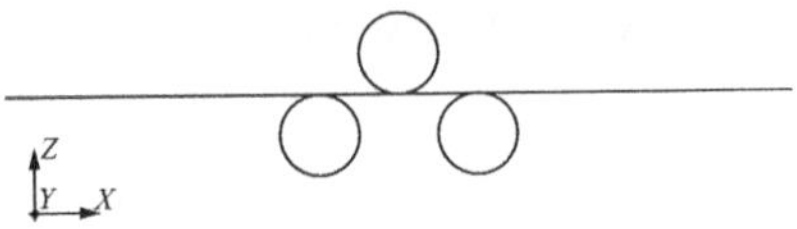

FIGURE 12.13 Three-point bending test.

equal to 3/4. Another group of curves are not so straight with an oscillatory pattern and have a slope asymptotically near one. It is worth noting that for the first group, the nodal displacement history curve of the analytical solution at these time points has discontinuous derivatives, that is, jump in velocity. But for the second group, the displacement has continuous derivative at these time points (node A and B have different timing). This type of error analysis is point wise, can be measured in $L^\infty(0, T; L^\infty)$-norm, cf. (2.29) for definition. The lower smoothness of solution may be responsible for the lower convergence rate at these discontinuity points. Theoretical assessment is yet to be developed.

Example 12.3 The simulation of a three-point bending test A steel strip of size 100 mm × 8 mm is placed in the test apparatus with three rigid rollers, depicted in Figure 12.13. This test is performed for very large bending deformation. The radii of the rollers are 5 mm. Two lower rollers are fixed at their centers. Acting as the loading mechanism, the upper roller presses down and forces the specimen to bend and slide through the gap between the two lower rollers.

We use cylindrical rigid wall to model the rollers. Here, we do not consider the thickness of the shell element for the contact with the rigid rollers. The friction is ignored. The strip specimen is uniformly graded with 100 × 2 quadrilateral elements. The two lower rollers are placed 20.02 mm apart from center to center, and 10.02 mm lower from the upper roller, allowing 0.01 mm gap to avoid possible confusion for penetration check.

For a high-speed test with constant loading velocity $v = 10$ mm/ms, we exercise 1.5 ms loading or 15 mm traveling of the upper roller. Computation is performed by using LS-DYNA V971. The deformation of the specimen calculated by using B-T element is depicted in Figure 12.14a. The mesh is then refined to 200 × 4 and 400 × 8, respectively. The deformation at 1.5 ms calculated by using 400 × 8

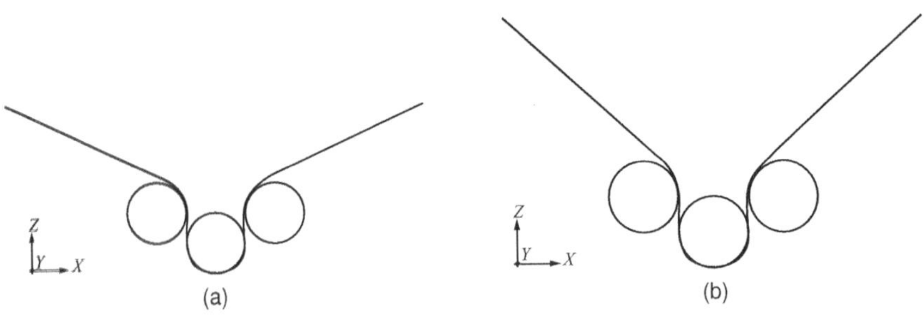

FIGURE 12.14 Deformation at 1.5 ms, by B-T element: (a) with 100 × 2 mesh; (b) with 400 × 8 mesh.

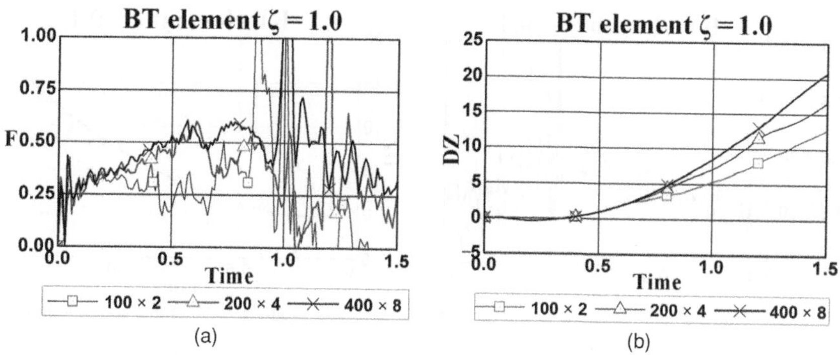

FIGURE 12.15 Results of B-T element: (a) time history of the applied force; (b) time history of z-displacement of an end node.

mesh is presented in Figure 12.14b, quite different from that in Figure 12.14a. The time history of the force exerted on the roller, calculated from the three meshes, is depicted in Figure 12.15a. The z-displacement of an end node is depicted in Figure 12.15b. Large deformation occurs in the area of contact with the rollers. There is large variation in the force calculated from the three meshes, developed after around 0.3 ms. The displacement from the three meshes before 0.6 ms are close to each other. Results of B-D element are shown in Figure 12.16 for comparison. The difference in

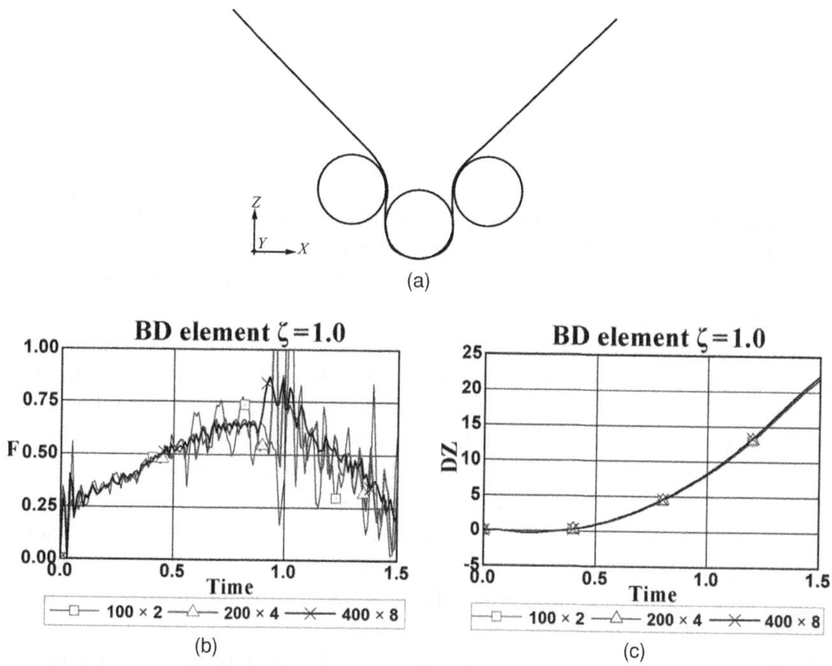

FIGURE 12.16 Results of B-D element: (a) deformation at 1.5 ms, with 400 × 8 mesh; (b) time history of the applied force; (c) time history of z-displacement of an end node.

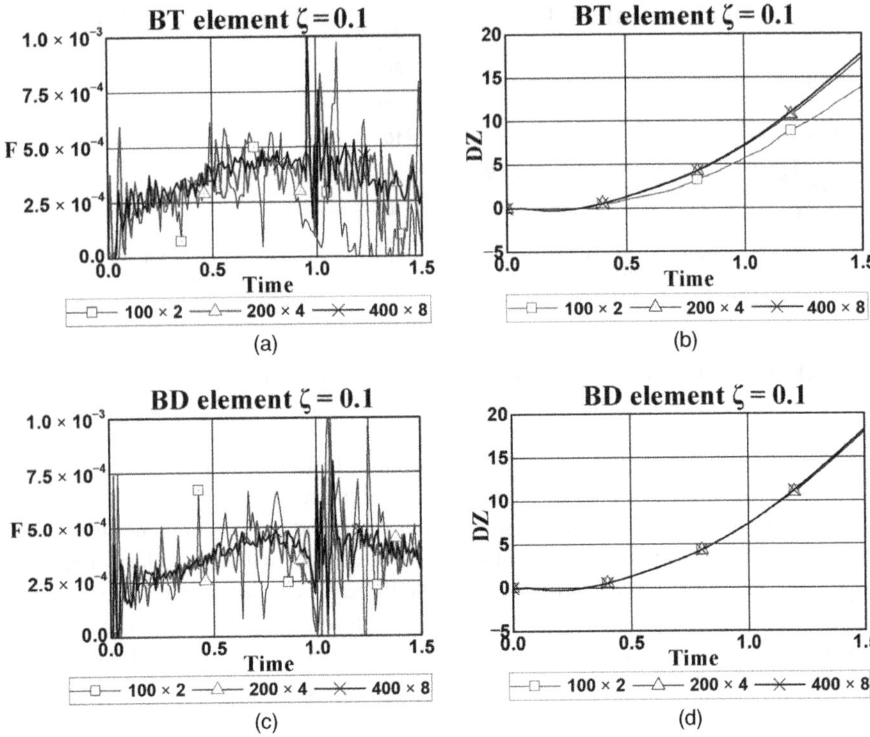

FIGURE 12.17 Results of reduced thickness: (a) applied force by B-T element; (b) z-displacement of an end node by B-T element; (c) applied force by B-D element; (d) z-displacement of an end node by B-D element.

results of B-D element by the three meshes is much smaller than that of B-T element. In fact, the result of B-T element with 400×8 mesh is close to that of B-D element with visible difference in the applied force in later time. For all these cases, there is essentially no penetration observed.

We continue the study with reduced thickness $\zeta = 0.1$. The load scaling and mass scaling are applied as discussed in Sections 3.8.1, 4.4.1, etc. For reduced thickness, with the same bending deformation, the strain in the thickness is scaled down. This results in the reduced stress. A scaling in yield stress is proposed here: when the thickness reduces with a factor of δ, the yield stress also reduces with a factor of δ.

The deformation at 1.5 ms is similar to that shown in Figure 12.14b. The time history of the applied load and the z-displacement of the end node are presented in Figure 12.17. The results from the three meshes are close for both B-T and B-D elements. Also, the results of B-T and B-D elements are closer than what are observed in the case of original thickness. Overall, both elements demonstrate a convergence tendency. Particularly, B-T element seems to have better convergence behavior than what it does in the case of the original thickness.

REFERENCES

CHAPTER 1

Anghileri M, Castelletti L-ML, Lanzi L, Mentuccia F (2005). Composite materials and bird-strike analysis using explicit finite element commercial codes. Proc. 8th Int. Conf. Structures under Shock and Impact (SUSI), Crete, Greece.

Argyris JH (1954). Energy theorems and structural analysis. *Aircraft Eng*. 26, 347–356.

Argyris JH (1957). The matrix theory of statics. *Ingr. Arch*. 25, 174–192.

Babuska I, Oden JT (2004). Verification and validation in computational engineering and science: basic concepts. *Comp. Meth. Appl. Mech. Eng*. 193, 4057–4066.

Belytschko T (1996). On difficulty levels in nonlinear finite element analysis of solids. *Bull. Int. Assoc. Comp. Mech*. 2, 6–8.

Belytschko T, Liu WK, Moran B (2000). *Nonlinear Finite Elements for Continua and Structures*. John Wiley & Sons, New York.

Chow CL, Tai (2000). Damage based formability analysis of sheet metal with LS-DYNA. *Int. J. Damage Mech*. 9, 241–254.

Ciarlet PG, Lions JL (1991). *Handbook of Numerical Analysis*. North-Holland, Amsterdam.

Clough RW (1960). The finite element method in plane stress analysis. J. Struct. Div., ASCE, Proc. 2nd Conf. Elect. Comp. pp. 345–378.

Constantino CJ (1967). Finite element approach to stress wave problems. *J. Eng. Mech. Div. ASCE* 93, 153–166.

Introduction to the Explicit Finite Element Method for Nonlinear Transient Dynamics, First Edition.
Shen R. Wu and Lei Gu.

Courant R (1943). Variational method for the solution of problems of equilibrium and vibration. *Bull. Am. Math. Soc.* 49, 1–23.

Ho CH, Smith MR (2006). Bird strike simulation for BA609 spinner and rotor controls. Proc. 9th Int. LS-DYNA Users Conf., pp. 18-21–18-22, Dearborn, MI.

Houssini SE (2006). Front bumper crashworthiness optimization. Proc. 9th Int. LS-DYNA Users Conf., pp. 11-57–11-61, Dearborn, MI.

Lu HS, Wu CT (2006). A grid-based adaptive scheme for the three-dimensional forging and extrusion problems with the EFG method. Proc. 9th Int. LS-DYNA Users Conf., pp. 17-33–17-44, Dearborn, MI.

Medvedev SV (2002). Computer modeling of residual welding strains in technological design of welded structures. *Weld. Int.* 16, 59–65.

Neumayer D, Chatiri M, Höermann M (2006). Drop test simulation of a cooker including foam packaging and pre-stressed plastic foil wrapping. Proc. 9th Int. LS-DYNA Users Conf., pp. 18-33–18-40, Dearborn, MI.

Oberkampf WL, Barone MF (2004). Measures of agreement between computation and experiment: validation metrics. AIAA 34th Fluid Dynamics Conf., AIAA-2004-2626, Portland, Oregon, June 28–July 1, 2004.

Oden JT (1972). *Finite Element of Nonlinear Continua*. McGraw-Hill, New York.

Oden JT, Belytschko T, Babuska I, Hughes TJR (2003). Research directions in computational mechanics. *Comp. Meth. Appl. Mech. Eng.* 192, 913–922.

Saha NK, Calso SM, Prasad P, Asjad MU (1995). Simulation of frontal offset impacts and comparison of intrusions and decelerations. SAE Paper 950647.

Turner MJ, Clough RW, Martin HC, Topp LJ (1956). Stiffness and deflection analysis of complex structures. *J. Aero. Sci.* 23, 805–823.

Wilkins ML (1964). *Calculation of elastic-plastic flow*. In Alder G, Fernbach S, Rotenberg M (editors), *Methods of Computational Physics*, Vol. 3. Academic Press, New York.

Xue X, Schmid F (2005). Modeling study to validate finite element simulation of railway vehicle behaviour in collisions. Proc. 5th Euro. LS-DYNA Users Conf., Paper 7b-30, Birmingham, UK.

CHAPTER 2

Ainsworth M, Oden JT (2000). *A Posteriori Error Estimation in Finite Element Analysis*. John Wiley & Sons, New York.

Andreev AB, Kascieva VA, Vanmaele M (1992). Some results in lumped mass finite-element approximation of eigenvalue problems using numerical quadrature formulas. *J. Comp. Appl. Math.* 43, 291–311.

Argyris JH (1964). *Recent Advances in Matrix Methods in Structural Analysis*. Pergamon Press, Oxford.

Argyris JH (1965). Continua and discontinua. In Proc. 1st Conf. Matrix Method in Structural Mech., Wright-Patterson Air Force Base, Ohio, October 26–28.

Babuška I, Osborn J (1991). Eigenvalue problems. In Ciarlet PG, Lions JL (editors), *Handbook of Numerical Analysis*, Vol. 2, pp. 641–787. North-Holland, Amsterdam.

Babuška I, Strouboulis T (2001). *The Finite Element Method and Its Reliability*. Oxford University Press, Oxford.

Belytschko T, Holmes N, Mullen R (1975). Explicit integration—Stability, solution properties, cost. *ASME AMD* 14, 1–21.

Belytschko T, Hsieh BJ (1973). Nonlinear transient finite element analysis with convected co-ordinates. *Int. J. Numer. Meth. Eng.* 7, 255–271.

Belytschko T, Liu WK, Moran B (2000). *Nonlinear Finite Elements for Continua and Structures*. John Wiley & Sons, New York.

Belytschko T, Yen HR, Mullen R (1979). Mixed methods for time integration. *Comp. Meth. Appl. Mech. Eng.* 17/18, 259–275.

Brenner SC, Scott LR (2002). *The Mathematical Theory of Finite Element Methods*, 2nd ed. Springer-Verlag, New York.

Ciarlet PG (1988). *Mathematical Elasticity: Three-dimensional Elasticity*, Vol. 1. Elsevier Science Publishing, New York.

Ciarlet PG, Lions JL (1991). *Handbook of Numerical Analysis*. North-Holland, Amsterdam.

Clough RW (1960). The finite element method in plane stress analysis. J. Struct. Div., ASCE, Proc. 2nd Conf. Electro. Comp. pp. 345–378.

Cohen G, Joly P, Tordjman N (1994). Higher-order finite elements with mass-lumping for the 1D wave equation. *Finite Elem. Anal. Des.* 16, 329–336.

Courant R, Friedrichs KO, Lewy H (1928). Über die partiellen Differenzensleichungen der Mathematischen Physik. *Math Ann.* 100, 32 (in German).

Dupont T (1973). L^2-estimate for Galerkin methods for second order hyperbolic equations. *SIAM J. Numer. Anal.* 10, 880–889.

Evans LC (1998). *Partial Differential Equations*. American Mathematical Society, Providence, Rhode Island.

Fix GJ (1973). Eigenvalue approximation by the finite element method. *Adv. Math.* 10, 300–316.

Flanagan DP, Belytschko T (1981a). Simultaneous relaxation in structural dynamics. *ASCE J. Eng. Mech. Div.* 107, 1039–1055.

Flanagan DP, Belytschko T (1981b). A uniform strain hexahedron and quadrilateral with orthogonal hourglass control. *Int. J. Numer. Meth. Eng.* 17, 679–706.

Fujii H (1972). Finite element schemes: Stability and convergence. In Oden JT, Yamamoto Y, Clough RW (editors), *Advances in Computational Methods in Structural Mechanics and Design*, pp. 201–218. UAH Press, Huntsville, Alabama.

Hughes TJR, Pister KS, Taylor RL (1979). Implicit-explicit finite elements in nonlinear transient analysis. *Comp. Meth. Appl. Mech. Eng.* 17–18, 159–182.

Hulbert GM, Hughes TJR (1988). Numerical evaluation and comparison of subcycling algorithms for structural dynamics. Technical Report, DNA-TR-88-8, Defense Nuclear Agency, Washington, DC.

Lions JL, Magenes E (1972). *Non-homogeneous Boundary Value Problems and Applications*. Springer-Verlag, New York (French ed., 1968).

Meirovitch L (1980). *Computational Methods in Structural Dynamics*. Sijthoff & Noordhoff, Alphen an den Rijn, The Netherlands.

Newmark NM (1959). A method of computation for structural dynamics. *J. Eng. Mech. Div. ASCE* 85 (EM3), 67–94.

Oden JT, Fost RB (1973). Convergence, accuracy and stability of finite element approximations of a class of non-linear hyperbolic equations. *Int. J. Numer. Meth. Eng.* 6, 357–365.

Oden JT, Reddy JN (1976). *An Introduction to the Mathematical Theory of Finite Elements.* John Wiley & Sons, New York.

Song SC, Duan ZP (1998). Iterative method using consistent mass matrix in axisymmetrical finite element analysis of hypervelocity impact. *Int. J. Impact Eng.* 21, 817–825.

Strang G, Fix GJ (1973). *An Analysis of the Finite Element Method.* Prentice-Hall, Englewood Cliffs, New Jersey.

Tong P, Pian THH, Bucciarelli LL (1971). Mode shapes and frequencies by finite element method using consistent and lumped masses. *Comp. Struct.* 1, 623–638.

Wu SR (2003). A-priori error estimates for explicit finite element for linear elasto-dynamics by Galerkin method and central difference method. *Comp. Meth. Appl. Mech. Eng.* 192, 5329–5353.

Wu SR (2006). Lumped mass matrix in explicit finite element method for transient dynamics of elasticity. *Comp. Meth. Appl. Mech. Eng.* 195, 5983–5994.

Wu SR (2009). A variational principle for dynamic contact with large deformation. *Comp. Meth. Appl. Mech. Eng.* 198, 2009–2015.

Wu SR, Gu L (2003). The effect of lumped mass matrix in explicit FEM. In Christon MA, Haber R (editors), *USNCCM VII Abstracts.* Proc. 7th US Nat. Cong. Comp. Mech., p. 938, 7/27–31/2003. Albuquerque, New Mexico.

Wu SR, Qiu W (2009). Nonlinear transient dynamic analysis by explicit finite element with iterative consistent mass matrix. *Commun. Numer. Meth. Eng.* 25, 201–217.

Yuan YR, Wang H (1985). The error estimate for the finite element method of a nonlinear hyperbolic equation. *J. Sys. Sci. Math. Sci.* 5, 161–171.

ZlÁmal M (1968). On the finite element method. *Numer. Math.* 12, 394–409.

CHAPTER 3

Argyris JH (1964). *Recent Advances in Matrix Methods in Structural Analysis.* Pergamon Press, Oxford.

Arnold D, Bolfi D, Falk R (2002). Remarks on quadrilateral Reissner-Mindlin plate element. In Mang HA, Rammerstorfer FG, Eberhardsteiner J (editors). Proc. WCCM V (5th World Cong. Comp. Mech.), Vienna, Austria, July 7–12.

Arnold D, Falk R (1989a). A uniformly accurate finite element method for the Reissner–Mindlin plate. *SIAM J. Numer. Anal.* 26, 1276–1290.

Arnold D, Falk R (1989b). Edge effect in the Reissner–Mindlin plate theory. In Noor AK, Belytschko T, Simo JC (editors), *Analytic and Computational Models of Shells*, pp. 71–90. ASME, New York.

Arnold D, Falk R (1990). The boundary layer for Reissner–Mindlin plate model. *SIAM J. Math. Anal.* 21, 281–312.

Arnold D, Falk R (1997). Analysis of a linear–linear finite element for the Reissner–Mindlin plate model. *Math. Models Meth. Appl. Sci.* 7, 217–238.

Babuška I, Li L (1992). The problem of plate modeling: Theoretical and computational results. *Comp. Meth. Appl. Mech. Eng.* 100, 249–273.

Babuška I, Oden JT, Lee JK (1975). Mixed-hybrid finite element approximations of second-order elliptic boundary-value problems. TICOM Report 75–7, The Texas Institute of Computational Mechanics, University of Texas at Austin.

Barlow L (1976). Optimal stress locations in finite element models. *Int. J. Numer. Meth. Eng.* 10, 243–251.

Bathe KJ (1982). *Finite Element Procedures in Engineering Analysis*. Prentice-Hall, Englewood Cliffs, New Jersey.

Bathe KJ, Brezzi F (1985). On the convergence of a four-node plate-bending element based on Mindlin-Reissner plate theory and a mixed interpolation. In Whiteman JR (editor), *The Mathematics of Finite Elements and Applications V (MAFELAP)*, pp. 491–503. Academic Press, London.

Bathe KJ, Dvorkin EN (1985). A four-node plate-bending element based on Mindlin-Reissner plate theory and a mixed interpolation. *Int. J. Numer. Meth. Eng.* 21, 367–383.

Bathe KJ, Wilson EL (1976). *Numerical Methods in Finite Element Analysis*. Prentice-Hall, Englewood Cliffs, New Jersey.

Belytschko T (1974). Finite element approach to hydrodynamics and mesh stabilization. In Oden JT et al. (editors), *Computation Methods in Nonlinear Mechanics*, pp. 231–238. The Texas Institute for Computational Mechanics, Texas.

Belytschko T, Bachrach WE (1986). Efficient implementation of quadrilaterals with high coarse-mesh accuracy. *Comp. Meth. Appl. Mech. Eng.* 54, 279–301.

Belytschko T, Hsieh BJ (1973). Nonlinear transient finite element analysis with convected co-ordinates. *Int. J. Numer. Meth. Eng.* 7, 255–271.

Belytschko T, Leviathan I (1994a). Physical stabilization of 4-node shell element with one-point quadrature. *Comp. Meth. Appl. Mech. Eng.* 113, 321–350.

Belytschko T, Leviathan I (1994b). Projection scheme for one-point quadrature shell element. *Comp. Meth. Appl. Mech. Eng.* 115, 277–286.

Belytschko T, Lin JL, Tsay CS (1984). Explicit algorithms for the nonlinear dynamics of shells. *Comp. Meth. Appl. Mech. Eng.* 42, 225–251.

Belytschko T, Liu WK, Moran B (2000). *Nonlinear Finite Element for Continua and Structures*. John Wiley & Sons, West Sussex, England.

Belytschko T, Tsay CS (1983). A stabilization procedure for the quadrilateral plate element with one-point quadrature. *Int. J. Numer. Meth. Eng.* 19, 405–419.

Belytschko T, Tsay CS, Liu WK (1981). A stabilization matrix for the bilinear Mindlin plate element. *Comp. Meth. Appl. Mech. Eng.* 29, 313–327.

Belytschko T, Wong BL, Chiang HY (1992). Advances in one-point quadrature shell elements. *Comp. Meth. Appl. Mech. Eng.* 96, 93–107.

Bernadou M (1993). C^1-curved finite elements with numerical integration for thin plate and thin shell problems, Part 2: Approximation of thin plate and thin shell problems. *Comp. Meth. Appl. Mech. Eng.* 102, 389–421.

Bernadou M (1996). *Finite Element Methods for Thin Shell Problems*. John Wiley & Sons, New York.

Brezzi F (2003). *On Scaling of Reissner-Mindlin Plate*. Private communication.

Brezzi F, Bathe KJ, Fortin M (1989). Mixed-interpolated elements for Reissner–Mindlin plates. *Int. J. Numer. Meth. Eng.* 28, 1787–1801.

Brezzi F, Fortin M (1986). Numerical approximation of Mindlin-Reissner plate. *Math. Comp.* 47, 151–158.

Brezzi F, Fortin M, Stenberg R (1991). Error analysis of mixed-interpolated elements for Reissner–Mindlin plate. *Math. Models Meth. Appl. Sci.* 1, 125–151.

Chapelle D, Bathe KJ (2003). *The Finite Element Analysis of Shells—Fundamentals*. Springer, Berlin.

Chien WZ (1944). The intrinsic theory of thin shells and plates. *Q. Appl. Math.* (a) 1, 297–327; (b) 2, 43–59; (c) 2, 120–135.

Ciarlet PG (2000). Mathematical Elasticity: *Theory of Shells*, Vol. 3. North-Holland, Amsterdam.

Doherty WP, Wilson EL, Taylor RL (1969). Stress analysis of axisymmetric solids utilizing higher order quadrilateral finite elements. SESM Report 69-3, Struct. Eng. Lab., University of California, Berkeley.

Durán R, Liberman E (1992). On mixed finite element methods for Reissner-Mindlin plate models. *Math. Comp.* 58, 561–573.

Dvorkin EN, Bathe KJ (1984). A continuum mechanics based four-node shell element for general nonlinear analysis. *J. Eng. Comp.* 1, 77–88.

Flanagan DP, Belytschko T (1981). A uniform strain hexahedron and quadrilateral with orthogonal hourglass control. *Int. J. Numer. Meth. Eng.* 17, 679–706.

Flügge W (1960). *Stresses in Shells*. Springer-Verlag, Berlin.

Gallagher RH (1975). *Finite Element Analysis Fundamentals*. Prentice-Hall, Englewood Cliffs, New Jersey.

Häggblad B, Bathe KJ (1990). Specifications of boundary conditions for Reissner/Mindlin plate bending finite elements. *Int. J. Numer. Meth. Eng.* 30, 981–1011.

Hu HC (1963). On problems of anti-symmetric small deflection of isotropic sandwich plate. *Mechanics* 6, 53–60 (in Chinese).

Hu HC (1981). *Variational Principles of Elasticity and their Applications*. Science Press, Beijing (in Chinese).

Hughes TJR (1987). *The Finite Element Method: Linear Static and Dynamic Finite Element Analysis*. Prentice-Hall, Englewood Cliffs, New Jersey.

Hughes TJR, Cohen M, Haroun M (1978). Reduced and selective integration techniques in finite element analysis of plates. *Nucl. Eng. Des.* 46, 203–222.

Hughes TJR, Liu WK (1981). Nonlinear finite element analysis of shells: Part I. Three-dimensional shells. *Comp. Meth. Appl. Mech. Eng.* 26, 331–362.

Hughes TJR, Taylor RL, Kanoknukulchai W (1977). A simple and efficient finite element for plate bending. *Int. J. Numer. Meth. Eng.* 11, 1529–1543.

Jacquotte OP, Oden JT (1984). Analysis of hourglass instabilities and control in underintegrated finite element methods. *Comp. Meth. Appl. Mech. Eng.* 44, 339–363.

Jacquotte OP, Oden JT (1986). An accurate and efficient a posteriori control of hourglass instabilities in underintegrated linear and nonlinear elasticity. *Comp. Meth. Appl. Mech. Eng.* 55, 105–128.

Kavanagh KT, Key SW (1972). A note on selective and reduced integration techniques in the finite element method. *Int. J. Numer. Meth. Eng.* 4, 148–150.

Kirchhoff G (1850). Über das Gleichgewicht und die Bewegung einer elastischen Scheibe. *J. Reine Angew. Math.* (Crelle) 40, 51–88 (in German).

Kosloff D, Frazier GA (1978). Treatment of hourglass patterns in low order finite element codes. *Int. J. Num. Anal. Meth. Geomech.* 2, 57–72.

Liew KM, Wang CM, Xiang Y, Kitipornchai S (1998). *Vibration of Mindlin Plates, Programming the P-Version Ritz Method.* Elsevier, Oxford.

Love AEH (1927). *A Treatise on the Mathematical Theory of Elasticity*, 4th ed. Dover Publications, New York.

Lyly M, Sternberg R, Vihinen T (1993). A stable bilinear element for the Reissner-Mindlin plate model. *Comp. Meth. Appl. Mech. Eng.* 110, 343–357.

MacNeal RH (1982). Derivation of element stiffness matrices by assumed strain distributions. *Nucl. Eng. Des.* 70, 3–12.

MacNeal RH, Harder RL (1985). A proposed standard set of problems to test finite element accuracy. *Finite Elem. Anal. Des.* 1, 3–20.

Maenchen G, Sack S (1964). The TENSOR code. In Alder B et al. (editors), *Methods in Computational Physics*, Vol. 3, pp. 181–210. Academic Press, London.

Malkus DS, Hughes TJR (1978). Mixed finite element methods—reduced and selective integration techniques: A unification of concepts. *Comp. Meth. Appl. Mech. Eng.* 15, 63–81.

Mindle W, Belytschko T (1983). A study of shear factors in reduced-selective integration Mindlin beam elements. *Comp. Struct.* 17, 334–344.

Mindlin RD (1951). Influence of rotatory inertia and shear on flexural motions of isotropic elastic plates. *J. Appl. Mech.* 18, 31–38.

Petschek AG, Hanson ME (1968). Difference equations for two-dimensional elastic flow. *J. Comp. Phys.* 3, 307–320.

Rankin CC, Nour-Omid B (1988). The use of projectors to improve finite element performance. *Comp. Struct.* 30, 257–267.

Reissner E (1945). The effect of transverse shear deformation on the bending of elastic plates. *J. Appl. Mech.* 12, A69–A77.

Reissner E (1947). On bending of elastic plates. *Q. Appl. Math.* 5, 55–68.

Stolarski H, Belytschko T, Lee SH (1995). A review of shell finite element and corotational theories. *Comp. Mech. Advances* 2, 125–212.

Suri M, Babuška I, Schwab C (1995). Locking effects in the finite element approximation of plate models. *Math. Comp.* 64, 461–482.

Taylor RL (1979). Finite elements for general shell analysis. 5th Int. Seminar Comp. Aspects Finite Element Method, Berlin, August.

Timoshenko SP, Woinowsky-Krieger S (1959). *Theory of Plates and Shells.* McGraw-Hill, New York.

Volmir AS (1956). *Flexible Plates and Shells.* Gosudarstv, Moscow (in Russian; English translation published by Air Force Flight Dynamics Laboratory, 1967).

Wempner GA, Oden JT, Kross DA (1968). Finite element analysis of thin shells. *ASCE J. Eng. Mech. Div.* 94, 1273–1294.

Wempner GA, Talaslidis D, Hwang CM (1982). A simple and efficient approximation of shells via finite quadrilateral elements. *J. Appl. Mech.* 26, 331–362.

Wilson EL, Taylor RL, Doherty WP, Ghaboussi J (1973). Incompatible displacement models. In Fenves SJ et al. (editors), *Numerical and Computer Methods in Structural Mechanics*, pp. 43–57. Academic Press, New York.

Wu SR (2004). Reissner-Mindlin plate theory for elasto-dynamics. *J. Appl. Math.* 2004, 179–189.

Wu SR (2005). A priori error estimation of a 4-node Reissner-Mindlin plate element for elasto-dynamics. *Comp. Meth. Appl. Mech. Eng.* 194, 2257–2281.

Wu SR (2006). Lumped mass matrix in explicit finite element method for transient dynamics of elasticity. *Comp. Meth. Appl. Mech. Eng.* 195, 5983–5994.

Wu SR, Chai L, Felitta B (2006). Investigation on the effect of hourglass control to the accuracy of component crashworthiness analysis. Presented at the 7th World Cong. on Comp. Mech., 7/17-21/2006, Los Angeles, California.

Wu SR, Gu L (2003). The effect of lumped mass matrix in explicit FEM. In Christon MA, Haber R (editors). *USNCCM VII Abstracts.* Proc. 7th US Nat. Cong. Comp. Mech., 7/27-31/2003, held in Albuquerque, New Mexico. p. 938; www.esc.sandia.gov/usnccm.html.

Wu SR, Qiu W (2009). Nonlinear transient dynamic analysis by explicit finite element with iterative consistent mass matrix. *Cummu. Numer. Meth. Eng.* 25, 201–217.

Wu SR, Saha N, Chen P (2004). Shell elements performance in crashworthiness analysis. *ASME Paper IMECE* 2004-61535.

Wu SR, Xie S, Wang Y (2007). Study on effect of warping to the accuracy of crash simulation. Proc. 5th INFATS (Int. Forum of Automotive Traffic Safety), pp. 164–168, 12/6-8/2007, Changsha, China.

Zeng Q, Combescure A (1998). A new one-point quadrature, general non-linear quadrilateral shell element with physical stabilization. *Int. J. Numer. Meth. Eng.* 42, 1307–1338.

Zhang Z, Zhang S (1994). Wilson's element for the Reissner-Mindlin plate. *Comp. Meth. Appl. Mech. Eng.* 113, 55–65.

Zienkiewicz OC (1977). *The Finite Element Method*, 3rd ed. McGraw-Hill, London.

Zienkiewicz OC, Taylor RL, Too JM (1971). Reduced integration technique in general analysis of plates and shells. *Int. J. Numer. Meth. Eng.* 3, 275–290.

CHAPTER 4

Arnold D, Falk R (1989). A uniformly accurate finite element method for the Reissner-Mindlin plate. *SIAM J. Numer. Anal.* 26, 1276–1290.

Arnold DN, Bolfi D, Falk RS (2002). Remarks on quadrilateral Reissner-Mindlin plate elements. In Mang FG, Rammerstorfer FG, Eberhardsteiner J (editors). Proc. WCCM V (5th World Cong. Comp. Mech.), July 7–12, Vienna, Austria.

Bathe KJ, Dvorkin E, Ho LW (1983). Our discrete Kirchhoff and isoparametric shell elements for nonlinear analysis—An assessment. *Comp. Struct.* 16, 89–98.

Batoz JL, Bathe KJ, Ho LW (1980). A study of three-node triangular plate bending element. *Int. J. Numer. Meth. Eng.* 15, 1771–1812.

Belytschko T, Stolarski H, Carpenter N (1984). A C_0 triangular plate element with one-point quadrature. *Int. J. Numer. Meth. Eng.* 20, 787–802.

Chapelle D, Bathe KJ (2003). *The Finite Element Analysis of Shells—Fundamentals.* Springer, Berlin.

Dhatt G (1969). Numerical analysis of thin shells by curved triangular elements based on discrete Kirchhoff hypothesis. Proc. ASCE Symp. Appl. FEM in Civil Eng., Vanderbilt University, Nashville, Tennessee, pp. 13–14.

Dhatt G (1970). An efficient triangular shell element. *AIAA J.* 8, 2100–2102.

Durán R, Liberman E (1992). On mixed finite element methods for Reissner-Mindlin plate models. *Math. Comp.* 58, 561–573.

Falk RS, Tu T (2000). Locking-free finite elements for the Reissner-Mindlin plate model. *Math. Comp.* 69, 911–928.

Kennedy JM, Belytschko T, Lin JI (1986). Recent developments in explicit finite element techniques and their application to reactor structures. *Nucl. Eng. Des.* 97, 1–24.

Li G, Belytschko T, Wu SR (2001). Performance of DKT thin shell element with 2 point integration scheme for large deformation analysis. Proc. 6th US Nat. Cong. Comp. Mech.

MacNeal RH, Harder RL (1985). A proposed standard set of problems to test finite element accuracy. *Finite Elem. Anal. Des.* 1, 3–20.

Stricklin JA, Haisler WE, Tisdale PR, Gunderson R (1969). A rapidly converging triangular plate element. *AIAA J.* 7, 180–181.

Wempner GA, Oden JT, Kross DA (1968). Finite element analysis of thin shells. *ASCE J. Eng. Mech. Div.* 94 (EM6), 1273–1294.

Wenzel T, Schoop H (2004). A non-linear triangular curved shell element. *Commun. Numer. Meth. Eng.* 20, 251–264.

Wu S, Li G, Belytschko T (2005). A DKT shell element for dynamic large deformation analysis. *Cummu. Numer. Meth. Eng.* 21, 651–674.

CHAPTER 5

Belytschko T (1983). Correction of article by Flanagan DP, Belytschko T. *Int. J. Numer. Meth. Eng.* 19, 467–468.

Belytschko T, Bachrach WE (1986). Efficient implementation of quadrilaterals with high coarse-mesh accuracy. *Comp. Meth. Appl. Mech. Eng.* 54, 279–301.

Belytschko T, Bindeman LP (1993). Assumed strain stabilization of the eight-node hexahedral element. *Comp. Meth. Appl. Mech. Eng.* 105, 225–260.

Fish J, Belytschko T (1988). Elements with embedded localization zones for large deformation problems. *Comp. Struct.* 30, 247–256.

Flanagan DP, Belytschko T (1981). A uniform strain hexahedron and quadrilateral with orthogonal hourglass control. *Int. J. Numer. Meth. Eng.* 17, 679–706.

Flanagan DP, Belytschko T (1984). Eigenvalues and stable time steps for the uniform strain hexahedron and quadrilateral. *J. Appl. Mech.* 51, 35–40.

Hu YK, Nagy LI (1997). Brake squeal analysis by using nonlinear transient finite element method. SAE Paper 971510.

Hughes TJR (1980). Generalization of selective integration procedures to anisotropic and nonlinear media. *Int. J. Numer. Meth. Eng.* 15, 1413–1418.

Kavanagh KT, Key SW (1972). A note on selective and reduced integration techniques in the finite element method. *Int. J. Numer. Meth. Eng.* 4, 148–150.

Koh BC, Kikuchi N (1987). New improved hourglass control for bilinear and trilinear elements in anisotropic linear elasticity. *Comp. Meth. Appl. Mech. Eng.* 65, 1–46.

Liu WK, Belytschko T, Ong JSJ, Law SE (1984). The use of stabilization matrices in nonlinear analysis. In Liu WK, Belytschko T, Park KC (editors), *Innovative Methods for Nonlinear Problems*, pp. 233–258, Pineridge, Swansea; Also *J. Eng. Comp.* (1985).

Liu WK, Hu YK, Belytschko T (1994). Multiple quadrature underintegrated finite elements. *Int. J. Numer. Meth. Eng.* 37, 3263–3289.

Liu WK, Ong JSJ, Uras RA (1985). Finite element stabilization matrices—A uniform approach. *Comp. Meth. Appl. Mech. Eng.* 53, 13–46.

Markus DS, Hughes TJR (1978). Mixed finite element methods-reduced and selective integration techniques: a unification of concepts. *Comp. Meth. Appl. Mech. Eng.* 15, 63–81.

Nagtegaal JC, Parks DM, Rice JR (1974). On numerical accurate finite element solutions in the fully plastic range. *Comp. Meth. Appl. Mech. Eng.* 4, 153–178.

Puso MA (2000). A highly efficient enhanced assumed strain physically stabilized hexahedral element. *Int. J. Numer. Meth. Eng.* 49, 1029–1064.

Schulz JC (1985). Finite element hourglassing control. *Int. J. Numer. Meth. Eng.* 21, 1039–1048.

Simo J, Hughes TJR (1986). On the variational foundations of assumed strain methods. *J. Appl. Mech.* 53, 51–54.

Zienkiewicz OC, Taylor RL, Too JM (1971). Reduced integration techniques in general analysis of plates and shells. *Int. J. Numer. Meth. Eng.* 3, 275–290.

CHAPTER 6

ALTAIR (2009). *RADIOSS Theory Manual*, 10.0 Version. Altair Engineering Inc., Troy, Michigan.

Chen X, Hang W, Huang H (1997). Finite element method for Timshenko beam, circular arch and Reissner-Mindlin plate problems. *J. Comp. Appl. Math.* 79, 215–324.

Ewing KM, Cheresh M, Thompson R, Kukuchek P (1982). Static and impact strengths of spot-welded HSLA and low carbon steel joints. SAE Paper 820281.

Hallquist JO (2006). *LS-DYNA Theory Manual*. Livermore Software Technology Corporation, Livermore, California.

Kaga S, Hotta T, Yamamoto Y, Ogawa K (1976). On the directionality of the strength of spot weld. SAE Paper 760024.

Lin SH, Pan J, Tyan T, Wu S, Prasad P (2002a). Modeling and testing of spot welds under dynamic impact loading conditions. SAE Paper 2002-01-0149.

Lin SH, Pan J, Wu SR, Tyan T, Wung P (2001). Spot weld failure loads under combined mode loading conditions. SAE Paper 2001-01-0428.

Lin SH, Pan J, Wu SR, Tyan T, Wung P (2002b). Failure loads of spot welds under combined opening and shear static loading conditions. *Int. J. Solids Struct.* 39 (2002) 19–39.

Riesner M, Sun X, Wu S, Hwang HY, Low E (2000). Modeling and optimization of structural joints in automotive applications. In Chirwa EC , Otte D (editors). Proc. ICRASH-2000 (International Crashworthiness Conf., 9/6-8/2000, London.), pp. 493–506.

Xiang Y, Wang Q, Fan Z, Fang H (2006). Optimal crashworthiness design of a spot-welded thin-walled hat section. *Finite Elem. Anal. Des.* 42, 846–855.

CHAPTER 7

Bader W (1927). Dissertation. University of Göttingen.

Bauschinger J (1886). On the change of the elastic limit and strength of iron and steel. Mittheilungen aus dem Mechanischen Technischen Laboratorium der Könglich Hochschule in Münich, pp. 1–115.

Budiansky B (1959). A reassessment of deformation theories of plasticity. *J. Appl. Mech.* 26, 259–264.

Chou CH, Pan J, Tang SC (1993). A hardening rule between stress resultants and generalized plastic strains for thin plates of power-law hardening materials. *J. Appl. Mech.* 60, 548–554.

Chou CH, Pan J, Tang SC (1994). Analysis of sheet metal forming operations by a stress resultant constitutive law. *Int. J. Numer. Meth. Eng.* 37, 717–735.

Chou CH, Pan J, Tang SC (1996). An anisotropic stress resultant constitutive law for sheet metal forming. *Int. J. Numer. Meth. Eng.* 39, 435–449.

Clark DS, Duwez PE (1950). The influence of strain rate on some tensile properties of steel. *Proc. ASTM.* 50, 560–575.

Cowper GR, Symonds PS (1957). Strain-hardening and strain-rate effects in the impact loading of cantilever beams. *Brown Univ. Dept. Appl. Math. T.R.* 28.

Drucker DC (1951). A more fundamental approach to plastic stress-strain relations. Proc. 1st US Natl. Cong. Appl. Mech. (Chicago 1951); Also ASME, New York, 1952, pp. 481–491.

Drucker DC (1956). On uniqueness in the theory of plasticity. *Q. Appl. Math.* 14, 35–42.

Han W, Reddy BD (1999). *Plasticity: Mathematical Theory and Numerical Analysis.* Springer-Verlag, New York.

Hencky H (1924). Zur Theorie plastischer Deformationen und der hierdurch im Material herrorgerufenen Nachspannungen. *ZAMM.* 4, 323–334 (in German).

Hill R (1950). *The Mathematical Theory of Plasticity.* Oxford University Press, Oxford.

Hill R (1958). A general theory of uniqueness and stability in elasto-plastic solids. *J. Mech. Phys. Solids* 6, 236–249.

Hodge PG Jr. (1959). *Plastic Analysis of Structures.* McGraw Hill, New York.

Hopkinson B (1905). The effects of momentary stress in metals. *Proc. Roy. Soc.* A74, 498–506.

Hopkinson B (1914). A method of measuring the pressure produced in the detonation of high explosives or by the impact of bullets. *Philos. Trans. Roy. Soc.* (London) A213, 437–456.

Huber MT (1904). Wlasciva praca odksztalcenia jako miara witezenia materialu. *Czas. Techn. Lwów* 22, 38–81 (in Polish).

Il'yushin AA (1948). *Plasticity.* Gostekhizdart, Moscow (in Russian).

Il'yushin AA (1961). On the postulate of plasticity. *Prik. Mat. Mekh.* 25, 503 (in Russian).

Johnson GR, Cook WH (1983). A constitutive model and data for metal subjected to large strains, high strain rates and high temperatures. In Proc. 7th Int. Symp. Ballistics, The Hague, The Netherlands. pp. 541–547.

Jones RM (2009). *Deformation Theory of Plasticity.* Bull Ridge Publisher, Blacksburg, Virginia.

Kachanov LM (1956). *Foundations of the Theory of Plasticity.* MIR, Moscow (English ed., 1971, North-Holland, Amsterdam).

Kim SJ, Oden JT (1984). Generalized potentials in finite elastoplasticity. *Int. J. Eng. Sci.* 22, 1235–1257.

Koiter WT (1953). Stress-strain relations, uniqueness and variational theorems for elastic-plastic materials with s singular yield surface. *Q. Appl. Math.* 11, 350–354.

Koiter WT (1960). General Theorems for Elastic-Plastic Solids. In Sneddon IN, Hill R (editors), *Progress in Solid Mechanics*, Vol. 1, Chapter 4, pp. 167–221. North-Holland, Amsterdam.

Krieg RD (1977). An efficient numerical method for time independent plasticity. SAND Rept. 77-0943, Sandia National Laboratories, Albuquerque, New Mexico.

Larmor J (1937). Origins of Clark Maxwell's electric ideas as described in familiar letters to William Thompson, Cambridge at the University Press, New York and Cambridge; See also Proc. Cambridge Phil. Soc. Vol. 32, Part 5, pp. 725–727.

Lemaitre J, Chaboche JL (1990). *Mechanics of Solid Materials*. Cambridge University Press, Cambridge.

Lévy M (1870). Mémoire sur les équations générales des mouvements intérieurs des corps solids ductiles au delà des limites où l'élasticité pourrait les ramener à leur premier état. *Comp. Rend. Acad. Sci.* (Paris) 70, 1323–1325; Also *J. Math. Pures Appl.* 16 (1871), 369–372 (in French).

Lubarda VA (2002). *Elastoplasticity Theory*. CRC Press, Boca Raton, Florida.

Manjoine M (1944). Influence of rate of strain and temperature on yield stresses of mild steel. *J. Appl. Mech.* 11, 211–219.

Martin JB (1975). *Plasticity: Fundamentals and General Results*. MIT Press, Cambridge, Massachusetts.

Melan E (1938a). Zur Plastizitat des raumlichen Kontinuum. *Ing. Arch.* 9, 116–126 (in German).

Melan E (1938b). Der Spannungszustand eines Hencky-Mises'schen Kontinuums bei veranderlicher Belastung. *Sitz. Ber. Ak. Wiss. Wien,* IIa 147, 73 (in German).

Mikhlin SG (1934). *Basic Equations of the Mathematical Theory of Plasticity*. Akad. Nauk., USSR (in Russian).

Mroz Z (1967). On the description of anisotropic workhardening. *J. Mech. Phys. Solids* 15, 163–175.

Nádai A (1931). *Plasticity*. McGraw-Hill, New York.

Perzyna P (1963). The constitutive equations for work-hardening and rate sensitive materials. *Proc. Vibr. Prob.* 4, 281–290.

Perzyna P (1966). Fundamental problems in viscoplasticity. *Adv. Appl. Mech.* 9, 244–377.

Perzyna P (1971). Thermodynamics theory of viscoplasticity. *Adv. Appl. Mech.* 11, 313–354.

Prager W (1953). On the use of singular yield conditions and associated flow rules. *J. Appl. Mech.* 20, 317–320.

Prager W (1955a). The theory of plasticity; a survey of recent achievements (James Clayton Lecture). *Proc. Inst. Mech. Eng.* 169, 41–57.

Prager W (1955b). *Introduction to Plasticity*. Addison Wesley (English ed. 1959), Reading, Massachusetts.

Prager W (1956). A new method of analyzing stress and strains in work-hardening plastic solids. *J. Appl. Mech.* 23, 493–496.

Prager W, Hodge PG Jr. (1951). *Theory of Perfectly Plastic Solids*. John Wiley & Sons, New York.

Prandtl LT (1924). Spannungsverteilüng in plastischen Körper. In Walman J Jr. (editor). Proc. 1st Intern. Congr. Mech. pp. 43–54. Delft, Technische Boekhandel en Druckerij (in German).

Reuss A (1930). Berücksichtigung der elastischen Formänderungen in der Plastizitätstheorie. *ZAMM*. 10, 266–274 (in German).

Rice JR (1970). On the structure of stress-strain relations for time-dependent plastic deformation in metals. *J. Appl. Mech.* 137, 728–737.

St. Vénant B (1870). Mémoire sur l'établissement des équations differéntielle des mouvements intérieurs opérés dans les corps solides ductiles au delà des limites où l'élasticité pourrait les ramener à leur premier état. *Comp. Rend. Acad. Sci.* (Paris) 70, 473–480 (in French).

Sanders JL (1954). Plastic stress-strain relations based on linear loading functions. Proc. 2nd US Nat. Cong. Appl. Mech. pp. 455–460.

Schreyer HL, Kulak RF, Kramer JM (1979). Accurate numerical solutions for elastic-plastic models. *J. Press. Vess. Tech.* 101, 226–234.

Shield RT, Ziegler H (1958). On Prager's hardening rule. *Z. Ang. Math. Phys.* 9a, 260–276.

Simo JC, Hughes TJR (1997). *Computational Inelasticity*. Springer-Verlag, New York.

Sokolovskii VV (1945). *Theory of Plasticity*. Gostekhizdat, Moscow (in Russion).

Tresca HE (1864). Mémoire sur l'écoulement des corps solides soumis a de fortes pression. *Comp. Rend. Acad. Sci.* (Paris) 59, 754 (in French).

von Mises R (1913). Mechanik der festen Körper im plastisch-deformablen Zustand. *Göttinger Nachrichten Math. Phys. Klasse*, 582–592 (in German).

Wilkins ML (1964). Calculation of elastic-plastic flow. In Alter B et al. (editors). *Methods of Computational Physics*, Vol. 3. Academic Press, New York.

Хуан КГ (1958). П. М. М. 22, 544–547 (in Russian).

Zerilli FJ, Armstrong RW (1987). Dislocation-mechanics-based constitutive relations for material dynamics calculations. *J. Appl. Phys.* 61, 1816–1825.

Zhao H (1997). A constitutive model for metals over a large range of strain rates, identification for mild-steel and aluminium sheets. *Material Sci. Eng.* A230, 95–99.

Ziegler H (1959). A modification of Prager's hardening rule. *Q. Appl. Math.* 17, 55–65.

CHAPTER 8

Chow CL, Wei Y (1999). Constitutive modeling of material damage for fatigue failure prediction. *Int. J. Damage Mech.* 8, 355–375.

Chow CL, Wei Y (2001). Anisotropic damage. In Lemaitre J (editor), *Handbook of Materials Behavior Models*, Vol. 2, pp. 430–435. Academic Press, San Diego, CA.

Chow CL, Wei Y, Wu S, Subbaraman B, Borchelt J, Frederick G (2000). Analysis of cumulative damage in a bumper due to multiple low-speed impacts. SAE Paper 2000-01-0631.

Chu CC, Needleman A (1980). Void nucleation effects in biaxially stretched sheets. *J. Eng. Mat. Tech.* 102, 249–256.

Gurson AL (1975). Plastic flow and fracture behavior of ductile materials incorporating void nucleation, growth and interaction. Ph.D. Dissertation, Brown University.

Gurson AL (1977). Continuum theory of ductile rupture by void nucleation and growth, I. Yield criteria and flow rules for porous ductile media. *J. Eng. Mat. Tech.* 99, 2–13.

Hill R (1950). *The Mathematical Theory of Plasticity*. Oxford University Press, Oxford.

Janson J, Hult J (1977). Fracture mechanics and damage mechanics: A combined approach. *J. Mech. Appl.* 1, 59–64.

Kachanov LM (1958). On the creep fracture time. *Ivz. Acad. Nauk U.S.S.R. Otd. Tech. Nauk.* 8, 26–31.

Lemaitre J (1971). Evaluation of dissipation and damage in metals submitted to dynamic loading. Mechanical behavior of materials; Proc. ICM-1, Kyoto, Japan, August 15–20, pp. 540–549.

Lemaitre J (editor) (2001), *Handbook of Materials Behavior Models*, Vol. 2. Academic Press, San Diego, CA.

Lemaitre J, Chaboche JL (1990). *Mechanics of Solid Materials*. Cambridge University Press, Cambridge.

McClintock FM (1968). A criterion for ductile fracture by the growth of holes. *J. Appl. Mech.* 35, 363–371.

Needleman A, Tvergaard D, Hutchinson JW (1992). Void growth in plastic solids. In Argon AS (editor), *Topics in Fracture and Fatigue*, pp. 145–178. Springer-Verlag, Berlin.

Rice JR, Tracey DM (1969). On the ductile enlargement of voids in triaxial stress fields. *J. Mech. Phy. Solids* 17, 201–217.

Robotnov YN (1963). On the equations of state of creep. ASME/ASTM Proceedings Joint International Conference on Creep, New York, Aug. 25–29, 1963, London, Sept. 30–Oct. 4, 1963. The Institution of Mechanical Engineers, Proceedings 1963–64, Vol. 178, Part 3A,

Robotnov YN (1969). Creep rupture. In Hetenyi M, Vincenti WG (editors), *Applied Mechanics*, Proc. 12th Int. Cong. Appl. Mech., pp. 342–349. Standford, Springer-Verlag, Berlin.

Tvergaard V (1990). Material failure by void growth to coalescence. *Adv. Appl. Mech.* 27, 83–151.

Tvergaard V, Needleman A (2001). The modified Gurson model. In Lemaitre J (editor), *Handbook of Materials Behavior Models*, Vol. 2, pp. 430–435. Academic Press, San Diego, CA.

Wu SR, Chen X, Hu W (2001). Global characterization of failure modes under impact loading in structural components made of low-ductility materials. ASME Paper IMECE2001/AMD-25437.

CHAPTER 9

Arruda E, Boyce M (1993). A three-dimensional constitutive model of the large stretch behavior of rubber elastic materials. *J. Mech. Phys. Solids* 41, 389–412.

Bansal NP, Doremus RH (1986). *Handbook of Glass Properties*. Academic Press, Inc., Orlando, Florida.

Blatz PJ, Ko WL (1962). Application of finite elastic theory to the deformation of rubbery materials. *Trans. Soc. Rheol.* 6, 223–251.

Boyce MC, Arruda EM (2000). Constitutive models of rubber elasticity: A review. *Rubber Chem. Tech.* 73, 504–523.

Chang FS (1995). Constitutive equation development of foam materials. Ph.D. Dissertation. Wayne State University, Detroit, Michigan.

Chang FS, Hallquist JO, Lu DX, Shahidi BK, Kudelko CM, Tekelly JP (1994). Finite element analysis of low-density high-hysteresis foam materials and the application in the automotive industry. SAE Paper 940908.

Chou CC, Zhao Y, Chai L, Co J, Lim GG, Lin TL (1994). Development of foam models using LS-DYNA3D. Proc. 2nd Int. LS-DYNA3D Conf., 9/20-21/1994, San Francisco.

Dharani LR, Mettu SR, Zhao S, Barbat SD, Chai L (2003). Modeling fracture in laminated automotive glazing impacted by spherical featureless headform. SAE Paper 2003-01-1225.

Du Bois PA, Kolling S, Fassnacht W (2003). Modeling of safety glass for crash simulation. *Comp. Mat. Sci.* 28, 675–683.

Ferry JD (1980). *Viscoelastic Properties of Polymers*, 3rd ed. Wiley, New York.

Fung YC (1965). *Foundations of Solid Mechanics*. Prentice-Hall, Englewood Cliff, N.J.

Gibson LJ, Ashby MF (1988). *Cellular Solids—Structures and Properties,* 1st ed. Pergamon Press, Oxford; 2nd ed. Cambridge University Press, Cambridge, 1997.

Green AE, Rivlin RS (1957). The mechanics of nonlinear materials with memory, Part 1. *Arch. Rat. Mech. Anal.* 1, 1.

Hill R (1950). *The Mathematical Theory of Plasticity*. Oxford University Press, Oxford.

Hinnerichs TD, Carme TG, Lu WY, Stasiunas EC, Neilsen MK, Scherzinger W, Rogillio BR (2006). Characterization of Aluminum Honeycomb and Experimentation for Model Development and Validation, Vol. II: Honeycomb Experimentation for Model Development and Validation. SAND2006-4455, Sandia National Laboratories, Albuquerque, New Mexico.

Hong ST, Pan J, Tyan T, Prasad P (2003). Crush strength of aluminum 5052-H38 honeycomb materials under combined compressive and shear loads. SAE Paper 2003-01-0331.

Kobayashi A, Wang LL (2001). *Quest for Dynamic Deformation and Fracture of Viscoelastic Solids*. Ryoin, Japan.

Krieg RD (1972). A simple constitutive description for soils and crushable foams. SC-DR-72-0883, Sandia National Laboratories, Albuquerque, New Mexico.

McFarland RK (1963). Hexagonal cell structure under post-buckling load. *AIAA J.* 1, 1380–1385.

Meyers MA, Chawla KK (1999). *Mechanical Behavior of Materials*. Prentice Hall, New Jersey.

Mooney M (1940). A theory of large elastic deformations. *J. Appl. Phys.* 11, 582–592.

Mulliken AD, Boyce MC (2006). Mechanics of the rate-dependent elastic-plastic deformation of glassy polymers from low to high strain rates. *Int. J. Solids Struct.* 43, 1231–1356.

Neilsen MK, Morgan HS, Krieg RD (1987). A phenomenological constitutive model for low-density polyurethane foams. SAND86-2927, Sandia National Laboratories, Albuquerque, New Mexico.

Neilsen MK, Morgan HS, Krieg RD, Yoshimura HR (1989). Development of a phenomenological constitutive model for polyurethane foams. SAND89-0941C, Sandia National Laboratories, Albuquerque, New Mexico.

Ogden RW (1972). Large deformation isotropic elasticity: On the correlation of theory and experiment for incompressible rubber like solids. *Proc. Roy. Soc. London* 326, 565–584.

Ogden RW (1984). *Non-linear Elastic Deformations.* Ellis Horwood Ltd., Chichester, Great Britain.

Ree T, Eyring H (1955). Theory for non-Newtonian flow I. Solid plastic system. *J. Appl. Phys.* 26, 793.

Rivlin RS (1948). Large elastic deformations of isotropic materials. *Proc. Roy. Soc. London* 241, 379–397.

Tran VX, Hong ST, Pan J, Tyan T, Prasad P (2006). Crush behavior of aluminum honeycomb of different cell geometries under compression dominant combined loads. SAE Paper 2006-01-0122.

Truesdell C, Toupin RA (1960). *The Classical Field Theories. Encyclopedia of Physics*, Vol. III/1. Springer-Verlag, Berlin.

Viana JC, Billon N, Cunha AM (2004). The thermomechanical environment and the mechanical properties of injection moldings. *Polymer Eng. Sci.* 44, 1522–1533.

Ward IM, Sweeney J (2004). *An Introduction to the Mechanical Properties of Solid Polymers*, 2nd ed. John Wiley & Sons, Chichester, UK.

Wierzbicki T (1983). A crushing analysis of metal honeycomb. *Int. J. Impact Eng.* 1, 157–174.

Zhang J, Ashby MF (1992a). The out-of-plane properties of honeycombs. *Int. J. Mech. Sci.* 34, 475–489.

Zhang J, Ashby MF (1992b). Buckling of honeycombs under in-plane biaxial stresses. *Int. J. Mech. Sci.* 34, 491–509.

Zhao S, Dharani LR, Chai L, Barbat SD (2006). Analysis of damage in laminated automotive glazing subjected to simulated head impact. *Eng. Failure Anal.* 13, 582–597.

CHAPTER 10

Amerio L, Prouse G (1975). Study of the motion of a string vibrating against an obstacle. *Rendiconti Di Matematica*, 8, 563–585.

Andress K, Auerbach F, Hort W (editors) (1922). *Handbuch der Physikalischen und Technischen Mechanik*, 3, S. 120. Leipzig, Barth.

Belytschko T, Liu WK, Moran B (2000). *Nonlinear Finite Elements for Continua and Structures.* John Wiley & Sons, New York.

Bourago NG (2002). A survey on contact algorithms. In Ivanenko SA, Garanzha VA (editors). Proc. Int. Workshop on Grid Generation and Indus. Appl., Moscow, pp. 42–59.

Boussinesq J (1885). *Applications des Potentiels.* Gauthier-Villars, Paris (in French).

Brezis H, Lions JL (1967). *Inéquations Hyperboliques.* C.R. Acad. Sci. Paris (in French).

Cabannes H (1985). Motion of a string vibrating against a rigid fixed obstacle. *Int. Conf. Nonlinear Mech. (Shanghai)*, pp. 879–884.

Campos LT, Oden JT, Kikuchi N (1982). A numerical analysis of a class of contact problems with friction in elastostatics. *Comp. Meth. Appl. Mech. Eng.* 34, 821–845.

Chien WZ, Ye KY (1956). *Elasticity.* Science Publisher, Beijing (in Chinese).

de Coulomb CA (1781). Théorie des machines simples, en ayant égard au frottement de leurs parties et la raideur des cordages. *Prix double Acad. Sci.* (in French)

Demkowicz L, Oden JT (1982). On some existence and uniqueness results in contact problems with nonlocal friction. *Nonlinear Anal.* 6, 1075–1093.

Duvaut G, Lions JL (1976). *Inequalities in Mechanics and Physics*. Springer-Verlag, Berlin Heidelberg (French ed., 1972).

Eck C, Jarušek J, Krbec M (2005). *Unilateral Contact Problems, Variational Methods and Existence Theorems*. Taylor & Friends Group, Boca Raton, Florida.

Glowinski R, Lions JL, Tremolieres R (1981). *Numerical Analysis of Variational Inequalities*. North-Holland, New York (French ed., 1976).

Han W, Reddy BD (1999). *Plasticity, Mathematical Theory and Numerical Analysis*. Springer, New York.

Han W, Sofonea M (2002). *Quasistatic Contact Problems in Viscoelasticity and Viscoplasticity*. American Mathematical Society and International Press, Providence, R.I.

Heinstein MW, Mello FJ, Attaway SW, Laursen TA (2000). Contact-impact modeling in explicit transient dynamics. *Comp. Meth. Appl. Mech. Eng.* 187, 621–640.

Hertz H (1881). Über die Berührung fester elastischer körper. *J. reine angewandte Math.* (Celle), 92, 156–171 (in German).

Hlasváčdk I, Haslinger J, Nečas J, Lovíšek J (1988). *Solution of Variational Inequalities in Mechanics*, Springer-Verlag, New York (Slovak ed., 1982).

Huber MT (1904). Zur Theorie der Berührung Fester Elasticher Körper. *Ann. Phys.* (Leipzig) (in German).

Johnson C (1976). Existence theorems for plasticity problems. *J. Math. Pures Appl.* 55, 431–444.

Johnson KL (1985). *Contact Mechanics*. Cambridge University Press, Cambridge, London.

Jones RE, Papadopoulos P (2001). A novel three-dimensional contact finite element based on smooth pressure interpolations. *Int. J. Numer. Meth. Eng.* 51, 791–811.

Kikuchi N, Oden JT (1988). *Contact Problems in Elasticity: A Study of Variational Inequalities and Finite Element Methods*. SIAM, Philadelphia.

Kloosterman G, van Damme RMJ, van den Boogaard AH, Huétink J (2001). A geometrical-based contact algorithm using a barrier method. *Int. J. Numer. Meth. Eng.* 51, 865–882.

Lebeau G, Schatzman M (1984). A wave problem in a half-space with a unilateral constraint at the boundary. *J. Diff. Eq.* 53, 309–361.

Li Y, Babuška I (1996). A convergent analysis of a p-version finite element method for one-dimensional elastoplasticity problem with constitutive laws based on the gauge function method. *SIAM J. Numer. Anal.* 33, 809–842.

Lions JL (1969). *Quelques Méthodes de Résolution des Problèmes aux Limites non Linéaires*. Dudon, Paris (in French).

Lions JL, Magenes E (1972). *Non-homogeneous Boundary Problems and Applications*. Springer-Verlag, New York (French ed., 1968).

Lions JL, Stampacchia G (1967). Variational inequalities. *Commun. Pure Appl. Math.* 20, 493–519.

Love AEH (1927). *A Treatise on the Mathematical Theory of Elasticity*, 4th ed. Dover Publications, New York (1st ed., 1892).

Martins JAC, Oden JT (1983). A numerical analysis of a class of problems in elastodynamics with friction. *Comp. Meth. Appl. Mech. Eng.* 40, 327–360.

Martins JAC, Oden JT (1987). Existence and uniqueness results for dynamic contact problems with nonlinear normal and friction interface laws. *Nonlinear Anal. Theo. Meth. Appl.* 11, 407–428.

Oden JT (1985). *Qualitative Methods in Nonlinear Mechanics*. Prentice-Hall, Englewood Cliffs, New Jersey.

Oden JT, Kikuchi N (1980). Theory of variational inequalities with applications to problems of flow through porous media. *Int. J. Eng. Sci.* 18, 1173–1284.

Oden JT, Martins JAC (1985). Models and computational methods for dynamic friction phenomena. *Comp. Meth. Appl. Mech. Eng.* 52, 527–634.

Oden JT, Pires EB (1981). Contact problems in elastostatics with non-local friction laws. TICOM Rept. 81–12, University of Texas at Austin.

Oden JT, Pires EB (1983a). Algorithms and numerical results for finite element approximations of contact problems with non-classical friction laws. *Comp. Struct.* 19, 137–147.

Oden JT, Pires EB (1983b). Nonlocal and nonlinear friction laws and variational principles for contact problems in elasticity. *J. Appl. Mech.* 50, 67–76.

Oden JT, Wu SR (1985). Existence of solutions to the Reynolds equation of elastohydrodynamic lubrication. *Int. J. Eng. Sci.* 23, 207–215.

Pires EB (1982). Analysis of nonclassical friction laws for contact problems in elastostatics. Ph.D. Dissertation, University of Texas at Austin.

Schatzman M, Bercovier M (1989). Numerical approximation of a wave equation with unilateral constraints. *Math. Comp.* 53, 55–79.

Shi P (1998a). The restitution coefficient for a linear elastic rod. *Math. Comp. Modeling* 28, 427–435.

Shi P (1998b). Simulation of impact involving an elastic rod. *Comp. Meth. Appl. Mech. Eng.* 151, 497–499.

Shillor M, Sofonea M, Telega JJ (2004). *Models and Analysis of Quasistatic Contact*. Springer, New York.

Signorini A (1933). Sopra alcune questioni di elastostatica. *Atti Soc. Ital. per il Progesso delle Scienze* (in Italian).

Stampacchia G (1964). Formes bilinéaires coercitives sur les ensemble convexes. *C.R. Acad. Sci.* (Paris) 258, 4413–4416 (in French).

Taylor RL, Papadopoulos P (1993). On a finite element method for dynamic contact/impact problems. *Int. J. Numer. Meth. Eng.* 36, 2123–2140.

Timoshenko SP, Goodier JN (1970). *Theory of Elasticity*, 3rd ed. McGraw-Hill, New York (1st ed., 1934).

Wriggers P (1995). Finite element algorithms for contact problems. *Arch. Comp. Meth. Eng.* 2, 1–49.

Wriggers P (2002). *Computational Contact Mechanics*, John Wiley & Sons, West Sussex, England.

Wu SR (1986). A penalty formulation and numerical approximation of the Reynolds Hertz problem of elastohydrodynamic lubrication. *Int. J. Eng. Sci.* 24, 1001–1013.

Wu SR (2001). Case study on convergence and accuracy of explicit finite element for contact-impact problems. In Dominguez VJ, Brebbia CA (editors), *Computational Methods in Contact Mechanics* (5th Inter. Conf. Comp. Meth. Contact Mech., 6/18-20/2001, Seville, Spain), pp. 273–282. WIT Press, Southampton, UK.

Wu SR (2009). A variational principle for dynamic contact with large deformation. *Comp. Meth. Appl. Mech. Eng.* 198, 2009–2015.

Zhong ZH (1993). *Finite Element Procedure for Contact-Impact Problems*. Oxford University Press, Oxford.

Zhong ZH, Mackerie J (1994). Contact-impact problems: a review with bibliography. *Appl. Mech. Rev.* 47, 55–76.

Беляев НМ (1924). Сборник Статей. Ленинград (in Russian).

CHAPTER 11

Belytschko T, Neal MO (1991). Contact-impact by the pinball algorithm with penalty and Lagrangian methods. *Int. J. Numer. Meth. Eng.* 31, 547–572.

Belytschko T, Yeh IS (1992). Adaptivity in nonlinear structural dynamics with contact-impact. *ASME AMD*. 157, 165–202.

Belytschko T, Yeh IS (1993). The splitting pinball method for contact-impact problems. *Comp. Meth. Appl. Mech. Eng.* 105, 375–393.

Benson DJ, Hallquist JO (1990). A single surface contact algorithm for the post-buckling analysis of shell structures. *Comp. Meth. Appl. Mech. Eng.* 78, 141–163.

Bourago NG (2002). A survey on contact algorithms. In Ivanenko SA, Garanzha VA (editors). Proc. Int. Workshop on Grid Generation and Indus. Appl., Moscow, pp. 42–59.

Chan SK, Tuba IS (1971) A finite element method for contact problems of solid bodies—Part I. Theory and validation. *Int. J. Mech. Sci.* 13, 615–625.

Conry TF, Seireg A (1971). A mathematical programming method for design of elastic bodies in Contact. *J. Appl. Mech.* 38, 387–392.

Hallquist JO, Goudreau GL, Benson DJ (1985). Sliding interfaces with contact-impact in large-scale Lagrangian computations. *Comp. Meth. Appl. Mech. Eng.* 51, 107–137.

Heinstein MW, Mello FJ, Attaway SW, Laursen TA (2000). Contact-impact modeling in explicit transient dynamics. *Comp. Meth. Appl. Mech. Eng.* 187, 621–640.

Hughes TJR, Taylor RL, Kanoknukulchai W (1977). A finite element method for large displacement contact and impact problems. In Bathe K et al. (editors), *Formulations and Computational Algorithms in Finite Element Analysis*, pp. 468–495. MIT Press, Massachusetts.

Hughes TJR, Taylor RL, Sackman JL, Curnier A, Kanoknukulchai W (1976). A finite element method for a class of contact-impact problems. *Comp. Meth. Appl. Mech. Eng.* 8, 249–276.

Wilson EA, Parsons B (1970). Finite element analysis of elastic contact problems using differential displacements. *Int. J. Numer. Meth. Eng.* 2, 387–395.

Wriggers P (1995). Finite element algorithms for contact problems. *Arch. Comp. Meth. Eng.* 2, 1–49.

Wu SR (2001). Case study on convergence and accuracy of explicit finite element for contact-impact problems. In Dominguez VJ, Brebbia CA (editors). *Computational Methods in Contact Mechanics* (5th Inter. Conf. Comp. Meth. Contact Mech., 6/18-20/2001, Seville, Spain), pp. 273–282, WIT Press, Southampton, UK.

Wu SR, Saha N, Chen P (2004). Shell elements performance in crashworthiness analysis. *ASME Paper IMECE* 2004-61535.

Zhong ZH (1993). *Finite Element Procedure for Contact-Impact Problems*. Oxford University Press, Oxford.

Zhong ZH, Mackerie J (1994). Contact-impact problems: a review with bibliography. *Appl. Mech. Rev.* 47, 55–76.

Zhong ZH, Nilsson L (1996). A unified contact algorithm based on the territory concept. *Comp. Meth. Appl. Mech. Eng.* 130, 1–16.

CHAPTER 12

ALTAIR (2009). *RADIOSS Theory Manual,* 10.0 Version. Altair Engineering Inc., Troy, Michigan.

Duvaut G, Lions JL (1976). *Inequalities in Mechanics and Physics.* Springer-Verlag, Berlin, Heidelberg (French ed., 1972).

Hallquist JO (2006). *LS-DYNA Theory Manual.* Livermore Software Technology Corporation, Livermore, California.

Hughes TJR, Winget J (1980). Finite rotation effects in numerical integration of rate constitutive equations arising in large-deformation analysis. *Int. J. Numer. Meth. Eng.* 15, 1862–1867.

INDEX

Acceleration, 6, 18–20, 21, 23, 30, 41, 45, 53, 55, 108, 160, 228, 231, 259, 275, 298, 299
Accuracy, 22, 27, 30, 56, 60, 87, 92, 174, 175, 209, 215, 268, 282, 301
Algorithm
  contact, 32, 221, 262, 272, 277, 279, 282–284, 287, 288, 290
  numerical, 25, 47, 237
  node-surface, 277
  of assumed strain, 120
  of explicit finite element method, 21
  pinball contact, 277–279
  search, 272
  single surface contact, 262
  sliding interface, 262
  splitting pinball, 278
  stabilization, 120
Angular
  velocity, 69, 297
Assembly, 17, 18
Assessment, 24, 26–30, 49–51, 69, 70, 78–80, 82, 102, 130, 150, 237, 302
Assumed deviatoric strain, 118, 121
Assumed strain, 56, 74, 77, 115–127
Assumption of constant pressure, 62
Assumption of K-L theory, 46
Assumption of straight normal, 47
Asymptotically, 29, 194, 255, 283, 301, 302

B-matrix, 73, 110, 115, 120, 121
Balance, 162, 222, 240, 241, 268, 269
Base vector, 61, 108, 113, 114
Basis, 23, 108, 198
Bauschinger effect, 144
Beam, 57, 62, 68, 70, 74, 80–83, 102–104, 123–125, 129–132, 136–138, 159, 170–172, 242, 243, 279
Bilinear
  form, 199
  interpolation, 52–55, 63, 86, 267
  shape function, 64
Body force, 12, 14, 27, 43, 52, 242, 243, 250
Boundary
  contact, 243, 244, 250, 268
  of the contact zone, 225
  layer, 51, 79

Introduction to the Explicit Finite Element Method for Nonlinear Transient Dynamics, First Edition.
Shen R. Wu and Lei Gu.

Boundary conditions, 14, 15, 24, 25, 27, 28, 43, 47–49, 51, 79, 222, 229, 244, 245, 289
Boundary value, 51, 232, 245
Boundedness, 25
Buckling, 5, 132, 141, 172, 210–212, 286
Bulk, 9, 196, 199–203, 205, 269

Cauchy stress, 24, 176, 178, 181, 183, 202, 203
Central difference, 19, 20, 26, 27, 259, 277, 301
Characteristic, 15, 40, 42, 50, 123, 153, 159, 189, 199, 204, 212, 213, 232, 269
Complete polynomials, 25
Component, 6, 7, 9–12, 13, 16, 18, 32, 33, 43, 44, 46, 50, 56, 57, 60, 63, 66, 73, 77–79, 99, 108, 110, 112, 114, 115, 117–121, 123, 128–130, 132, 134, 137, 147, 148, 154, 163–165, 172, 173, 176, 181, 187, 190, 196, 198, 199, 207, 212, 215, 221, 234–236, 241, 245, 247, 250, 257, 260, 262, 269, 282–284, 286, 292, 294, 295
Computation, 23, 28, 66, 70, 84, 122, 131, 160, 162–164, 169, 191, 197, 205, 259, 266, 273, 275, 277, 282, 288, 290, 294, 297, 298, 300, 302
Computational mechanics, 3, 221, 237
Condition
  boundary, 14, 15, 24, 25, 27, 28, 43, 47–49, 51, 79, 222, 229, 244, 245, 289
  initial, 24, 27, 28, 44, 47, 49, 51, 79, 228, 246, 249, 251, 252, 257, 258, 297
  Kuhn-Tucker, 234, 245
  of impenetrability, 233
  stability, 22–23
Configuration, 39, 40, 69, 93, 96, 108, 134, 242
Conjugate, 30, 65, 185
Connectivity, 208
Conservation
  of momentum, 290, 292, 293, 297
Conservative systems, 13
Consistent mass matrix, 24, 30, 32–35
Consistency, 136, 155, 159, 167, 179, 184, 186, 189
Constitutive
  equation, 133, 153–160, 168, 179–180, 183, 194, 195, 204, 206, 212
  law, 73, 116, 121
  model, 203, 205, 209
  relation, 8, 13, 74, 76, 121, 141, 154, 155, 159, 164, 167, 180, 181, 189, 198–199, 207, 209, 213
Constant strain, 58, 98, 153, 199
Constant stress, 194
Constraint, 11–13, 21, 27, 99, 138, 163, 172, 200, 222, 229, 232–235, 237, 242, 243, 252, 253, 259, 261, 262, 275, 276, 289, 290, 293, 296, 298–301
Constraint condition, 11, 21, 27, 138, 163, 172, 200, 229, 233, 234, 237, 243, 252, 253, 261, 276, 289, 293, 296, 298–301
Constraint method, 259, 275, 289
Contact
  domain, 224, 233, 242–243, 250–252, 258
  dynamic, 5, 237, 243, 247, 254, 256–258, 261, 262
  frictional, 233, 247
  frictionless, 243, 244, 246
  line, 279, 282
  node, 261
  penalty method of, 252
  point, 222–225, 231, 232, 235, 237, 238, 254–257, 261, 266, 267, 269, 270, 282, 294
  pin ball, 277, 279
  surface, 221, 235, 261, 262, 272, 279, 281, 282, 290
Contact-impact, 11
Continuity, 99, 105, 176, 223, 231, 255
Continuation, 142, 232, 256
Continuum mechanics, 24, 60, 131, 175, 177, 180, 208
Contradiction, 224, 247
Convergence
  of B-D element, 78–79
  rate, 26–27, 30, 56, 79, 86–88, 302
  tendency of, 84, 85
  trend, 283
  uniform, 55, 79
Corotational stress, 24, 68, 117
Corotational system, 66, 122
Coulomb's friction model, 248

Coulomb's law, 240, 241, 249, 250, 261, 270, 271, 291, 292
Criteria
yield, 148, 150
rupture, 175, 213
failure, 135–138
Critical
condition, 176
time step, 22, 282
value, 22, 146, 147, 175, 178, 211, 215, 301
Curvature, 16, 40, 83, 92–94, 97, 166, 172
Curvilinear, 73
Cycle, 23, 150, 151, 158, 231, 298
Cyclic, 20, 21, 298, 299
Cylindrical, 147, 159, 177, 205, 302

Damper, 66, 132, 192–195
Damping, 114, 115, 132, 133
Decomposition, 48, 49, 74, 92, 93, 95, 96, 145, 150, 154, 199, 241, 245, 247, 270, 291, 294
Decomposition method, 92
Helmholtz, 48, 49
orthogonal, 23, 61, 63, 74–75, 108, 113–114, 239
Deformation
axisymmetric, 159
bending, 39, 44, 46, 130, 302, 304
elastic, 55, 169, 200, 215, 254, 261
incremental, 20
initial, 27, 194
large, 5, 11, 12, 24, 27, 34, 39, 41, 63, 68–70, 129, 131, 132, 135, 141, 159, 170, 172, 173, 197, 203, 213, 221, 224, 235, 236–238, 242, 243, 256, 262, 269, 284, 285, 287, 303
lateral, 204
mode, 61, 71, 108, 112, 210, 283
pattern of, 57, 111
permanent, 142
plastic, 122, 141, 142, 153, 158, 159, 169, 170, 177, 197–199, 204, 283, 284
rate of, 71, 76, 192
shear, 24, 112, 181
small, 9, 41, 44, 46, 131, 222, 233, 235, 256, 260
Deformation energy, 66, 93
Deformation gradient, 200, 202
Degree of freedom, 71, 74, 75, 100, 128, 129, 131, 132, 136, 138
Density, 169, 181, 200, 203–205, 211
Diagonal mass matrix, 23–24, 30, 33, 35
Directional derivatives, 99
Discontinuity, 224, 233, 249, 302
Discrete, 4, 19, 56, 89, 97–100, 131, 137, 224, 279
Discrete kirchhoff condition, 99, 100
Discrete kirchhoff theory, 89, 97
Discretization, 16, 19, 23, 26, 68, 279
Displacement, 6, 7, 9, 12–15, 18–20, 22–30, 35, 40–44, 53, 55, 56, 60–63, 66–68, 74, 77, 80–85, 102, 108, 109, 123, 125, 128, 130, 132, 170, 171, 210, 228, 229, 231, 233–238, 241, 243, 244, 256, 257, 261, 283, 289, 300–304
Diverge, 22, 23, 102
Divergence, 7
Drill projection, 74–76
Drucker's postulate, 150–152

Effective strain, 182
Effective stress, 176, 181–183, 189
Eigenvalue, 30, 57, 61, 123
Elastic materials, 9, 21, 180, 233
Elastic modulus, 185
Elastic-plastic, 122, 141, 152, 159, 197
Elasticity constants, 6, 9, 112, 182, 188, 199
Elasticity tensor, 9, 74, 151, 155, 156, 164, 167, 182, 215
Elasto-plasticity, 158
Element
B-D, 73, 77–86, 102, 103, 170, 171, 174, 285, 286, 300, 303, 304
B-L, 71–74, 80–82, 84, 85, 102, 103, 170, 172–174, 286, 300
B-T, 66, 68–74, 80, 81, 84–86, 92, 170, 172–174, 284–287, 300, 302–304
B-W-C, 70
DKT, 97, 102–104, 106, 170, 171, 285–287
H-L, 58, 60
R-M, 56, 83, 86, 103, 105
R-M plate, 52, 55–58, 64, 70, 80, 85, 86, 102, 105, 112, 130

Element
  quadrilateral, 56, 58, 62, 63, 87, 88, 91, 92, 102, 103, 105, 169, 170, 172, 277, 278, 302
  triangular, 4, 88–92, 94, 96–98, 102–106, 170, 172, 277, 278, 286
Element
  beam, 57, 129–131, 137, 279
  plate, 52, 55, 56–58, 64, 70, 80, 85, 86, 102, 105, 111, 112, 130
  rod, 22, 128
  shell, 33, 39, 40, 50, 56, 61–63, 65, 68, 80, 87, 88, 92, 97, 107, 109, 112–114, 123–125, 129–131, 137, 138, 142, 169, 172, 174, 187, 216, 260, 271, 278, 280, 287, 300, 302
  solid, 62, 107, 111, 112, 123–125, 137, 138, 204, 271, 277
  truss, 22, 28, 129, 231, 282
Element
  coordinate system, 78, 117
  performance, 72
  properties, 4
  shape functions, 17, 276
  technology, 3, 21, 32, 37
Empirical, 68, 135, 153, 198, 207, 208, 211
Energy
  strain, 39, 56, 62, 66, 75, 92, 93, 95, 112, 114, 116, 122, 130, 173–175, 200, 202, 203, 284
  internal, 29, 173, 286, 287
  kinetic, 13–15, 23, 29, 227, 284, 297
Equation of motion, 14, 19, 47, 259, 298
Equilibrium, 27, 46, 48, 80, 194, 224–226, 240
Equilibrium equation, 43, 45, 48
Equivalent plastic strain rate, 154, 166, 167, 183, 186
Equivalent stress, 148, 154, 189, 190
Error(s)
  boundedness of, 25
  curves, 29, 283, 301
  of the displacement, 25, 28, 231, 283, 301
Error estimate
  a-priori, 25
Experiment, 34, 82, 123, 137, 168, 180, 189, 207, 217, 226, 240
Explicit
  integration, 5, 26, 298
  method, 5, 23
  scheme, 19, 21, 23, 257, 259, 270
Explicit finite element, 4–6, 11, 13, 20, 21, 23–25, 27, 28, 30, 40, 56–58, 60, 61, 63, 68, 79, 87, 123, 142, 160, 237, 252, 258, 259, 262, 298, 299
Explicit finite element procedure, 21, 160, 258, 259, 298, 299
External load, 51, 269
External forces, 20, 289, 297

Fiber, 59, 60
Finite element
  approximation, 283
  discretization, 68, 279
  equation, 15, 20, 54, 55, 64, 258
Flow rule, 141, 158, 166, 189
Flow chart, 299
Force
  body, 12, 14, 27, 43, 52, 242, 243, 250
  contact, 226, 240–243, 248, 250, 253, 254, 256, 258, 259, 261, 262, 268, 269, 275–279, 281, 295
  damping, 114, 132, 133
  driving, 231, 240
  external, 20, 289, 297
  friction, 240, 241, 248, 258, 270, 271, 282, 291
  hourglass, 74
  impact, 32, 33, 35, 231, 259, 270, 277, 281, 282, 290, 292
  inertia, 47
  internal, 20, 289
  membrane, 164, 165
  nodal, 65, 69, 75, 114, 116–118, 120, 122, 128, 130, 131, 133, 160, 183
  penalty, 256, 269
  shear, 43, 44, 47, 129, 137, 164
Force-deflection, 132, 211, 212
Formulation
  element, 6, 74, 97, 102
  incremental, 58, 95
  variational, 48, 243, 247, 250, 258
Friction, 123, 233, 240, 241, 247–250, 252, 256, 258, 261, 262, 270, 271, 282, 291, 292, 295, 302
Friction models, 241, 248, 250

Frictionless, 243–246, 251, 295
Functional
  energy, 252, 253
  friction, 249, 252
  penalty, 257

Galerkin method, 15, 48, 244
Gap, 137, 229, 231, 235, 269, 271, 273, 279–281, 302
Gap function, 235–237, 257, 260, 268, 269
Gauss quadrature, 56–58, 91, 120, 170
Gauss-green theorem, 14, 243, 244, 246
Gaussian curvature, 83
Generalized, 9, 10, 13, 44, 65, 66, 69, 74, 76, 117, 150, 163, 166, 167, 182, 187, 196
Geometric nonlinearity, 224
Global search, 263–264, 267, 272, 274
Governing equations, 11, 28, 47, 224, 228, 250
Gradient, 15, 30, 151, 200, 202
Gurson's model, 177–180, 189

Hardening
  isotropic, 144, 145f, 148, 149, 158, 163, 178
  kinematic, 145, 148, 149, 158
Helmholtz free energy, 181, 182, 186
Hexagonal, 146, 209, 210, 212
Hooke's law, 9, 150
Hourglass
  force, 74
  strain, 65, 69, 74
  stress, 65, 66, 69, 74, 117
  velocity, 70, 114, 117
Hourglass control
  elastic, 66
  perturbed, 68, 113
  plastic, 66, 76
Hourglass modes, 62, 63, 65, 71, 108, 113
Hyperbolic equation, 24, 29
Hyperbolic paraboloid, 83–86, 103–106, 125–127
Hyperelastic, 200, 202

Impact
  condition, 212, 214, 216, 231
  elastic, 226, 231, 232
  energy, 205, 212, 214
  engineering, 5, 34, 56, 141, 203, 209, 212
  force, 33, 231, 259, 270, 277, 281, 282, 290, 292
  load, 27, 123, 175, 214
  plastic, 289–291, 294, 297
  test, 137, 212, 289
Impactor, 212, 262, 263, 289, 291
Impenetrability, 233, 235
Imperfection, 176
Implicit method, 5
Implicit scheme, 23, 26
Incompressibility, 154, 156, 164, 179, 185, 203, 204
Incompressible, 57, 111, 112, 123, 128, 200, 201, 217
Incremental
  constitutive relation, 74, 76
  form, 13, 58, 60, 95, 114, 144, 151, 152, 154, 161, 183, 259
  method, 5
Index, 7, 24–27, 53, 108, 166
Inequality
  variational, 246, 247, 249, 251–253, 256, 257
Inertia
  moment of, 55, 68, 131, 296
  force, 47
Initial
  conditions, 24, 27, 28, 44, 47, 49, 51, 79, 228, 246, 249, 251, 252, 257, 258, 297
  deformation, 27, 194
  velocity, 12, 21, 22, 27, 28, 66, 79, 169, 172, 227, 228, 297, 300
Interface, 240, 262
Integration
  by parts, 14, 15
  explicit, 5, 26, 298
  full, 56, 74, 216
  one-point, 58, 63, 71, 73, 76, 92, 112, 114, 117, 118, 123
  reduced, 56–58, 60–64, 66, 70, 74, 77, 92, 112, 113, 115, 118, 123, 130
  selective reduced, 56–58, 115, 118, 123
  surface, 226
  through thickness, 131
  2x2 integration, 58, 60, 78
Internal energy, 173, 287
Internal force, 20, 289

Internal nodal force, 75, 120, 128, 160
Internal variable, 158, 177, 178
Interpenetration, 284, 287
Interpolation, 4, 7, 17–19, 21, 23–27, 47, 52, 53, 55, 59–61, 63, 70, 77, 86, 88, 89, 91, 92, 94, 95, 96, 98–100, 102, 105–108, 113, 128, 267, 270, 275, 276, 279, 282
Invariance, 58, 75
Invariant, 57, 147, 200, 202, 205, 207
Inverse, 9, 30, 32, 90, 111, 120, 166, 167, 180, 182, 186
Isoparametric elements, 18, 89
Isotropic hardening, 144, 145, 148, 149, 158, 163, 178
Isotropic material, 189
Isotropic model, 180, 181
Iterative process, 33, 34, 261, 295

J-integral, 214–215
J2, 147, 205
Jacobian, 63, 70, 87–90, 93, 111, 120, 122

Kinematic constraint condition, 234, 289–304
Kinematic hardening, 145, 148, 149, 158
Kirchhoff-Love theory, 46–47
Kuhn-Tucker conditions, 234, 245

Lagrange multiplier, 200, 259, 260, 295
Lamé elasticity constants, 6, 9, 112
Large deformation, 5, 11, 12, 24, 27, 34, 39, 41, 63, 68–70, 129, 131, 132, 135, 141, 159, 170, 172, 173, 197, 203, 213, 221, 224, 235–238, 242, 243, 256, 262, 269, 284, 285, 287, 303
Large rotation, 5, 11, 235, 236, 238
Linear application, 95, 105
Linear elasticity, 13, 44–48, 93, 117, 128, 131, 164, 165, 216, 260, 261
Linear interpolation, 26, 88, 89, 92, 94, 96, 98, 105, 106, 128, 279
Linear problem, 4, 224
Local search, 266–268, 273, 274
Local system, 24, 63, 70, 90, 93, 123, 128, 208, 210, 280
Locking
  membrane, 121
  shear, 50, 55, 56, 64, 72, 74, 76, 77, 79, 81, 85, 88, 91–93, 98, 99, 105, 111, 112, 115, 118, 119, 121–123, 127, 130
  volumetric, 56, 111, 112, 115, 118, 119, 121, 123
Lower order, 23, 56
Lumped mass, 24, 30, 32, 68

Mass matrix
  consistent, 24, 30, 32–35
  diagonal, 23–24, 30, 33, 35
  lumped, 24
  mixed, 30, 33–35
Master node, 262–267, 272–274
Master segment, 262–264, 266, 268, 270–275
Material model, 20, 21, 32, 141, 160, 169, 177, 199, 200, 203, 207, 208, 211–213, 217
Membrane
  force, 164, 165
  strain, 96, 166
  stress, 98
Mesh
  quadrilateral, 88, 102, 103, 123, 125, 170
  solid, 125
  triangular, 88, 102, 103, 170
  uniform, 28, 282, 283, 301
  warped, 80, 102
Mesh refinement, 56, 70, 87, 103, 138, 279, 286
Mises stress, 148, 154, 163, 164, 170, 175, 189
Mixed finite element, 57
Mixed method, 49, 56, 57, 66, 86
Modeling, 88, 128, 129, 131, 132, 169, 212, 216, 217, 289
Modulus
  elasticity, 185
  plastic, 144
  tangent, 13, 76, 144, 204
Moment, 42, 48, 55, 68, 75, 129–133, 136, 164, 165, 183, 212, 238, 290, 292, 293, 296, 297

Momentum, 290, 292, 293, 296, 297
Motion
  decelerating, 227
  equation, 15, 42, 43, 45, 47, 227
  equation of, 14, 19, 47, 259, 298
  infinitesimal, 239
  relative, 241, 271, 291
  rigid body, 24, 57, 62, 74, 75, 298
  rotational, 55, 300
  small, 238
  sliding, 240, 291
  tangential, 271, 291–293
  vibrating, 249

Natural frequency, 22
Node
  eight-, 56, 63, 107, 109, 111–112, 123
  four-, 16, 39–40, 52, 53, 57, 58, 61, 63, 65, 66, 69, 75, 80–85, 87, 88, 105, 107, 109, 112, 215, 263, 277, 278
  three-, 16, 88, 89, 91, 94, 98, 102, 105, 133, 264, 277, 278
  two-, 18, 22, 57, 78, 99, 128–129, 131, 137, 138, 265, 279
Node
  contact, 261
  master, 262–267, 272–274
  slave, 262–277, 289–298
Nonconservative, 13
Nonlinear
  analysis, 4, 13, 47, 262
  dynamic contact, 256–257
  frictional dynamic problem, 257
  material, 5, 13, 21, 27, 76, 128, 131, 192–196
  spring, 199
  transient dynamics, 4, 5, 11–15, 27, 60, 61, 63, 78, 98, 106
  viscoelastic model, 199–200
Nonlinearity
  geometry, 11, 224
  material, 11, 224
Norm
  Sobolev, 25
  L^2, 26, 30, 79
  L^∞, 26, 82, 302
  H^1, 27, 29–30, 78, 79

Normal
  component, 43, 234–236, 241, 247, 257, 292, 295
  distance, 257, 273, 275, 276
  direction, 44, 63, 70, 138, 146, 161, 162, 213, 235, 240–243, 250, 266, 274, 277, 281, 282, 291, 294, 295
  vector, 13, 75, 234, 238, 264, 267, 268, 273, 274
  velocity, 269, 292
Normal contact force, 240–241, 248
Normal rotation, 92–93, 96–98
Normal tension, 134–137
Normality, 46, 150, 152, 154–155, 158–159, 179, 180, 183, 267
Numerical
  algorithm, 25, 47, 237
  approach, 3, 5, 221, 233
  example, 79, 231
  experiment, 34, 123, 168
  integration, 91
  method, 4, 6, 39, 141, 177, 221, 233, 262
  model, 216
  procedure, 215, 233, 257, 261–262, 290
  quadrature, 91
  scheme, 4, 68
  solution, 5–6, 231

Objective, 6, 48, 135, 142
Objectivity, 24, 60, 208
One-dimensional, 17, 76, 224, 265
One-point
  integration, 58, 63, 71, 73, 76, 92, 112, 114, 117, 118, 123
  quadrature, 56, 71, 91, 114, 117, 122
  reduced integration, 57, 58, 60, 62, 66, 74, 92, 112, 113
One-sided, 272
One-to-four splitting, 82, 278
One-to-two splitting, 102–103
Operator, 7–8, 24–26, 47–48, 77, 93, 96
Optimal convergence rate, 26, 30, 79
Order, 7, 23, 25, 28, 52, 56–58, 60, 61, 78, 79, 85–88, 91, 102, 105–107, 134, 141, 181, 189, 200, 241, 282, 301
Orthogonal, 23, 61, 63, 74, 108, 239, 300
Out of plane, 67, 96, 210
Outer normal, 234–235, 238, 242

Parabolic, 25, 222, 224
Parameter, 6, 9, 18, 28, 30, 50, 55, 59, 68–69, 71, 73, 80, 83, 99, 102–103, 108, 115, 119–120, 135, 137, 149, 151, 153, 165–166, 169, 172, 177, 180–182, 185, 186, 188, 199–200, 202, 203, 207–209, 215, 216, 231, 270, 281
Parametric, 98
Path
  loading, 132
  reloading, 143
  unloading, 132, 142
Penalty
  coefficient, 282–283, 287, 288
  contact force, 254, 256, 268–272, 278–279, 281
  contact method, 282, 289
  force, 256, 269
  function, 57, 257
  functional, 257
  method, 200, 237, 252–260, 262, 268, 269, 278, 281–283, 295, 301
  parameter, 68, 270, 281
  term, 203, 253, 256–258, 260, 269
Perfect plasticity, 142, 144, 150, 165
Perturbation hourglass control, 60–61, 63, 66, 70, 73, 115
Perturbation method, 69
Perturbed lagrangian method, 260
Plane stress, 4, 10, 44, 62, 158, 163–165, 167, 187
Plastic
  damage, 141, 186–187, 189
  deformation, 141–142, 153, 158, 159, 169–173, 177, 197–199, 204, 283, 284
  flow, 166, 183
  hardening, 150, 162, 184, 186
  incompressibility, 154, 156, 164, 179, 185
  loading, 144, 154, 155, 159, 179
  material, 158
  modulus, 144
  perfectly, 135, 158–159
  power, 154
  rigid-, 158, 211
  strain, 143, 151, 154, 158, 162, 165–167, 175, 178, 179, 181, 183, 185, 186, 188, 189
  tangent modulus, 13, 76, 144, 204
  unloading, 132
  work, 150–151, 154, 166, 179
  zone, 149, 165
Plastic impact, 289–291, 294, 297
Plasticity
  hardening, 148, 165
  perfect-, 142, 144, 150, 165
  rigid-, 211
  rigid-perfect-, 142
Plate
  bending, 57, 63, 77, 92, 97–98, 130
  element, 52, 55–58, 64, 70, 80, 85, 86, 102, 105, 111, 112, 130
  K-L, 50–51, 56
  moderately thick, 47
  R-M, 47–52, 54–58, 63, 64, 70, 80, 85, 86, 102, 105, 112, 129, 130, 296f
  theory, 39–40, 47, 88, 164, 296
  thin, 46
Plateau, 212
Point wise, 258, 302
Procedure
  contact, 270, 288
  cyclic, 20, 299
  explicit, 15, 23
  finite element, 160, 183, 258, 259, 298, 299
  iterative, 30, 32, 163, 261
  Newton-Raphson, 4
  numerical, 215, 233, 257, 261, 290
Product
  cross, 264
  scalar, 264
  tensor, 88, 91, 107
Projection
  drill, 74–76
  method, 70, 72, 75, 76, 89
  operator, 77
  static, 26
  shear, 76–78

Quadrature
  2x2, 56, 58, 61
  eight-point, 116
  four-point, 116, 120, 122
  Gauss, 56–58, 91, 120, 170
  one-point, 56, 71, 91, 114, 117, 122
  point, 56, 63, 76, 114
  reduced, 92, 120

three-point, 92
two-point, 102
Quadrilateral, 16, 56–58, 62, 63, 65, 68, 70, 87, 88, 91–92, 97, 102, 103, 105, 109, 123–125, 169, 170, 172, 264, 266, 268, 277–278, 286, 302
Quasi-static, 169

Radius, 172, 225–226, 278, 280, 293
Rate
convergence, 26, 27, 30, 56, 79, 86–88, 302
damage energy release, 185–186
damage growth, 185
deformation, 194–195
dependent, 169, 198, 200
independent, 55, 200
form, 13, 41, 74, 151, 154, 155, 160, 161, 168, 185, 256
of convergence, 25–26
of deformation, 71, 76, 192
of strain, 114–115, 121–123
of stress, 121–122
of work, 237
optimal, 28, 30, 70, 78–79, 86
strain, 65, 69, 73, 74, 76, 110, 114, 119, 152–154, 158, 166, 167, 169, 170, 172, 176, 183, 186, 192, 198, 199, 200, 209
stress, 24, 69, 74, 76, 121, 128
work, 65, 95, 114, 154, 166
Rectangular element, 55, 62, 66, 78, 87
Rectangular mesh, 78, 87–88
Reference
coordinate, 107, 116–117, 120, 122
domain, 18, 53, 108, 267
element, 18, 71
system, 6, 53, 77, 78, 119
Refined mesh, 33–34, 83, 123, 174, 215, 283, 286, 301
Relation
constitutive, 8–10, 13, 74, 76, 121, 141, 154, 155, 159, 164, 167, 180, 181, 189, 198–199, 207, 209, 213
geometry, 44, 59
incremental, 65
rate form, 185
stress-strain, 153, 169, 193–194, 212
Relaxation, 192–193, 195–200
Residual, 15, 55
Resultant
of stress components, 164
form, 43
force, 136, 169
stress, 42, 43, 45, 131, 164, 165, 168
Rod, 22, 28–30, 128, 228, 229, 231, 243, 282, 301
Rotation
large, 5, 11, 235–236, 238
nodal, 132
normal, 92, 93, 96, 97, 98
of the normal, 40–42, 46, 72, 164
rigid, 69–70, 300, 301
rigid body, 75–76, 78, 96, 133, 300
Rotational motion, 55, 300

Searching process, 264–265, 268, 270
Segment
line, 263, 265, 279–281
master, 262–264, 266, 268, 270–275
quadrilateral, 266
slave, 262
surface, 242, 262–263, 273, 279
triangular, 264, 268
Shear
locking, 50, 55–56, 64, 72, 74, 76, 77, 79, 81, 85, 88, 91–93, 98, 99, 105, 111, 112, 115, 118, 119, 121–123, 127, 130
transverse, 44, 46, 47, 49, 51, 55–58, 60, 71–72, 76–79, 86, 92, 97, 99, 105, 106, 164, 211
Shear correction factor, 68, 130
Shell element
B-L, 123
B-T, 63, 113, 114
Bathe-Dvorkin (B-D), 123
Belytschko-Tsay (B-T), 63, 113, 114
quadrilateral, 63, 65, 92, 97, 109, 124
four-node, 61, 80, 87, 112
Reissner-Mindlin, 47, 111, 164
three-node, 88–106
Shell structure, 39–40, 46, 242, 260, 262, 268, 272, 279
Shell theory, 39
Sliding, 262, 266, 271, 293
Sliding motion, 240, 291
Sobolev norm, 25
Spot weld, 134–138
Spurious, 57, 122

Stability, 22–23, 28, 35, 39, 152, 271, 288, 291, 298
Strain
assumed, 56, 74, 77, 115–127
bending, 92, 93, 95
deviatoric, 118, 121, 207
effective, 182
elastic, 165, 182
energy, 39, 56, 62, 66, 75, 92, 93, 95, 112, 114, 116, 122, 130, 173–175, 200, 202, 203, 284
hardening, 181
hourglass, 65, 69, 74
increment, 58, 60, 63, 117, 150, 153, 155, 158, 161, 199
membrane, 96, 166
plastic, 143, 151, 154, 158, 162, 165–167, 169, 175, 178, 179, 181, 183, 185, 186, 188, 189
rate, 65, 69, 73, 74, 76, 110, 114, 119, 152–154, 158, 166, 167, 169, 170, 172, 176, 183, 186, 192, 198–200, 209
shear, 44, 46, 48–49, 51, 55–58, 60, 72–74, 76–79, 86, 92, 93, 95, 97, 99, 105, 106, 116, 121, 130, 202
tensile, 121, 152
velocity, 64–65, 71, 74, 128, 130
volumetric, 116, 118, 205–206
Stress
back-, 145, 148–149, 159
bending, 214
Cauchy, 24, 176, 178, 181, 183, 202–203
compressive, 176, 204, 209, 214, 226, 234
contact, 229, 234, 235
corotational, 24, 68, 117
deviatoric, 147, 158, 162, 205
effective, 176, 181–183, 189
equivalent, 148, 154, 189–190
hourglass, 65–66, 69, 74, 117
increment, 20, 60, 153, 155, 199
membrane, 98
Mises, 148, 154, 163, 164, 170, 171, 175, 189
nominal, 176, 182, 183, 189, 190, 198, 211
objectivity, 24, 60, 208
plane, 4, 10, 44, 62, 158, 163, 167, 187
principle, 158
rate, 24, 69, 74, 76, 121, 128
resultant, 42, 43, 45, 131, 164, 165, 168
shear, 44, 78, 146, 191, 202, 212, 216
-strain curve, 142–143, 145, 151, 153, 170, 205
-strain relation, 153, 169, 193–194, 212
tensile, 135, 205, 214, 215
tensor, 147, 178, 181, 241
trial, 161–162
true, 176, 198
yield, 135–137, 143–144, 147–149, 162, 167, 170, 172, 176, 178, 183, 185, 186, 198, 199, 207, 208, 304

Tangent modulus, 13, 76, 144, 204
Tensor
compliance, 9, 151, 182, 187
damage effect, 181–183, 187, 190
elasticity, 9, 74, 151, 155–156, 164, 167, 182, 215
permutation, 8
strain, 71, 74, 119, 200, 202
stress, 147, 178, 181, 241
Tensor product, 88, 91, 107
Theory
discrete Kirchhoff, 89, 97
elasticity, 180
K-L, 44, 46–48, 50, 94
Kirchhoff-Love (K-L), 46–47
mathematical, 40, 57, 237
plate, 40, 47
of continuum mechanics, 24, 60
of damage mechanics, 176, 177
R-M, 47, 48, 50, 55, 68, 88, 91, 130
R-M plate, 47, 296
Reissner-Mindlin (R-M) plate, 47
shell, 39
small deformation, 41
Three-dimensional, 6, 24, 39, 107, 145, 180, 196, 221–222, 233, 261, 262
Time step, 22–23, 26, 28, 35, 60, 61, 63, 68, 82–83, 102, 125, 138, 161, 183, 202, 215, 216, 257, 259, 269–271, 273, 274, 277, 282, 288, 291, 297, 298, 301
Tolerance, 32–35, 279
Total
deformation, 195, 227
mass, 296
momentum, 296

strain, 142, 159, 165, 175
strain energy, 175
Traction, 12–14, 52, 155, 234, 241, 243
Transformation, 24, 53, 60, 63–64, 66, 89–90, 101, 108, 110, 111, 202
Transient dynamics, 4, 5, 11–13, 23, 26–27, 30, 34, 39, 48, 60, 61, 63, 78–79, 82, 87, 88, 98, 103, 107, 128, 141, 175, 192, 207, 221, 238
Transient structural dynamics, 11–13
Two-dimensional, 7, 39, 109, 215, 233, 262

Unilateral contact, 224, 229, 232, 269, 289, 290, 295
Unity property, 17, 24, 276
Unique solution, 21, 49
Uniqueness, 24, 49, 152, 223–224, 229, 232, 237

Variation, 14, 15, 85, 107, 303
Variational
equation, 15, 19, 21, 25, 26, 52, 243, 257–258
formulation, 48, 243–247, 250, 258
inequality, 246–247, 249, 251–253, 256, 257
principle, 11, 13–15, 47, 55, 243–244, 246, 251, 253, 254, 257, 258
Velocity, 6, 12–13, 15, 18–22, 26–30, 41, 52, 53, 55, 61–67, 69–71, 73–76, 79, 96, 108, 113, 114, 116, 117, 119, 128–130, 160, 169, 172, 227–231, 236–242, 244, 246, 248, 249, 251, 256–259, 269–271, 273, 291–293, 295, 297–300, 302
Velocity strain, 64–65, 71, 74, 128, 130
Vector
base, 61, 108, 113, 114
form, 20, 50, 61, 86, 93, 94, 295
normal, 13, 75, 234, 238, 264, 267, 268, 273, 274
position, 256, 266, 296
rotation, 298
Virtual
displacement, 15
power, 15, 120, 248
velocity, 15
work, 74
Volume
fraction, 176, 178–179, 204–207
integration, 5, 14–15, 19–20, 23, 26, 56–60, 62, 66, 70
Volumetric locking, 56, 111–112, 115, 118–119, 121, 123
Volumetric strain, 116, 118, 205–206

Weak form, 19
Weak solution, 15, 21, 249
Weight function, 244–245
Weighted residual, 15, 55
Work
plastic, 150–151, 154, 166, 179
rate, 65, 95, 114, 154, 166
virtual, 74

Yield
condition, 135, 146–150, 161, 162, 165, 208
criterion, 141, 145, 150
function, 146, 148, 154, 158, 163, 165, 166, 178, 183, 205, 207, 213
point, 135, 143, 144, 150, 161, 198, 205
surface, 145–152, 155, 158–159, 161–162, 186, 204
initial, 143–144, 149, 208
stress, 135–137, 143, 144, 147–149, 162, 167, 170, 172, 176, 178, 183, 185, 186, 198, 199, 207, 208, 304
subsequent, 143–144, 159

Zone
contact, 214, 224–226, 228, 233, 243, 250
elastic, 149, 152, 165
heat affected, 134–135, 137–138
impact, 259
non-contact, 224, 250
plastic, 149, 165

Printed and bound by CPI Group (UK) Ltd, Croydon, CR0 4YY
16/09/2026
14963417-0003